CONSTRUCTION SAFETY HANDBOOK

A Practical Guide to OSHA Compliance and Injury Prevention

2nd Edition

Mark McGuire Moran

ABS Consulting

Government Institutes
Rockville, Maryland

ABS Consulting

4 Research Place, Rockville, Maryland 20850, USA
Phone: (301) 921-2300
Fax: (301) 921-0373
Email: giinfo@govinst.com
Internet: http://www.govinst.com

06   05   04   03   4   3   2   1

Library of Congress Cataloging-in-Publication Data

Construction safety handbook : a practical guide to OSHA compliance and injury
    prevention / Mark McGuire Moran.
      p. cm.

   Includes bibliographical references and index.
   ISBN 0-86587-813-7
   1. Building--Safety measures--Standards--United States. 2. Construction industry--
Safety regulations--United States.

TH443.C683 2003
363.11'969'00973--dc21

                                                    2003050349

Printed in the United States of America

Summary Contents

Chapter 1
The Occupational Safety and Health Act 1

Chapter 2
Recordkeeping: Employers Subject to OSHA
Recordkeeping 13

Chapter 3
Construction Industry Standards 67

Chapter 4
Hazard Communication: Communicating the Hazards
of Chemicals to Workers 123

Chapter 5
Training Requirments in OSHA: Construction
Standards Guidelines 149

Chapter 6
Noise Control 183

Chapter 7
Signs, Signals, and Barricades: Safety Standards for
Signs, Signals, and Barricades 197

Chapter 8
Ergonomics: Preventing Musculoskeletal Injuries
at Your Workplace 211

Chapter 9
Personal Protective Equipment: Protecting Your
Workers from Head to Toe 237

Chapter 10
Fall Protection: Preventing Falls at Your Worksite 255

Chapter 11
Excavations, Trenching, and Shoring: Avoiding
Cave-Ins and Other Accidents 273

Chapter 12
Demolition Operations: Using Safe Blasting Procedures
at Your Worksite 293

Chapter 13
Cranes and Crane Operation: Avoiding Crane Hazards 299

Chapter 14
Forklift Safety: Training Your Forklift Operators in
Safe Driving Techniques 333

Chapter 15
Controlling Lead Exposures: Working with Lead
in the Construction Industry 347

Chapter 16
Asbestos: Controlling Asbestos Exposures at Your
Worksite 367

Chapter 17
Process Safety Management: Conducting a Process
Safety Analysis at Your Worksite 379

Chapter 18
Heat Stress: Setting up an Effective Program to
Reduce Heat Exhaustion 399

Chapter 19
OSHA Inspections: How to Prepare for and Respond
to an Inspection 405

Glossary 433

Index 445

Contents

List of Figures xxii

List of Tables xxiv

Preface xxvii

About the Author xxix

Chapter 1
The Occupational Safety and Health Act 1

Introduction 1

Coverage of the Act 2

State OSHA Plans 2

Duties of Employers and Employees 4

    Multi-Employer Job Sites 5

Labor–Management Relations 6

Who Is Obligated to Observe the Law? 8

Chapter 2
Recordkeeping: Employers Subject to OSHA Recordkeeping 13

Introduction 13

Functions of the Recordkeeping System 13

What Are the Reasons for OSHA Recordkeeping? 14

OSHA Recordkeeping Requirements 16

Who Is Covered by the Recordkeeping Rule? 16

Employers Exempt from Recording Injuries and illnesses 17

Comparison of the Old and New Rules 19

Questions About the New Rule 22

    Which Employers Are Covered by the New Rule? 22

    What Are the New Forms for Recording Injuries and Illnesses? ... 22

    Is the Injury or Illness Work-Related? 22

    Is the Injury or Illness a New Case? 23

    What Is the General Recording Criteria? 23

    What Is Restricted Work? 23

    What Is Medical Treatment? 24

    Diagnosis of a Significant Injury or Illness 24

    Recording Injuries and Illnesses to Soft Tissues 24

    Employee Privacy 24

    Certification, Summarization, and Posting 24

    Employee Involvement 25

Reporting Fatalities and Multiple Hospitalizations 25
Clarification of Recordkeeping Citation Policy in the Construction Industry 25

OSHA Recordkeeping Forms 26

What Recordkeeping Forms Are Required by OSHA? 26
How Often Are the Forms Updated and/or Maintained? 26
What Is the OSHA Form 300 (Log of Work-Related Injuries and Illnesses)? 26
What Do We Need to Do if an Injury/Illness Occurs? 28
How Should the OSHA 300 Form Be Filled Out? 29
What Is the OSHA 300-A (Summary of Work-Related Injuries and Illnesses)? 32
How Should the OSHA 300-A Summary Form Be Filled Out? 32
What Is the OSHA 301 (Injury and Illness Incident Report)? 36
How Should the OSHA 301 Form Be Filled Out? 36

Maintaining Your Records 39

Retention of Records 39
Establishments Required to Maintain Records 39
Location of Records 40
Access to the Records 40

Injury and Illness Incidence Rates 41

What Is an Incidence Rate? 41
How Do You Calculate an Incidence Rate? 41
How Are Incidence Rates Used? 42
Reports by the Bureau of Labor Statistics 43

Definition of Terms 46

Death 46
Days Away From Work 46
Restricted Work or Transfer to Another Job 46
Medical Treatment and First Aid 47
Loss of Consciousness 48
Significant Work-Related Injuries and Illnesses 48
Significant Criteria 48
Additional Criteria 49

Determining if a Case is Recordable 50

Is This Injury or Illness Recordable? 50

Reporting Occupational Hearing Loss Cases 54

Is the Hearing Loss Requirement Similar to OSHA's Noise Standard? 55
Age Correction 56
Record Consistently 56
Baseline Reference and Revision of Baseline 57
Work Relationship 57
Adding a Column to the 300 Log 58

Enforcement of the Recordkeeping Rule 60

Inspection Procedures 60
Inspecting the OSHA 300 Log 60
Records Review 61
Incomplete Recorded Cases on the OSHA 300 or 301 61

Penalties 62
Failure to Post Annual Summary 62
Failure to Report Fatalities and Hospitalizations 62
Access to Records for Employees 63
Willful, Significant, and Egregious Cases 63
Enforcement Procedures for Occupational Exposure to Bloodborne Pathogens 63
Enforcement Procedures for Occupational Exposure to Tuberculosis 64

Compliance Officer Checklist 64
Check Data for the Establishment 64
Obtain Administrative Subpoena/Medical Access Orders 64
Verify the SIC Code 64
Ask for Information 64
Ask Your Regional Recordkeeping Coordinator for Assistance 65
Procedures for a Recordkeeping Inspection 65

Chapter 3
Construction Industry Standards 67

Introduction 67
The Remaining 24 Subparts Are Listed Below 67
Subpart C—General Safety and Health Provisions 68
General Safety and Health Provisions—§1926.20 69
Inspections 70
Personal Protective Equipment—§1926.28 71
Access to Employee Exposure and Medical Records—§1926.33 72
Means of Egress—§1926.34 75
Employee Emergency Action Plans—§1926.35 75

Subpart D—Occupational Health and Environmental Controls 76
OSHA Summary of Hazard Communication Standard 79

Subpart E—Personal Protective and LifeSaving Equipment 80

Subpart F—Fire Protection and Prevention 82

Subpart G—Signs, Signals, and Barricades 83
Safety Standards for Signs, Signals, and Barricades 83

Subpart H—Materials, Handling, Storage, Use, and Disposal 84
Materials Handling and Storing 84

Subpart I—Tools (Hand and Power) 85

Subpart J—Welding and Cutting 85

Subpart K—Electrical 86
Installation Safety Requirements 87
Safety-Related Work Practices 89
Safety-Related Maintenance and Environmental Considerations 90
Safety Requirements for Special Equipment 91
What are the most frequently cited standards in Subpart K? 91
What are some effective control measures that can be used to control electrical

hazards listed in Subpart K?.. 92
What work practices are needed for protection against electrical hazards? 92

Subpart L—Scaffolding 93
Specific Requirements for Scaffolds 93

Subpart M—Fall Protection 94
Introduction 94
Duty to Have Fall Protection 95
Fall Protection Systems, Criteria, and Practices 95
What are the most frequently cited violations of the Fall Protection Standard?....... 97
What are some control measures that can be used for the serious hazards
for which OSHA has most frequently cited employers? 98
What work practices can be used to provide protection from falls when
conventional systems are not feasible? 98

Subpart N—Cranes, Derricks, Hoists, Elevators, and Conveyors 98
Requirements for Mobile Cranes 99

Subpart O—Motor Vehicles, Mechanized Equipment, and Marine Operations ... 99

Subpart P—Excavations 100
What are some control measures that can be used for hazards in Subpart P? 101

Subpart Q—Concrete and Masonry Construction 101
What are the most frequently cited violations of the Concrete and Masonry
Construction Standard? 102
What are some of the effective control measures that can be used to eliminate
the hazard of being struck by flying brick and block in the event of a wall
collapse? 102
What are some of the ways to brace a wall over eight feet to provide protection
against the hazard of collapse? 102

Subpart R—Steel Erection 103

Subpart S—Underground Construction, Caissons, Cofferdams, and
Compressed Air 104
Introduction 105
Competent Person 105
Safety Instruction 105
Access and Egress 106
Check-In/Check-Out 106
Hazardous Classifications 106
Air Monitoring 107
Oxygen 107
Hydrogen Sulfide 107
Other Precautions 108
Ventilation 108
Illumination 109
Fire Prevention and Control 109
Hot Work 109
Cranes and Hoists 109
Emergencies 110

Subpart T—Demolition 111
   Preparatory Operations 111
   What are the most frequently cited Demolition Standards? 111
   What are some control measures that can be used to protect employees from
     these hazards? 112

Subpart U—Blasting and Use of Explosives 112
   General Provisions 113

Subpart V—Power Transmission and Distribution 115
   Initial Inspections, Tests, or Determinations 116
   Clearances 116

Subpart W—Rollover Protective Structures; Overhead Protection 117
   Remounting 118
   Labeling 118
   Machines Meeting Governmental Requirements 118

Subpart X—Stairways and Ladders 118

Subpart Y 120

Subpart Z—Toxic and Hazardous Substances 120

Chapter 4
Hazard Communication: Communicating the Hazards of Chemicals to Workers 123

Introduction 123

Scope 123

Hazard Determination 123

Container Labeling 123

Material Safety DAta Sheets (MSDS) 124

List of Hazardous Substances 125

Employee Information and Training 125

Refresher Training Shall Be Conducted Annually 126

Non-Routine Tasks 126

Informing Contractors 126

Explanation of Terms Used on Material Safety Data Sheets 134
   Section I 134
   Section II 134
   Section III 134
   Section IV 135
   Section V 135
   Section VI 135
   Section VII 136
   Section VIII 136

The Most Frequently Asked Questions on Hazard Communication 136

Chapter 5
Training Requirments in OSHA: Construction Standards Guidelines......................149

Introduction 149
Voluntary Training Guidelines 150
Training Model 150
    Review Commission Implications 151
Training Guidelines 151
    Determining If Training is Needed 152
    Identifying Training Needs 152
    Identifying Goals and Objectives..................................... 153
    Developing Learning Activities 154
    Conducting the Training... 154
    Evaluating Program Effectiveness 155
    Improving the Program 155
Matching Training to Employees 156
    Identifying Employees at Risk 156
    Training Employees at Risk 157
    Conclusion 158
Construction Training Requirements 158
    General Safety and Health Provisions 158
    Safety Training and Education 159
    Medical Services and First Aid 159
    Ionizing Radiation 160
    Non-ionizing Radiation 160
    Gases, Vapors, Fumes, Dusts, and Mists 160
    Asbestos 160
    Hazard Communication, Construction 161
    Lead in Construction 162
    Process Safety Management of Highly Hazardous Chemicals 163
    Hearing Protection 164
    Respiratory Protection 164
    Fire Protection 164
    Signaling 165
    Powder-Operated Hand Tools.. 165
    Woodworking Tools 165
    Gas Welding and Cutting 166
    Arc Welding and Cutting 168
    Fire Prevention 168
    Welding, Cutting, and Heating in Way of Preservative Coatings 168
    Ground-Fault Protection 169
    Scaffolding 169
    Guarding of Low-Pitched Roof Perimeters during the Performance of Built-Up
      Roofing Work 170
    Fall Protection 171

Cranes and Derricks 171
Material Hoists, Personnel Hoists, and Elevators 172
Material Handling Equipment 173
Site Clearing 173
Excavations—General Protection Requirements (Excavations, Trenching,
    and Shoring) 173
Concrete and Masonry Construction 174
Bolting, Riveting, Fitting-Up, and Plumbing-Up 175
Underground Construction 175
Compressed Air 177
Preparatory Operations 178
Chutes 178
Mechanical Demolition 178
General Provisions (Blasting and Use of Explosives) 178
Blaster Qualifications 179
Surface Transportation of Explosives 179
Firing the Blast 179
General Requirements (Power Transmission and Distribution) 180
Overhead Lines 181
Underground Lines 181
Construction in Energized Substations 181
Ladders 182

Chapter 6
Noise Control 183

Introduction 183
Noise Exposure In Construction 183
Occupational Noise Exposure Standards in Construction 184
Construction Noise 185
A Sample Hearing Conservation Program 185
Employers Covered 185
Monitoring 186
When Should Monitoring Be Done? 186
Where Should Monitoring Be Done? 187
How Should Monitoring Be Done? 187
Who Should Conduct the Monitoring? 187
Employee Involvement in Noise Measurements 187
Noise Measuring 188
Provide Options 188
Training 189
Fitting 189
Attenuation 189
Training 190
Potential Hearing Loss Cases 190
Employee Information 191
Records 191

Chapter 7
Signs, Signals, and Barricades: Safety Standards for Signs, Signals, and Barricades 197

Introduction 197
Background 197
  Temporary Traffic Control Elements 199
  Pedestrian and Worker Safety 199
  Hand Signaling or Flagger Control.......................................200
  Devices 200
  Temporary Traffic Control Zone Activities 202
Relationship to Existing DOT Regulations 202
Percentage of Roads Covered Under OSHA's Standard Versus the DOT Standard 203
Federal-Aid Highways...203
  State, Local, County, and Municipal Roads Not Receiving Federal Aid 203
  Private Roads 204
Incentives to Comply with the MUTCD 204
State Plan Standards 205
Accident Prevention Signs and Tags—§1926.200.......................206
  General 206
  Danger Signs 206
  Caution Signs 206
  Exit Signs 207
  Safety Instruction Signs 207
  Directional Signs 208
  Traffic Signs..208
  Accident Prevention Tags 208
Signaling—§1926.201 209
  Flagmen...209
  Crane and Hoist Signals 209
Barricades—§1926.202 210

Chapter 8
Ergonomics: Preventing Musculoskeletal Injuries at Your Workplace 211

Introduction 211
OSHA Activity Relating to Ergonomics 212
How OSHA Guidelines Can Reduce Injuries and Illnesses Related to MSDs 213
  Problem Recognition 213
Review and Analysis of Injury and Illness Records 214
  Step 1: Review Injury and Illness Records 214
  Step 2: Identify Potential MSD Cases 214

Step 3: Categorize 215

Analysis of Jobs or Work Tasks 215

Defining the Problem 215
Correcting the Problem 216
Disorders and Injuries Associated with MSDs 216

Introduction 220

Management Commitment and Employee Involvement 220
Job Hazard Analysis 222
Quick Fix Option 222
Hazard Prevention and Control 223
Engineering Controls 223
Administrative Controls 224
Personal Protective Equipment 224
Implementing Hazard Controls 225
Evaluating Hazard Controls 225
MSD Management 226
The HCP will provide the following: 226
Evaluations 226
Early Reporting 227
Training and Education 227

Program Evaluation 229

Construction Ergonomics Checklist 230

Materials Handling 230
Tools 231
Repetitive Work 231
Awkward Postures 232
Standing 232
Surfaces for Walking and Working 233
Seating 233
Lighting 233
Weather 233
Production Pressures 233
Training 234
Musculoskeletal Symptoms 234
Solutions 235

Chapter 9
Personal Protective Equipment: Protecting Your Workers from Head to Toe 237

Introduction 237

General Requirements 238

Hazard Assessment 239
Equipment Selection 239
Training 240

Eye and Face Protection 241

Inspection and Maintenance for Eye Protection 242

Head Protection 242

Inspection and Maintenance for Head Protection 244

Foot Protection 245

Hand Protection 246

Respiratory Protection 247

Respiratory Protection Inspections 248
Education and Training 248
Fitting and Maintenance of Respirators 249

Limitations of PPE 249

Personal Protective Equipment Safety Checklist 250

Responsibilites 250
Head Protection 251
Eye and Face Protection 251
Hand Protection 252
Foot and Leg Protection 252
Light Reflective Products 252
Safety Belts, Lifelines, and Lanyards 252

Chapter 10
Fall Protection: Preventing Falls at Your Worksite 255

Introduction 255

Fall Hazard Locations 255
Identifying Fall Hazards 256
Interviews 256
Job Site Inspections 256

Fall Prevention 256

Fall Prevention Measures 257
Engineering Controls 257
Determining Alternative Approaches to the Work 257
Providing Proper Access 257
Providing Guardrail Protection 257

Fall Protection 258

Fall Arrest Systems 258
Selecting Proper Equipment 258

Components of a Fall Arrest System 259

Anchorage 259
Anchoring Devices/Points 260
Non-Anchorages 260
Body Support 260
Connecting Means 261

Developing a Fall Protection Program 264

Policy 268

Controlling Fall Hazards 269

Chapter 11
Excavations, Trenching, and Shoring: Avoiding Cave-Ins and Other Accidents 273

Excavation, Trenching, and Shoring Program 273

Excavation Responsibilities 274
Competent Person 274
Training 275
Underground Installations 275
Access and Egress 275
Vehicular Traffic 276
Falling Loads 276
Mobile Equipment 276
Hazardous Atmospheres 276
Water Accumulation 276
Loose Rock or Soil 277
Fall Protection 278
Daily Inspections 278
Protection from Cave-ins 279
Support Systems 279
Protection Systems Decision Tree 280
Sloping and Benching 281
Sloping and Benching Systems 281
Requirements 282
Actual Slope 282
Configurations 283
Figures: Slope Configurations 283
Excavations Made in Type A Soil 283
Excavations Made in Type B Soil 284
Excavations Made in Type C Soil 284
Excavations Made in Layered Soils 285

Soil Classification 286

Scope 286
Definitions 286

Classification of Soil and Rock Deposits 288

Basis of Classification 288
Visual and Manual Analyses 288
Layered systems 288
Reclassification 288
Visual tests 289
Manual tests 289
Materials and Equipment 290
Installation and Removal of Support 291
Shield Systems 291

Chapter 12
Demolition Operations: Using Safe Blasting Procedures at Your Worksite 293

Introduction 293
Engineering Survey 293
Utility Location 294
Medical Services and First Aid 294
Police and Fire Contact 295
Safe Blasting Procedures 296
Fire Precautions 296
Personnel Selection 297
Proper Use of Explosives 297
Procedures After Blasting 298

Chapter 13
Cranes and Crane Operation: Avoiding Crane Hazards 299

Introduction 299
Supervisor of Equipment 300
Supervisor of Lift 300
Operator Requirements 301
Crane Inspections 302
Hazard Prevention Requirements 303
    Preconstruction Planning 303
    Job Hazard Analysis 304
    Hand Signals 304
    Signaling Devices 304
    Lifting Capabilities 304
    Rigging Practices 304
    Controlling the Load 304
    Wire Rope Requirements 305
    Annual Inspections 305
    Preventive Maintenance 305
Competent Person Requirements 305
    Operators 305
    Riggers, Signalers, and Others 306
Hazards Common to Most Cranes 306
Powerline Contact 306
    Definition 306
    Description 306
    Risks Presented by Hazard 307

Preventive Measures 307
OSHA Requirements 307

Overloading 308

Definition 308
Description 308
Risks Presented by Hazard 308
Preventive Measures 309
OSHA Requirements 309

Failure to Use Outriggers, Soft Ground, and Structural Failure 309

Definition 309
Description 309
Risks Presented by Hazard 310
Preventive Measures 310
OSHA Requirements 310

Two-Blocking 310

Definition 310
Description 311
Risks Presented by Hazard 311
Preventive Measures 311
OSHA Requirements 312

Pinchpoint 312

Definition 312
Description 312
Risks Presented by Hazard 313
Preventive Measures 313
OSHA Requirements 313

Unguarded Moving Parts 313

Definition 313
Description 313
Risks Presented by Hazard 314
Preventive Measures 314
OSHA Requirements 314

Unsafe Hooks 314

Definition 314
Description 314
Risks Presented by Hazard 315
Preventive Measures 315
OSHA Requirements 315

Obstruction of Vision 315

Definition 315
Description 315
Risks Presented by Hazard 316
Preventive Measures 316
OSHA Requirements 316

Sheave-Caused Cable Damage 316

Definition 316
Description 316
Preventive Measures 317
OSHA Requirements 317

Cable Kinking 317
Definition 317
Description 317
Risks Presented by Hazard 318
Preventive Measures 318
OSHA Requirements 318

Sidepull 318
Definition 318
Description 318
Risks Presented by Hazard 318
Preventive Measures 318
OSHA Requirements 319

Boombuckling 319
Definition 319
Description 319
Risks Presented by Hazard 319
Preventive Measures 319
OSHA Requirements 319

Access to Cabs, Bridges, and/or Runways 319
Definition 319
Description 320
Risks Presented by Hazard 320
Preventive Measures 320
OSHA Requirements 320

Control Confusion 320
Definition 320
Description 320
Risks Presented by Hazard 321
Preventive Measures 321
OSHA Requirements 321

Turntable Failure 321
Definition 321
Description 321
Risks Presented by Hazard 321
Preventive Measures 322
OSHA Requirements 322

Removable or Extendable Counterweight Systems 322
Definition 322
Description 322
Risks Presented by Hazard 322
Preventive Measures 323

OSHA Requirements 323

Chapter 14
Forklift Safety: Training Your Forklift Operators in Safe Driving Techniques 333

Powered Industrial Truck Hazards 333

    Qualifications of Machinery Operators 334

    Operator Training 335

    Testing, Retraining, and Enforcement 336

    Powered Industrial Truck Characteristics 337

    Introduction 338

    General Requirements 339

    Designations 339

    Designated Locations 340

    Safe Guards 340

    Fuel Handling and Storage 340

    Batteries 341

    Control of Noxious Gases and Fumes 341

    Trucks and Railroad Cars 341

    Truck Operation 342

    Traveling 342

    Loading 343

    Maintenance 343

    Required Operator Training 343

    Training Program 344

    Workplace-Related Topics 344

    Refresher Training and Evaluation 345

    Avoidance of Duplicate Training 345

    Training Certification 345

Powered Industrial Truck Training 346

Chapter 15
Controlling Lead Exposures: Working with Lead in the Construction Industry 347

Introduction 347

Sample Lead Exposure Control Program 348

    Health Effects of Lead 348

Definitions 349

The OSHA Permissible Exposure Limit (PEL) 350

Exposure Monitoring 351

Methods of Compliance 352

Respiratory Protection 353

Protective Work Clothing and Equipment 355

Hygiene Areas and Practices 357

Housekeeping 357
Medical Surveillance 358
Medical Removal Protection 360
Employee Information and Training 362
Recordkeeping 363

Chapter 16
Asbestos: Controlling Asbestos Exposures at Your Worksite..367

Introduction 367
Scope and Application 368
Permissible Exposure Limit (PEL) 368
Work Classification 368
Exposure Monitoring 369
Competent Person Requirements 369
Regulated Areas 369
Employee Information and Training 370
Respiratory Protection 370
Protective Clothing 371
Housekeeping 372
Medical Surveillance 372
Recordkeeping 373
    Monitoring Records 374
    Medical Surveillance Records 374
    Other Recordkeeping Requirements 374

Chapter 17
Process Safety Management: Conducting a Process Safety Analysis at Your Worksite..............379

Introduction 379
Sample Process Safety Management Program 380
    Scope and Application 380
    Process Safety Information 380
    Process Hazard Analysis 381
    Operating Procedures 383
    Training 384
    Contractors 385
    Pre-Startup Safety Review 385
    Mechanical Integrity 386
    Hot Work Permit 387
    Management of Change 387

Incident Investigation 388
Emergency Planning and Response 388
Compliance Audits 389
Trade Secrets 389

Chapter 18
Heat Stress: Setting up an Effective Program to Reduce Heat Exhaustion 399

Introduction 399
Causal Factors 399
Heat Stroke 399
Heat Exhaustion 400
Heat Cramps 400
Heat Collapse (Fainting) 400
Heat Rashes 401
Heat Fatigue 401
Preventing Heat Stress 401
Acclimatization 401
Fluid Replacement 402
Engineering Controls 402
Convection 402
Administrative Controls 403
Other Administrative Controls 403
Heat Stress Questionnaire 404

Chapter 19
OSHA Inspections: How to Prepare for and Respond to an Inspection 405

Introduction 405
How Construction Inspection Candidates are Chosen 406
Guidance for Focused Inspections in Construction 407
Background 407
General Guidelines 409
Specific Guidelines 410
Tips on What to Do When an OSHA Inspector Appears at Your Worksite 411
Understanding 411
Preparation 412
Responding to the Inspection 414

Glossary 433

Index 445

Figures

Figure 2.1 OSHA 300 Log 27
Figure 2.2 OSHA 300A Form 33
Figure 2.3 OSHA worksheet 34
Figure 2.4 OSHA 301 Form 37
Figure 2.5 Do I Record the Injury or Not? 51
Figure 4.1 Sample Request for Material Safety Data Sheets 127
Figure 4.2 Sample Training Acknowledgment 128
Figure 4.3 Sample Material Safety Data Sheets 129
Figure 4.4 Sample Material Safety Data Evaluation Checklist 132
Figure 5.1 Code of Blasting Signals 179
Figure 6.1 Hearing Conservation Program CheckList for a Mine Operation ... 193
Figure 6.2 Hearing Conservation Training Log 194
Figure 6.3 Record of Hearing Protection Needs 195
Figure 7.1 Danger Sign 206
Figure 7.2 Caution Sign 207
Figure 7.3 Exit Sign 207
Figure 7.4 Safety Instruction Sign 208
Figure 7.5 Traffic Sign 208
Figure 7.6 Accident Prevention Tags 209
Figure 7.7 Hand signaling 209
Figure 7.8 Barricade 210
Figure 8.1 Sample Ergonomics Program 220
Figure 8.2 Construction Ergonomics Checklist 230
Figure 9.1 PPE Checklist 250
Figure 10.1 Sample Fall Protection Program 268
Figure 11.1 Excavations made in Type A soil 283
Figure 11.2 Excavations Made in Type B Soil 284
Figure 11.3 Excavations Made in Type C Soil 284
Figure 11.4 Excavations Made in Layered Soils 285
Figure 13.1 Types of Cranes Generally Found in the Workplace 323
Figure 13.2 Sample Inspection Checklist 332
Figure 14.1 A Sample Forklift Safety Program 338
Figure 14.2 Sample Certificate of Training 346
Figure 15.1 Sample Airborne Lead Notification Form 365
Figure 15.2 Results of Airborne Lead Monitoring Form 366
Figure 15.3 Blood Lead Test Form 366
Figure 16.1 Sample Asbestos Medical Questionnaire 375
Figure 17.1 Process Safety Management Standard Requirements and
   Recommended Response 390
Figure 17.2 Process Safety Management Training Certification Document ... 392
Figure 17.3 List of Highly Hazardous Chemicals 393
Figure 17.4 Process Safety Managment Flow Chart 396
Figure 17.5 Examples of a Block Flow Diagram and a Process Flow Diagram ... 397
Figure 19.1 Sample Handout for Contractors and Subcontractors 429
Figure 19.2 Sample Notice to OSHA Inspectors on Policy Regarding
   Warrantless Searches 430
Figure 19.3 Sample Notice of Protest to OSHA Inspectors Who Have a Warrant ... 431

Tables

Table 2.1 Partially Exempt Industries 18
Table 2.2 Newly Covered Industries 19
Table 2.3 Comparison of Old and New Rules 20
Table 2.4 Incidence Rates 42
Table 2.5 Incidence Rates for Construction Machinery Manufacturing 43
Table 2.6 Recordable Work-Related Injuries and Illnesses 53
Table 3.1 Alternating Current—Minimum Distances 117
Table 6.1 Permissible Noise Exposures 184
Table 11.1 Maximum Allowable Slopes for Excavations Less Than 20 Feet 282

Preface

"What I like about construction is working outside and working with my hands. I can drive through town and say, 'Hey, I built that.' But it's a dangerous job. Since 1991, in our Local, at least twelve people have been killed. You walk onto a site and you don't know what you're breathing, and you don't know who's working with what."

Robert Welch

Many of the dangers of construction work are obvious. We're all familiar with safety hazards like unstable scaffolding, falling objects, electric shocks, and fires.

Other dangers, especially chemical hazards, are less obvious. Some are hidden. If you're doing demolition work and breathe in asbestos, you may not notice any effect at the time, but you can develop lung cancer many years later. Other chemicals have both short-term and long-term effects. You may feel sick or dizzy right away when you work with adhesives, paints, or other materials that contain solvents.

The *Construction Safety Handbook* is written to help employers and employees with all the many safety hazards that they deal with from day to day. This book is an effort to place everything that most employers in the construction industry need to know about OSHA in a single volume. It is a reference book, designed to be consulted frequently.

The new Second Edition includes all the latest changes in OSHA standards and includes the following:

- OSHA's new recordkeeping requirements. It covers who must comply with OSHA's recordkeeping rule. It covers which employers are exempt and which employers are newly covered. It has a comparison of the old and new rules and explains how to record an injury or illness under the new recording criteria.
- OSHA's new guidelines on ergonomics. It covers how OSHA will conduct ergonomic inspections.
- OSHA's new signs, signals, and barricades safety standards. It covers OSHA's new traffic control signs.
- The new requirements to the Steel Erection Standard.

The purpose of this book is to give you information you can use to protect yourself. This guide will teach employers and employees how to recognize chemical hazards on the job and how to plan strategies to make the workplace safer.

It is the author's hope that the second edition of this book will continue to help employers prevent countless construction injuries and fatalities while simplifying the OSHA compliance process.

Mark McGuire Moran

About the Author

Mark McGuire Moran is the president of Moran Associates, a management consulting firm that specializes in occupational safety and health. The firm is engaged in keeping its clients abreast of administrative and legislative developments in occupational safety and health and related environmental laws, analyzing their effect upon business and industry, preparing written compliance programs, and providing advice and guidance on the means and methods of obtaining compliance with regulatory requirements. The firms' clients include employers throughout the country engaged in all aspects of business and industry.

Mr. Moran is certified by the U.S. Department of Labor as an accredited instructor in voluntary compliance with OSHA requirements. Mr. Moran also worked on Capitol Hill as a legislative aide to Representative Bill McCollum (R-Florida). He recently wrote the scripts for two videos for the American Subcontractors Association on OSHA's new Fall Protection Standard and Ergonomics. Mr. Moran also represents various national trade associations including National Rifle Association, American Subcontractors Association, National Fire Sprinkler Association, Industrial Fasteners Association, Forging Industry Association, etc. He is the editor of a bi-monthly newsletter entitled *OSHA Casebook*. The newsletter focuses on important trends in the area of Occupational Safety and Health issues.

Mr. Moran authored two 20-page compliance manuals which complement the ASA videos. He is also the author of several books on occupational health and safety topics, including written programs for compliance with customizable computer disks (*OSHA's Electrical Safety and Lockout/Tagout Standards, OSHA's Hazard Communication Standard, OSHA's Process Safety Management Standard*, and *OSHA's Respiratory Protection Standard*); *OSHA Made Easy*, a Van Nostrand Reinhold publication on OSHA recordkeeping, reporting and compliance requirements; *The OSHA Answer Book*, a book that covers OSHA enforcement activities; and *The OSHA 500*, a subscription book that covers OSHA regulations concerning employee training, recordkeeping and written safety programs.

Mr. Moran is also a speaker who makes presentations at conferences sponsored by many trade associations, including the National Safety Council.

1

The Occupational Safety and Health Act

INTRODUCTION

The Occupational Safety and Health Act of 1970 endeavored to provide safe and healthful working conditions for all American workers. The method provided in the Act for accomplishing this goal is to require employers to furnish their employees with hazard-free employment conditions and to authorize stiff penalties for those employers who fail to do so.

The purpose of the Act is "to assure so far as possible every working man and woman in the nation safe and healthful working conditions and to preserve our human resources," a lofty goal with which few can disagree, but the means and methods for achieving that purpose produced sharp divisions of opinion when this legislation was being considered. Because much of this division of opinion continues more than 30 years after the law's enactment, it might be helpful to summarize it briefly.

Essentially, there were two competing bills pending in Congress in the autumn of 1970. One was fashioned on President Nixon's proposal which provided for safety standards to be promulgated by an independent board, inspection for compliance to be the province of the Secretary of Labor, and adjudication of alleged violations to be determined by an independent commission. The other bill had solid support from organized labor. That bill would have reposed all authority in the Secretary of Labor: standard-setting, inspection for compliance, and adjudication of violations. The two versions also contained differences over the authority of Labor Department inspectors to close unsafe work areas or operations, the right of employees to quit unsafe work, and the rights of union representatives to demand inspections and to accompany inspectors on their rounds.

There did not seem to be anyone who opposed "sound" job safety legislation. Everyone wanted a law that would reduce the increasing toll of on-the-job deaths and injuries, the annual figures for which were then estimated to be 14,000 deaths and two million injuries.[1] Nevertheless, in the heated debate over appropriate legislation, which Senator Javits of New York called "the most bitter labor-management fight in years," few thought a compromise could be reached which would enable passage during the 91st Congress. However, in the lame duck session that convened on November 16, 1970, Congressman Steiger led the Administration forces to victory in the House over the Labor-backed Daniels Bill. Senator Javits succeeded in amending the Williams Bill which had strong support from organized labor in order to bring it closer to what the Administration sought. A conference committee then reconciled remaining differences and

President Nixon pronounced himself satisfied with the result.[2] He signed it into law on December 29, 1970 and it took effect on April 28, 1971.

There were also 37 states that had job safety acts, but they suffered from ineffective enforcement and inadequate funding. The existence of these state laws, which varied in their requirements, was also troublesome to employers both because it was more costly to do business in one state than in another and also because the varying requirements was burdensome to those operating across state lines. It was thought that a comprehensive federal law would result in both a uniform standard of job safety and also equality in resulting costs.

COVERAGE OF THE ACT

The OSH Act strives for maximum effectiveness by covering virtually every person and business organization which employs people. Congress intended the coverage of the Act to be as broad as the scope of the commerce clause of the Constitution by declaring it to be applicable to every person "engaged in a business affecting commerce who has employees."[3] The Act does have some exclusions and modifications, however, for businesses regulated under the Atomic Energy Act of 1954,[4] under job safety rules of federal agencies other than the Department of Labor,[5] and for government employees.[6]

The meatpacking industry, railroads, airlines, water transportation employers and a number of others have sought to escape OSH Act coverage under the § 653(b)(1) exclusion because of safety rules adopted by the Department of Agriculture, Department of Transportation, the U.S. Coast Guard and other agencies. They have been unable to completely escape, however, because courts have held that the exclusion was only meant to prevent duplicate regulation and that, as a result, OSHA rules apply to such employers unless one of these other agencies enforces a safety rule regulating the same hazard for which OSHA seeks to cite them.[7]

A somewhat broader view of the § 653(b)(1) exclusion was applied in the case of *Organized Migrants in Community Action v. Brennan*.[8] That decision held that OSHA was precluded from adopting any standard regulating reentry time for fields treated with pesticides because EPA had ample statutory authority to do so.

STATE OSHA PLANS

The law was enacted at a time when decentralization of federal authority was very much in vogue in the nation's capitol, particularly at the White House. Despite the fact that one of the major reasons for its adoption was the many inadequacies with the job safety efforts of the states, the Act encouraged the states to assume full responsibility for OSHA administration and enforcement within their borders. As soon as it was enacted, the new Assistant Secretary of Labor for OSHA devoted considerable time and energy to a major effort to get all fifty states to do so. Each of the states responded affirmatively at the time, but less than half of them actually followed through with the process and obtained the required approval.[9]

OSHA administration and enforcement has the responsibility of the state government in 21 states and two territories: Alaska, Arizona, California, Hawaii, Indiana, Iowa, Kentucky, Maryland, Michigan, Minnesota, Nevada, New Mexico, North Carolina, Oregon, Puerto Rico, South Carolina, Tennessee, Utah, Vermont, Virgin Islands, Virginia, Washington and Wyoming.[10] These are known as "State Plan" jurisdictions and, for all intents and purposes, employers in these states are only subject to their state OSHA agency. Federal OSHA plays virtually no role at all there.

As a general rule, employers prefer state OSHA to federal OSHA while organized labor's views are exactly the opposite. Business people seem to feel that the states will not be as tough on them as the feds. The AFL-CIO, on the other hand, believes that the states provide only the lowest common denominator in job safety and health protection.

In addressing this subject, it bears repeating that one of the Act's principle objectives is to obtain a uniform level of job safety for all employees throughout the nation, and at the same time ensure that the employer's compliance burdens will be equal regardless of where a place of employment happens to be located. This objective does not seem to have been well served by the State Plan system. It was thought at the time this law was being considered that safeguards could be written into the Act to ensure uniformity and equality by including provisions which require that state OSHA standards must be "at least as effective" as the federal standards and by federal OSHA monitoring of the State programs. Provisions of this kind were written into the law[11] but, even though most State Plan states have adopted the OSHA standards verbatim, there is no escape from the fact that the same standard can be interpreted and applied differently from one state to another.

Differences in interpretation of the same text can also occur within a single agency and, indeed, have actually occurred within OSHA itself, but at least with federal enforcement there is some hope that a kind of fuzzy uniformity in applying such a standard may eventually evolve. The federal inspectors and their superiors operate out of one compliance operations manual under the aegis of one central policy-making body and with a single federal court system capable of making a uniform and binding adjudication at some point in the process. That cannot happen when there are a number of different states with OSHA authority because each of them is a separate sovereign entitled to interpret and apply its own law.[12]

The OSHA noise standard will be used to illustrate this phenomenon. It sets specific limits on noise and provides that when those limits are exceeded "feasible administrative or engineering controls shall be utilized" by the employer to reduce the noise to within those limits. "Feasible" has been interpreted to mean that the relative costs of such controls to the employer and their benefits to employees must be weighed and matched against the costs and benefits of attaining the permissible noise limits by personal ear protective devices worn by the affected employees. The State Plan states have adopted the same noise standard but are not bound to interpret the word "feasible" in the same way. A state can force a non-complying employer to install expensive engineering controls if he can afford them even though they would cost 10 thousand times more than ear protective devices and would not reduce noise exposures as much.

An employer who tells a State Plan official that he is in compliance with a particular standard because of a federal OSHA interpretation is certain to be told in reply that the federal interpretation is not binding in that state. That has occurred many times. Such a reply is truthful and legally correct.[13]

This point can be made in another way. A ruling by the highest court in State A as to the meaning of a safety and health standard adopted under the law of State A would apply within State A even if the highest court of State B had ruled in a similar case that an identically worded occupational safety and health standard adopted under the law of State B meant something different. This would also be true even if a federal court had ruled in a similar case that identically worded occupational safety and health standard meant something entirely different when it was in effect under the federal law. Multiply this potential difference in interpretation by ten, 15, 20, or more and you have a problem of Gordian complexity for multi-state employers.

There is no solution to this problem, short of canceling the State Plan authority. The States are regularly coming up with a diversity of interpretations and there is no way that can be avoided.

The State Plan states receive a considerable amount of money from the federal treasury to finance their OSHA programs.[14] In 1987, they received approximately $55 million—about one-fourth of OSHA's $229 million budget for that year. Nevertheless, the states must pay a substantial sum for the privilege of having their own OSHA because it wouldn't cost them a dime if they stayed out of it like the majority of states have done. Although there is no immediate prospect that any of the remaining 20 State Plans will be ended, the financial logic in favor of such a result is quite compelling.

DUTIES OF EMPLOYERS AND EMPLOYEES

Since safety precautions that are considered necessary and feasible at any one time are constantly changing, Congress did not include specific safety or health standards in the Act. It sets forth employer responsibilities only in general terms. The employer must maintain a workplace which is free of recognized hazards likely to cause death or serious physical harm;[15] observe those safety and health standards which may be promulgated by the Secretary of Labor; maintain accurate records of all work-related accidents and diseases; inform employees of their protection and duties under the Act; and permit the inspection of the premises.[16]

Congress felt that because employers are in control of the workplace, they should be made responsible for OSHA compliance therein. As a general rule, that principle has been followed throughout the Act's existence.[17] In view of the well-known maxim that ignorance of the law is no excuse, each employer is obligated to know and observe the OSHA requirements that apply to his or her business, a rather heavy and difficult responsibility. Employers, however, are not responsible for guaranteeing that their employees always observe OSHA requirements, but they must do whatever is reasonable to ensure that—so far as possible—their employees work under conditions that meet the Act's requirements.

The employer's duty to observe OSHA requirements is further limited by the scope of the Act's coverage and intent: the protection of employees from job hazards. For example, an employer has no duty to observe OSHA requirements where the beneficiary of compliance is the general public.

The Act is not a building code. Its purpose is to protect employees from harm, not to force employers to woodenly adhere to physical safety requirements even when no employee harm could result from the failure to do so. Consequently, a violation of the Act generally requires the presence of both these two conditions: (1) noncompliance with an OSHA standard and (2) the endangerment (exposure or access to hazard) of one or more employees as a result of such noncompliance.

In one of the earliest cases decided under the Act, an employer was charged with failure to provide a screen on a scaffold, a requirement of an OSHA construction safety standard designed to protect people beneath the scaffold from getting hit by falling objects that the workers on the scaffold might drop. It was admitted that the standard specifies a screen and that this employer had none, but the hearing produced no evidence that there were any employees working beneath the scaffold. The only people below it were passers-by on the street. Was there a failure to comply with the standard? Yes. A violation of the law? No. Since OSHA standards cannot be construed more broadly than the Act's coverage, the citation in this case was vacated.[18]

Multi-Employer Job Sites

In enacting this law, Congress gave little thought to the unique relationship that arises when employees of a number of different employers work in and around the same job site and are subject to the hazards which may exist at that site—hazards which may or may not have been created by their own employer—or someone else's—or by some other employees of a totally unrelated and unknown employer. The legislators focused their attention only upon the typical industrial or commercial workplace where a single employer has one or more employees engaged in furthering the employer's business purposes. They therefore framed the Act so as to place upon such an employer the duty of providing for the safety and health of his own employees.[19] A number of varying rules have since been applied in OSHA enforcement cases at multi-employer worksites, but each employer's duty at such places is still not completely clear.

In OSHA's early years the rule at multiple-contractor work sites was very simple: You were in violation of the Act if one or more of your employees was exposed to the hazard resulting from noncompliance with an OSHA standard. It did not matter who had created the hazard. If, for example, the carpenter subcontractor had failed to build proper railings on the stairs, any other contractor whose employees used those stairs was in violation of the Act. OSHA frequently issued citations to all employers at the work site when a situation such as this existed. OSHA insisted that it was each contractor's duty to keep his employees away from hazardous conditions even if the only way to accomplish it was to stop the work. Some OSHA inspectors continue to follow that rule.

The rule was successfully challenged in 1975 by a masonry subcontractor who was cited for an absent guardrail because his employees were working at a place where the general contractor had not placed guardrails around the floor openings, as required by an OSHA standard. The masons were clearly exposed to the resulting hazard but the court rules that—regardless of exposure—subcontractors should not be held in violation for non-serious violations which they "neither created nor were responsible for pursuant to their contractual duties." The court reasoned that "in relation to non-serious violations we do not believe that Congress intended to subvert the well-established craft jurisdiction concept or to impose burdensome expenses on subcontractors which do not have the appropriate employees to abate certain hazards."

The court criticized the OSHA rule holding that all employers must keep their employees away from hazards even by shutting down the job if necessary: "To the extent that ... [that rule] will lead to the removal of workers from construction sites because of non-serious violations we find that policy inconsistent with the Act. Correcting the hazard, not shutting down constructions sites is the desired result."

However, calling its determination a "narrow holding," the court specifically noted that" we have not held that the ... policy of imposing liability on employers for exposure to conditions that are serious violations of promulgated standards is invalid. Nor have we held that exposure by a subcontractor's employees to a non-serious standard violation which he created or is otherwise responsible for is impermissible."[20]

OSHA has generally tried to follow that rule even though the rule is by no means without problems. On multi-employer worksites, "citations shall normally be issued only to employers whose employees are exposed to hazards. When these employers have a valid defense to a citation, however, the employer with responsibility for correcting the hazard may be cited instead, in accordance with the policy set forth [herein]."[21] This rule obviously is not crystal clear and OSHA policy is subject to change without notice.

An employer on such a job site who is aware of an OSHA violation created by some other employer would be well advised to think along these lines. "Do I have the power to abate the violation?" If the answer is "no," either because your employees are of the wrong craft, it is beyond your expertise, or you are barred by your building contract from doing so, then you probably have a pretty good defense to any citation for the non-complying condition. Even though such a situation exists, however, you would still be well advised to call the hazardous condition to the attention of the responsible contractor when the hazard is readily apparent to you.

Although it provided no firm rules for responsibility in multi-employer situations, Congress not only recognized that employers and employees have separate but dependent responsibilities for achieving safe and healthful working conditions. It included a requirement in the Act that each employee shall comply with OSHA standards and all rules, regulations, and orders issued pursuant to this Act which are applicable to his or her own actions and conduct. However, no sanctions for failure to do so were adopted. Many people have therefore dismissed section 654 (b) as empty rhetoric, but there are others who believe it has meaning if employers adopt shop rules for their employees to follow.

LABOR–MANAGEMENT RELATIONS

There are a number of areas where OSHA has an impact upon labor-management relations. Unions have been granted a number of rights under the Act itself,[22] under some OSHA standards,[23] and in the procedural rules which apply when an employer contests an OSHA citation.[24]

Despite this official status granted to unions in the Act itself, OSHA takes the position that it should not interfere in labor-management relations. It has thus adopted policies designed to keep OSHA out of strike situations which provide as follows:

Programmed Inspections.[25] As a rule, programmed inspections will be deferred during a strike or labor dispute, either between a recognized union and the employer or between two unions competing for bargaining rights in the establishment.[26]

Unprogrammed Inspections. As a rule, unprogrammed inspections will be performed during strikes or labor disputes. However, the seriousness and reliability of any complaint shall be thoroughly investigated by the supervisor prior to scheduling an inspection to ensure as far as possible that the complaint reflects a good faith belief that a true hazard exists and is not merely an attempt to harass the employer or to gain a bargaining advantage for labor. If there is a picket line at the establishment, the CSHO (OSHA inspector) shall inform the appropriate union official of the reason for the inspection prior to initiating the inspection.[27]

In at least one case, a union petitioned to set aside a noise abatement program the company had implemented in order to comply with an OSHA citation because it allegedly would adversely affect employee job classifications, seniority, and rates of pay. The abatement procedure called for so-called "administrative controls" under which employee noise exposure would be reduced by rotating them in and out of high noise areas so that no single employee would be exposed to noise for a period of time which exceeded that permitted under the OSHA noise standard.[28] In rejecting the union's petition, the OSH Review Commission observed that:

> In this case the petition shows that the [employer] elected to abate the hazard by implementing feasible administrative controls. No showing is made that such method of abatement would not accomplish the desired result within the period for abatement prescribed in the citation. Accordingly, the employer, so far as we know, is doing all that is required of it by the standard and the Act. Under these circumstances we conclude that a decision to grant the relief requested by the petition would be contrary to the congressional purpose to ... assure so far as possible every working man and woman in the nation safe and healthful working conditions...

> Moreover, we could not grant this petition even if abatement has not been achieved. The only reasons given for the requested relief relate to problems of labor-management relations and not to problems of occupational safety and health. Labor relations matters are not within our jurisdiction. Petitioner is undoubtedly aware of the fact that another agency of the Federal Government [the National Labor Relations Board] is statutorily empowered to deal with labor relation matters.

Some OSHA standards provide for removal of certain employees from work areas where OSHA-regulated substances are present and require that their employer pay the removed employees the same wages and benefits they would otherwise have received, a provision which is known as Medical Removal Protection (MRP). An appellate court upheld the validity of such a provision in the OSHA occupational lead exposure standard in the face of industry challenges that it interfered with national labor policy.[29] "We do not doubt that MRP will have a noticeable effect on future collective bargaining," the court conceded, "but such an effect hardly proves that

OSHA has misconstrued the legislative intent [of the National Labor Relations Act]."[30] The court added this explanation:

> When an issue related to earnings protection not wholly covered by OSHA regulation arises between labor and management, it will remain a mandatory subject of collective bargaining. Meanwhile we find nothing in the OSH Act or other labor legislation to suggest that Congress could remove from OSHA the power to create a program instrumental to achieving worker safety simply because such a program could otherwise be vented through collective bargaining."

In view of the many areas where OSHA requirements and labor-management relations interrelate, those companies utilizing a safety department which is organizationally separate from its employee relations department to handle OSHA matters may not be properly positioned to act appropriately in such areas.

WHO IS OBLIGATED TO OBSERVE THE LAW?

If you have one or more employees, you are an employer. If you are an employer, you are obligated to comply with the Occupational Safety and Health Act (OSH Act). There are very few exceptions.

It doesn't just apply to construction sites, or only to hazardous jobs. It applies to all employers. Office work is covered. Retail stores are covered. If you employ anyone to do any kind of work, you must observe the OSH Act.

OSH Act requirements apply even if there are no accidents or injuries. The purpose of the OSH Act is to prevent the first injury. Thousands of employers with perfect safety records have been penalized by OSHA. Other employers who have never been inspected by OSHA have been found liable in civil liability court cases because they had failed to observe an OSHA requirement.

You must observe the OSHA requirements even if you are unaware of any hazardous conditions. It can be compared with the income tax rule that requires you to file a return even if you don't owe any taxes.

Employers are cited even when the OSHA violation is an employee's fault such as failure to wear a hard hat, safety shoes, or goggles. OSHA places the onus on employers.

The law provides that each employer is supposed to make sure that his employees observe all safety requirements.

There is no small business exception. Employers with ten or fewer employees don't have to keep injury/illness records and are exempted from some OSHA inspections, but they must comply with all OSHA safety and health standards and requirements.

There are thousands of OSHA requirements, including the following:

- Adopting written compliance programs
- Electricity rules

- Employee training and qualifications
- Fire protection
- Keeping detailed records
- Machine use, maintenance, and repair
- Posting notices and warnings
- Reporting accidents and illnesses

OSHA violations occur when you don't observe all the requirements that apply. Simply running a safe operation is not enough. And ignorance of the law is no excuse. The penalties for non-compliance are severe. Million dollar fines are not unusual. But that's not all! You could got to jail or be sued for millions even if you have never received an OSHA inspection.

Here's an example:

> William Boss hired Joe Workman to do some painting. Boss didn't know that Workman had a slight asthma condition. Working with the paint severely aggravated Workman's condition. He had to be hospitalized and nearly died. He can never work again. Boss was sued for ten million dollars because he failed to heed the material safety data sheet (MSDS) warnings for the paint which provided that no one should use the paint unless first given a pulmonary function test. The fact that Boss didn't know about the MSDS warnings or Workman's asthma condition was no defense. There is an OSHA requirement that an MSDS must be obtained for all products that contain hazardous chemicals and that each employer must provide training to his employees on the MSDS provisions. Boss hadn't done that.

Everyone can learn a lesson from that example. If you are required to provide protection, give a warning, provide employee training, or use particular compliance methods, and you have not complied because you didn't know about it, the consequences could be severe. If someone is hurt or becomes disabled as a result, you could face criminal charges, a multi-million dollar liability suit, or both. And your defenses may be weak because ignorance of the law is no excuse. Of course, there could also be sanctions imposed by OSHA even if no one was hurt.

In conclusion, it's very important for the employer to understand what the compliance responsibilities are under the OSHA Standards. There are many requirements that employees must understand while doing their jobs and to avoid any legal consequences.

Endnotes:

[1] The provisions of the Act are directed not only at the promotion of job safety, but at occupational health problems as well. The problems in the health field were thought at the time to be at least as widespread as those in the area of safety. The U.S. Public Health Service estimated that there were 390,000 occurrences of occupational diseases each year. Some said this was a conservative estimate. For example, Dr. William J. Nicholson of the Environmental Science Laboratory at Mt. Sinai Hospital in New York, reported a study made of 643 New York and New Jersey asbestos workers over the preceding 30 years. About 60 percent died from asbestos-related diseases, most of them from cancer or respiratory diseases traced to the inhalation of tiny asbestos fibers. (Spencer, "Nader Pushes Job Safety," *The Washington Star*, 15 November 1970).

[2] 7 W.H. *Compilation of Presidential Docs*. 5 (1971). The Conference committee report is H.R. Rep. No. 91-1765, 91st Cong., 2d Sess. (1970).

[3] 29 U.S.C. § 652(5).

[4] 42 U.S.C. § 2021

[5] 29 U.S.C. § 653 (b)(1)

[6] Employees of state and local governments are not covered by the Act but must be covered if a state chooses to avail itself of the opportunity to operate its own OSHA program (29 U.S.C. § 667 (c)(6)). Executive branch employees of the federal government are covered under a program that is somewhat different from that imposed upon the private sector. 29 U.S.C. § 668, Executive Order No. 12196, Feb. 26, 1980; 45 Fed. Reg. 12769; and the implementing rules adopted by various federal agencies.)

[7] See, for example, *In Re-Inspection of Norfolk Dredging Company*, 783 F. 2d 1526 (11th Cir. 1986)(company regulated by Coast Guard is subject to OSHA citation because Coast Guard rules do not cover the condition for which company was cited); *Southern Railway Company v. OSHARC*, 539 F.2d 335 (4th Cir. 1976)(same for railroads); *U.S. Air, Inc. v. OSHRC*, 689 F.2d 1191 (4th Cir. 1982)(airline cannot be cited under OSHA standard where FAA has rule covering the cited condition); and *Secretary of Labor v. Mushroom Transportation Co., Inc.*, 1 BNA OSHC 1390 (1973)(trucking company cannot be cited under OSHA standard where Department of Transportation (DOT) has rule covering the cited condition even though DOT rule doesn't regulate the condition in the same manner or in an equally stringent manner).

[8] 520 F.2d 1161, 1169 (D.C. Cir. 1975).

[9] A state must satisfy the Secretary of Labor that it meets a number of criteria established by the Act before it can assume OSHA authority (Act § 18, 29 U.S.C. § 667). To do so, it enacts its own Occupational Safety and Health Act and submits to the Secretary a "State Plan" for administration and enforcement thereof. When he approves a state's plan, the Secretary causes the terms and conditions thereof to be published in the *Federal Register* (example: 29 CFR § 1952.240 *et seq.* for the Alaska State Plan).

[10] These jurisdictions currently have the authority to inspect private employers, cite them, adjudicate contested enforcement actions and adopt binding occupational safety and health standards and regulations. For the most part, OSHA itself does not do any of this in these places. There are a few states where only state and local government agencies are within the state's OSHA authority. In all of the remaining states and territories, the state government plays no part in OSHA administration or enforcement except to provide consultative services upon request.

[11] OSH Act §§ 18(c) and (f), 29 U.S.C. §§ 667 (c) and (f).

[12] When a state, for example, adopts a federal rule verbatim there then exist two rules—one federal and one state. From that point on, the federal rule is interpreted and applied by federal authorities and the state rule by state authorities. Neither is bound by the interpretation of the other and neither can overrule the other.

[13] This is even true in Michigan despite the fact that the Michigan Occupational Safety and Health Act created a Board of Health and Safety Compliance and Appeals with functions which parallel the federal Occupational Safety and Health Review Commission and that Sec. 46(6) of the Michigan Act provides that: "In construing or applying any state occupational safety or health standard which is identical to a federal occupational safety and health standard... the board shall construe and apply the state standard in a manner which is consistent with any federal construction or application by the occupational safety and health review commission."

[14] Act § 23(g) U.S.C. § 672(g) authorizes OSHA to pay half of the states' costs.

[15] Act § 5(a)(1), 29 U.S.C. § 654(a)(1). This is the so-called "General Duty Clause."

[16] Act § (8)(a), 29 U.S.C. §657(a). The Supreme Court has held that this section of the Act is unconstitutional insofar as it authorizes inspection without a warrant or equivalent process. *Marshall v. Barlow's Inc.*, 436 U.S. 307 (1978).

[17] See, for example, *Atlantic & Gulf Stevedores v. OSAHRC*, 534 F.2d 541, 555 (3d Cir. 1976) ("the entire thrust of the Act is to place primary responsibility upon the employer").

[18] *Secretary v. City Wide Tuckpointing Service Co.*, 1 BNA OSHC 1232 (1973).

[19] The General Duty Clause specifically states that each employer shall furnish a hazard-free workplace to *his* employees. 29 U.S.C. § 654(a)(1). However, the specific duty clause, § 654 (a)(2), which requires observance of OSHA standards, does not include the underscored pronoun and the resulting case law has not been uniform. In some areas of the country, the employer's OSHA compliance duty under § 654(a)(2) is only for his own employees but in other places, courts have interpreted the duty more broadly.

[20] *Anning Johnson Co. v. OSHARC*, 516 F.2d 1081 (7th Cir. 1975).

[21] OSHA FOM, Chapter V, §F.

[22] For example, the right to file a complaint with OSHA that a violation exists and to obtain either an OSHA inspection of the employer or a written explanation if OSHA decides against the inspection. 29 U.S.C. § 657 (f).

[23] The noise standard requires employers in certain situations to monitor noise levels and to provide the employee's representatives with the opportunity to observe the measuring process. 29 CFR § 1910.95(i). Similar provisions are included in a number of other standards.

[24] Unions representing employees who are affected by a contested OSHA citation may elect party status and participate in the resulting litigation. (29 CFR § 2200.20). They may also contest the abatement period included in an OSHA citation on the grounds that too much time has been allowed the employer to correct the alleged violation. (29 U.S.C. § 659(c); 29 CFR §§ 2200.20 and 38).

[25] These are routinely scheduled inspections pursuant to an OSHA plan to check all employers engaged in a particular kind of activity or known to be using certain chemicals or similar substances, as contrasted with inspections made in response to a particular event like an accident or a complaint.

[26] Note the interesting phrasing in this last clause. It apparently excludes the typical case in which a single union is trying to organize a non-union plant.

[27] OSHA FOM, Chapter III.

[28] *Detroit Printing Pressmen, Local No. 13 v. Secretary of Labor*, 1 BNA OSHC 1071 (1972).

[29] The lead standard provides for medical removal of employees with *blood*-lead levels of 50 micrograms per 100 grams of whole blood from those work areas with *air* lead levels of 30 micrograms per cubic meter of air averaged over an 8-hour period. 29 CFR § 1910.1025(k).

[30] *United Steelworkers of America v. Marshall*, 647 F.2d 1189, 1236 (D.C. Cir. 1980).

2

Recordkeeping

Employers Subject to OSHA Recordkeeping

INTRODUCTION

The Occupational Safety and Health Act requires the Secretary of Labor to adopt regulations pertaining to two areas of recordkeeping. First, the Act requires the Secretary to issue regulations requiring employers to "maintain accurate records of, and to make periodic reports on, work-related deaths, injuries and illnesses other than minor injuries requiring only first aid treatment and which do not involve medical treatment, loss of consciousness, restriction of work or motion, or transfer to another job." The OSH Act also authorizes the Secretary of Labor to develop regulations requiring employers to keep and maintain records regarding the causes and prevention of occupational injuries and illnesses.

Second, the Act requires the Secretary to develop and maintain an effective program of collection, compilation, and analysis of occupational safety and health statistics. This also directs the Secretary to "compile accurate statistics on work injuries and illnesses which shall include all disabling, serious, or significant injuries and illnesses, whether or not involving loss of time from work, other than minor injuries requiring only first aid treatment and which do not involve medical treatment, loss of consciousness, restriction of work or motion, or transfer to another job."

After passage of the Act, OSHA issued the required occupational injury and illness recording and reporting regulations as 29 CFR Part 1904. The Bureau of Labor Statistics (BLS) is now responsible for conducting the nationwide statistical compilation of occupational illnesses and injuries (called the Annual Survey of Occupational Injuries and Illnesses), while OSHA administers the regulatory components of the recordkeeping system.

FUNCTIONS OF THE RECORDKEEPING SYSTEM

Occupational injury and illness records have several distinct functions or uses. One use is to provide information to employers whose employees are being injured or made ill by hazards in their workplace. The information in OSHA records makes employers more aware of the kinds of injuries and illnesses occurring in the workplace and the hazards that cause or contribute to them. When employers analyze and review the information in their records, they can identify and correct hazardous workplace conditions on their own. Injury and illness records are also an essential tool to help employers manage their company safety and health programs effectively.

13

Employees who have information about the occupational injuries and illnesses occurring in their workplace are also better informed about the hazards they face. They are therefore more likely to follow safe work practices and to report workplace hazards to their employers. When employees are aware of workplace hazards and participate in the identification and control of those hazards, the overall level of safety and health in the workplace improves.

The records required by the recordkeeping rule are also an important source of information for OSHA. During the initial stages of an inspection, an OSHA representative reviews the injury and illness data for the establishment as an aid to focusing the inspection effort on the safety and health hazards suggested by the injury and illness records. OSHA also uses establishment-specific injury and illness information to help target its intervention efforts on the most danger-ous worksites and the worst safety and health hazards. Injury and illness statistics help OSHA identify the scope of occupational safety and health problems and decide whether regulatory intervention, compliance assistance, or other measures are warranted.

Finally, the injury and illness records required by the OSHA recordkeeping rule are the source of the BLS-generated national statistics on workplace injuries and illnesses, as well as on the source, nature, and type of these injuries and illnesses. To obtain the data to develop national statistics, the BLS and participating state agencies conduct an annual survey of employers in almost all sectors of private industry. The BLS makes the aggregate survey results available both for research purposes and for public information. The BLS has published occupational safety and health statistics since 1971 which chart the magnitude and nature of the occupational injury and illness problem across the country. Congress, OSHA, and safety and health policy makers in federal, state and local governments use the BLS statistics to make decisions con-cerning safety and health legislation, programs, and standards. Employers and employees use them to compare their own injury and illness experience with the performance of other estab-lishments within their industry and in other industries.

The Occupational Safety and Health Administration (OSHA) is revising its rule addressing the recording and reporting of occupational injuries and illnesses (29 CFR parts 1904 and 1952), including the forms employers use to record those injuries and illnesses. The revisions to the final rule will produce more useful injury and illness records, collect better information about the incidence of occupational injuries and illnesses on a national basis, promote improved em-ployee awareness and involvement in the recording and reporting of job-related injuries and illnesses, simplify the injury and illness recordkeeping system for employers, and permit in-creased use of computers and telecommunications technology for OSHA recordkeeping pur-poses.

The final rule also revises 29 CFR 1952.4, Injury and Illness Recording and Reporting Require-ments, which prescribes the recordkeeping and reporting requirements for states that have an occupational safety and health program. This final rule became effective January 1, 2002.

WHAT ARE THE REASONS FOR OSHA RECORDKEEPING?

OSHA standards are binding regulations that employers must observe in order to maintain a safe and healthful workplace. They are intended to be the "heart" of the OSH Act. By the same token, a systematic recordkeeping program would be the "brains" of the Act. Congress recognized that the government must first know the particulars of the job injury and illness problem before it could begin to solve it. Records would be developed to provide that knowledge.

The Act granted OSHA full authority to do whatever was necessary in order to compile the needed information. OSHA can require employers to "make, keep and preserve" whatever records it may deem "necessary or appropriate for the enforcement of this Act or for developing information regarding the causes and prevention of occupational accidents and illnesses." In addition, Congress directed the Secretary of Labor to "develop and maintain an effective program of collection, compilation, and analysis of occupational safety and health statistics" and authorized money, private research grants, and cooperative programs with the states in order to do so.

The Act directs OSHA to prescribe regulations requiring employers to maintain accurate records of work-related deaths, injuries, and illnesses that involve loss of consciousness, restrictions of work or motion, transfer to another job, or medical treatment, except where such treatment is for minor injuries requiring only first aid. OSHA's implementing regulations appear in 29 CFR Part 1904. Both the relevant statutory requirement and OSHA's recordkeeping requirements are phrased so generally that additional guidance is essential if uniformity is ever to be achieved and employers are to know what should and should not be recorded.

In fulfilling the employer's obligation to maintain a log and summary of all "recordable occupational injuries and illnesses," OSHA recommends that employers utilize OSHA Form No. 300 and, when preparing the supplementary information, the OSHA Form No. 301. However, any document that provides the same information in an equally comprehensible manner will suffice, including worker compensation reports and insurance reports. Employers are obligated to make log and summary entries on form No. 300 within seven working days of receiving notice of the injury or illness in question. A similar time limit applies for the completion of the OSHA 301 supplementary record forms.

Only "recordable" injuries and illnesses have to be entered on the OSHA 300 or its equivalent. That term is defined as an occupational injury or illness which results in one or more of the following:

- A fatality
- One or more lost workdays
- Transfer to another job
- Termination of employment
- Medical treatment other than first aid
- Loss of consciousness
- Restrictions on performance of normal job functions
- Restriction of motion
- A diagnosed occupational illness which is reported to the employer

OSHA RECORDKEEPING REQUIREMENTS

The OSH Act itself provides that to be recordable, an injury or illness must be "work-related." A case is considered work related if an event in the work environment either causes or contributes to an injury or illness or aggravates a pre-existing condition. The work environment is primarily composed of the employer's premises and other locations where employees are engaged in work-related activities or are present as a condition of their employment. A work relationship is presumed when an employee is on the premises of the employer. That presumption may be rebutted by evidence that symptoms are exhibited which merely surfaced while the individual was on the premises and were not caused by any on-site experience. Work relationship must be proved when the employee is off premises.

Employers should initially determine whether in fact a recordable instance has occurred at all, that is, whether there was a death, an illness, or an injury. If the answer is "yes," the employer must then decide if the case occurred in the work environment. Upon a determination that it has, the employer must then determine if the case is an injury or an illness. If the employer concludes that the incident was an illness, he then must record it in the portion of the OSHA Form No. 300 devoted to illness reports. If the employer decides that the incident was a recordable injury, he must record it in the injury portion of the OSHA Form No. 300.

The types of decisions employers make in the recordkeeping process include the following:

1. The employer must distinguish between employees and other workers on site. Employers must record the recordable injuries and illnesses of all employees on their payroll. Employers must also record injuries and illnesses that occur to employees who are not on their payroll if they supervise these employees on a day-to-day basis (i.e., workers from a temporary help service that are supervised day-to-day by the employer).
2. An injury or illness is considered work-related if an event or exposure in the work environment caused or contributed to the condition or significantly aggravated a preexisting condition. Work-relatedness is presumed for injuries and illnesses resulting from events or exposures occurring in the workplace, unless an exception specifically applies. The work environment includes the establishment and other locations where one or more employees are working or are present as a condition of their employment.

WHO IS COVERED BY THE RECORDKEEPING RULE?

All employers covered by the Occupational Safety and Health Act of 1970 must report to OSHA any workplace incident resulting in a fatality or the in-patient hospitalization of three or more employees within eight hours. If you had ten or fewer employees during all of the last calendar year or your business is classified in a specific low-hazard retail, service, finance, insurance, or real estate industry, you do not have to keep injury and illness records unless the Bureau of Labor Statistics or OSHA informs you in writing that you must do so.

EMPLOYERS EXEMPT FROM RECORDING INJURIES AND ILLNESSES

1. **OSHA and BLS Surveys**. All employers who receive the OSHA annual survey form or the BLS Survey of Occupational Injuries and Illnesses Form are required to complete and return the survey forms. This requirement also applies to those establishments under the small establishment exemption and the low-hazard industry exemption.

2. **Small Employer Exemption.** The recordkeeping regulations have exempted from the regular recordkeeping requirements employers that had ten or fewer employees at all times during the last calendar year. The new rule continues this small employer exemption. Still follow record keeping

3. **Low-Hazard Industry Exemption.** OSHA has exempted some low-hazard industries from maintaining injury and illness records on a regular basis. The new rule updates the old rule's listing of partially exempted low-hazard industries, which are those Standard Industrial Classification (SIC) code industries within SICs 52–89 that have an average Days Away, Restricted, or Transferred (DART) rate at or below 75 percent of the national average DART rate. The new rule continues this low-hazard industry exemption.

See Table 2.1 below for the list of Partially Exempt Industries and Table 2.2 for a list of Newly Covered Industries.

OSHA uses the Standard Industrial Classification (SIC) Code to determine which establishments must keep records.

Table 2.1 Partially Exempt Industries[1]

SIC Code	Industry Description	SIC Code	Industry Description
592	Liquor stores	802	Offices and clinics of dentists
594	Miscellaneous shopping goods stores	803	Offices and clinics of doctors of osteopathy
599	Retail stores, not elsewhere classified	804	Offices and clinics of other health practitioners
60	Depository institutions	807	Medical and dental laboratories
61	Non-depository credit institutions	809	Miscellaneous health and allied services, not elsewhere classified
62	Security and commodity brokers, dealers, exchanges, and services	81	Legal services
63	Insurance carriers	82	Educational services
64	Insurance agents, brokers, and services	832	Individual and family social services
653	Real estate agents and managers	835	Child day care services
654	Title abstract offices	839	Social services, not elsewhere classified
67	Holding and other investment offices	841	Museums and art galleries
722	Photographic studios, portrait	86	Membership organizations
723	Beauty shops	87	Engineering, accounting, research, management, and related services
724	Barber shops	899	Services, not elsewhere classified
525	Hardware stores	792	Theatrical producers (except motion picture), bands, orchestras, and
542	Meat and fish (seafood) markets, including freezer	793	Bowling centers
544	Candy, nut, confectionary	801	Offices and clinics of medical doctors
545	Dairy products stores	802	Offices and clinics of dentists
546	Retail bakeries	803	Offices and clinics of doctors of osteopathy
549	Miscellaneous food stores	804	Offices and clinics of other health practitioners
764	Reupholstery furniture	807	Medical and dental laboratories
791	Dance studios, schools, and halls	809	Miscellaneous health and allied services, not elsewhere classified

Table 2.2 Newly Covered Industries[2]

SIC Code	Industry Description	SIC Code	Industry Description
553	Auto and home supply stores	651	Real estate operators (except developers) and lessors
555	Boat dealers	655	Land subdividers and developers
556	Recreational vehicle dealers	721	Laundry, cleaning, and garment services
559	Automotive dealers, not elsewhere classified	734	Services to dwellings and other buildings
571	Home furniture and furnishing stores	735	Miscellaneous equipment rental and leasing
572	Household appliance	736	Personnel supply services
593	Used merchandise stores	833	Job training and vocational rehabilitation services
596	Non-store retailers	836	Residential care
598	Fuel dealers	842	Arboreta and botanical or zoological gardens

COMPARISON OF THE OLD AND NEW RULES

Some of the specific changes in the new rule include (a) changes in coverage; (b) the OSHA Forms; (c) the Recording Criteria in determination of work-relationship, elimination of different recording criteria for injuries and illnesses, days away and job restriction/ transfer, definition of medical treatment and first aid, recording of needlestick and sharps injuries, and recording of tuberculosis; (d) change in ownership; (e) employee involvement; (f) privacy protections; and (g) computerized and centralized records.

This listing is not comprehensive of an employer's obligations under OSHA's recordkeeping rule. Please see 29 CFR Part 1904 for the details concerning all recordkeeping obligations.

Table 2.3 Comparison of Old and New Rules[3]

Old Rule	New Rule
Recordkeeping Forms §1904.29	
OSHA 200—Log and Summary OSHA 101—Supplemental Record	OSHA 300—Log OSHA 300A—Summary OSHA 301—Incident Report
Work-Related §1904.5	
Any aggravation of a pre-existing condition by a workplace event or exposure makes the case work-related	**Significant** aggravation of a pre-existing condition by a workplace event or exposure makes the case work-related
Exceptions to presumption of work relationship: 1) Member of the general public 2) Symptoms arising on premises totally due to outside factors 3) Parking lot/Recreational facility	Exceptions to presumption of work relationship: 1) Member of the general public 2) Symptoms arising on premises totally due to outside factors 3) Voluntary participation in wellness program 4) Eating, drinking, and preparing one's own food 5) Personal tasks outside working hours 6) Personal grooming, self-medication, self infliction 7) Motor vehicle accident in parking lot/access road during commute 8) Cold or flu 9) Mental illness unless employee voluntarily presents a medical opinion stating that the employee has a metal illness that is work-related.
2 doses prescription medicine—Medical Treatment (MT) Any dosage of OTC medicine—First Aid (FA) 2 or more hot/cold treatments—MT Drilling a nail—MT Butterfly bandage/Steri-Strip—MT	1 dose prescription medicine—MT OTC medicine at prescription strength—MT Any number of hot/cold treatments—FA Drilling a nail—FA Butterfly bandage/Steri-Strip—FA
Non-minor injuries recordable: 1) fractures 2) 2nd and 3rd degree burns	Significant diagnosed injury or illness recordable: 1) fracture 2) punctured ear drum 3) cancer 4) chronic irreversible disease

Table 2.3 Comparison of Old and New Rules, continued

Specific Disorders	
Hearing loss—Federal enforcement for 25dB shift in hearing from original baseline	Hearing loss—From 1/1/02 until 12/31/02, record shift in hearing averaging 25dB or more from the employee's original baseline
Needlesticks and 'sharps injuries'—Record only if case results in med treatment, days away, days restricted or sero-conversion	Needlesticks and 'sharps injuries'—Record all needlesticks and injuries that result from sharps potentially contaminated with another person's blood or other potentially infectious material
Medical removal under provisions of other OSHA standards—all medical removal cases recordable	Medical removal under provisions of other OSHA standards—all medical removal cases recordable
TB—Positive skin test recordable when known workplace exposure to active TB disease. Presumption of work relationship in five industries.	TB—Positive skin test recordable when known workplace exposure to active TB disease. No presumption of work relationship in any industry.

Other Recordkeeping Issues	
Must enter the employees name on all cases	Must enter 'Privacy Cases' rather than the employee's name, and keep a separate list of the case number and corresponding names
Access—employee access to entire log, including names; No access to supplementary form (OSHA 101)	Access—employee and authorized representative access to entire log, including names; Employee access to individual's Incident Report (OSHA 301); Authorized Representative access to portion of all OSHA 301s
Fatality reporting—Report all work-related fatalities to OSHA	Fatality reporting—do not need to report fatalities resulting from motor vehicle accident on public street or highway that do not occur in construction zone
Certification—the employer or the employee who supervised the preparation of the Log and Summary can certify the annual summary	Certification—company executive must certify annual summary
Posting—post annual summary during month of February	Posting—post annual summary from February 1 to April 30
No such requirement	You must inform each employee how he or she is to report an injury or illness

QUESTIONS ABOUT THE NEW RULE

Which Employers Are Covered by the New Rule?

The rule requires employers to keep records of occupational deaths, injuries, and illnesses, and to make certain reports to OSHA and the Bureau of Labor Statistics. Smaller employers (with ten or fewer workers) and employers who have establishments in certain retail, service, finance, real estate or insurance industries are not required to keep these records. However, they must report any occupational fatalities or catastrophes that occur in their establishments to OSHA, and they must participate in government surveys if they are asked to do so.

What Are the New Forms for Recording Injuries and Illnesses?

Employers who operate establishments that are required by the rule to keep injury and illness records are required to complete three forms: the OSHA 300 Log of Work-Related Injuries and Illnesses, the annual OSHA 300A Summary of Work-Related Injuries and Illnesses, and the OSHA 301 Injury and Illness Incident Report. Employers are required to keep separate 300 Logs for each establishment that they operate which is expected to be in operation for one year or longer.

The log must include injuries and illnesses to employees on the employer's payroll as well as injuries and illnesses of other employees the employer supervises on a day-to-day basis, such as temporary workers or contractor employees who are subject to daily supervision by the employer. Within seven calendar days of the time the fatality, injury, or illness occurred, the employer must enter any case that is a work-related new case and which meets one or more of the recording criteria in the rule on the Log and Form 301.

Is the Injury or Illness Work-Related?

The employer must consider an injury or illness to be work-related if an event or exposure in the work environment either caused or contributed to the resulting condition. Work-relatedness is presumed for injuries and illnesses resulting from events or exposures occurring in the work environment. A case is presumed work-related if, and only if, an event or exposure in the work environment is a discernable cause of the injury or illness or of a significant aggravation to a pre-existing condition. The work event or exposure need only be one of the discernable causes; it need not be the sole or predominant cause.

A case is not recordable if it "involves signs or symptoms that surface at work but result solely from a non-work-related event or exposure that occurs outside of the work environment." Regardless of where signs or symptoms surface, a case is recordable only if a work event or exposure is a discernable cause of the injury or illness or of a significant aggravation to a pre-existing condition.

If it is not obvious whether the precipitating event or exposure occurred in the work environment or elsewhere, the employer "must evaluate the employee's work duties and environment to decide whether or not one or more events or exposures in the work environment caused or contributed to the resulting condition or significantly aggravated a pre-existing condition." This means that the employer must make a determination whether it is more likely than not that work events or exposures were a cause of the injury or illness or of a significant aggravation to a pre-existing condition. If the employer decides the case is not work-related and OSHA subsequently issues a citation for failure to record, the government would have the burden of proving that the injury or illness was work-related.

Is the Injury or Illness a New Case?

Only new cases are recordable. Work-related injuries and illnesses are considered to be new cases when the employee has never reported similar signs or symptoms before, or when the employee has recovered completely from a previous injury or illness and workplace events or exposures have caused the signs or symptoms to reappear.

What Is the General Recording Criteria?

Employers must record new work-related injuries and illnesses that meet one or more of the general recording criteria or meet the recording criteria for specific types of conditions. Recordable work-related injuries and illnesses are those that result in one or more of the following:

1. Death
2. Days away from work
3. Restricted work
4. Transfer to another job
5. Medical treatment beyond first aid
6. Loss of consciousness
7. Diagnosis of a significant injury or illness

Employers must classify each case on the 300 Log in accordance with the most serious outcome associated with the case. The outcomes listed on the form are death, days away, restricted work/ transfer, and "other recordable." For cases resulting in days away or in a work restriction or transfer of the employee, the employer must count the number of calendar days involved and enter that total on the form. The employer may stop counting when the total number of days away, restricted or transferred, reaches 180.

What Is Restricted Work?

An employee's work is considered restricted when, as a result of a work-related injury or illness, (a) the employer keeps the employee from performing one or more of the routine functions of his or her job (job functions that the employee regularly performs at least once per week) or from working the full workday that he or she would otherwise have been scheduled to work, or (b) a physician or other licensed health care professional recommends that the employee not perform one or more of the routine functions of his or her job or not work the full workday that he or she would otherwise have been scheduled to worked.

The new rule continues the policy established under the old rule that a case is not recordable as a restricted work case if the employee experiences minor musculoskeletal discomfort, a health care professional determines that the employee is fully able to perform all of his or her routine job functions, and the employer assigns a work restriction to that employee for the purpose of preventing a more serious condition from developing.

What Is Medical Treatment?

Medical treatment means any treatment not contained in the list of first aid treatments. Medical treatment does not include visits to a healthcare professional for observation and counseling or diagnostic procedures. First aid means only those treatments specifically listed in § 1904.7. Examples of first aid include the use of non-prescription medications at non-prescription strength, the application of hot or cold therapy, eye patches or finger guards, and others.

Diagnosis of a Significant Injury or Illness

A work-related cancer, chronic irreversible disease such as silicosis or byssinosis, punctured eardrum, or fractured or cracked bone is a significant injury or illness that must be recorded when diagnosed by a physician or a licensed health care professional.

Recording Injuries and Illnesses to Soft Tissues

Work-related injuries and illnesses involving muscles, nerves, tendons, ligaments, joints, cartilage, and spinal discs are recordable under the same requirements applicable to any other type of injury or illness. There are no special rules for recording these cases. If the case is work-related and involves medical treatment, days away, job transfer, or restricted work, it is recordable.

Employee Privacy

The employer must protect the privacy of injured or ill employees when recording cases. In certain types of cases, such as those involving mental illness or sexual assault, the employer may not enter the injured or ill employee's name on the Log. Instead, the employer simply enters "privacy case," and keeps a separate, confidential list containing the identifying information. If the employer provides the OSHA records to anyone who is not entitled to access to the records under the rule, the names of all injured and ill employees generally must be removed before the records are turned over.

Certification, Summarization, and Posting

At each facility after the end of the year, employers must review the Log to verify its accuracy, summarize the 300 Log information in an OSHA 300-A form, and certify the summary (a company executive must sign the certification). This information must then be posted for three months, from February 1 to April 30. The employer must keep the records for five years following the calendar year covered by them, and if the employer sells the business, he or she must transfer the records to the new owner.

The summary must be posted in a manner that is visible and accessible for all employees. Employee or safety bulletin boards are the most common places for posting the annual summary.

Employee Involvement

Each employer must establish a way for employees to report work-related injuries and illnesses, and each employee must be informed about how he or she is to report an injury or illness. Employees, former employees, and employee representatives also have a right to access the records, and an employer must provide copies of certain records upon request.

Reporting Fatalities and Multiple Hospitalizations

Within eight hours of a work-related fatality or an incident involving the hospitalization of three or more employees, the employer must orally report to the nearest OSHA office or the OSHA Hotline at 1-800-321-OSHA. There is an exception for certain motor vehicle or public transportation accidents. An employer also must participate in an OSHA or BLS injury and illness survey if he or she receives a survey form from OSHA or the BLS.

The report is made to the nearest office of the Area Director of the Occupational Safety and Health Administration, U.S. Department of Labor, unless the state in which the accident occurred is administering an approved plan. The report must contain three pieces of information:

- The circumstances surrounding the incident
- The number of fatalities
- The extent of any injuries

Clarification of Recordkeeping Citation Policy in the Construction Industry

You need to keep OSHA injury and illness records for short-term establishments (i.e., establishments that will exist for less than a year on construction projects).

You must also keep a separate OSHA 300 Log for each such establishment. You may keep one OSHA 300 Log that covers all of your short-term establishments. You may also include the short-term establishments' recordable injuries and illnesses on an OSHA 300 Log that covers short-term establishments for individual company divisions or geographic regions.

You may keep the records for an establishment at your headquarters or other central location if you can transmit information about the injuries and illnesses from the establishment to the central location within seven (7) calendar days of receiving information that a recordable injury or illness has occurred.

OSHA RECORDKEEPING FORMS

What Recordkeeping Forms Are Required by OSHA?

The Occupational Safety and Health (OSH) Act of 1970 requires employers to prepare and maintain records of work-related injuries and illnesses.

The new injury/illness recordkeeping requirements apply to all employers with 11 or more employees, except for some employers engaged in certain retail trades and in certain service industries.

How Often Are the Forms Updated and/or Maintained?

Recordkeeping forms are kept on a calendar-year basis and must be maintained at the workplace for five years and available for inspection by OSHA or the designated state agency.

Three forms must be maintained to comply with OSHA Recordkeeping requirements:

1. The Log of Work-Related Injuries and Illnesses (Form 300) is used to classify work-related injuries and illnesses and to note the extent and severity of each case. When an incident occurs, use the form to record specific details about what happened and how it happened. (See Figure 2.1.)
2. The Summary is a separate form (Form 300-A) which shows the totals for the year in each category. At the end of the year, post this in a visible location so that your employees are aware of the injuries and illnesses occurring in their workplace. (See Figure 2.2.) There is a worksheet (see Figure 2.3) to help you fill out the summary.
3. The OSHA Form 301 is the Injury and Illness incident report. It includes more information about how the injury or illness occurred. (See Figure 2.4.)

Employers must keep a separate OSHA 300 Log for each establishment or site. If you have more than one establishment, you must keep a separate log for each physical location that is expected to be in operation for one year or longer.

Employees have the right to review your injury and illness records. Cases listed on the Log of Work-Related Injuries and Illnesses, the OSHA 300 form, are not necessarily eligible for workers' compensation or other insurance benefits. Listing a case on the log does not mean that the employer or worker was at fault or that an OSHA standard was violated.

What Is the OSHA Form 300 (Log of Work-Related Injuries and Illnesses)?

The OSHA 300 log is used for recording and classifying recordable occupational injuries and illnesses and for noting the extent and outcome of each case. The log shows when the occupational injury or illness occurred, to whom, what the injured or ill person's regular job was at the time of the injury or illness exposure, the kind of injury or illness, how much time was lost, and whether the case resulted in a fatality, etc.

OSHA's Form 300

Log of Work-Related Injuries and Illnesses

Attention: This form contains information relating to employee health and must be used in a manner that protects the confidentiality of employees to the extent possible while the information is being used for occupational safety and health purposes.

Year 20___

U.S. Department of Labor
Occupational Safety and Health Administration

Form approved OMB no. 1218-0176

You must record information about every work-related death and about every work-related injury or illness that involves loss of consciousness, restricted work activity or job transfer, days away from work, or medical treatment beyond first aid. You must also record significant work-related injuries and illnesses that are diagnosed by a physician or licensed health care professional. You must also record work-related injuries and illnesses that meet any of the specific recording criteria listed in 29 CFR Part 1904.8 through 1904.12. Feel free to use two lines for a single case if you need to. You must complete an Injury and Illness Incident Report (OSHA Form 301) or equivalent form for each injury or illness recorded on this form. If you 're not sure whether a case is recordable, call your local OSHA office for help.

Establishment name _____

City _____ State _____

Identify the person

(A) Case no.	(B) Employee's name	(C) Job title (e.g., Welder)

Describe the case

(D) Date of injury or onset of illness	(E) Where the event occurred (e.g., Loading dock north end)	(F) Describe injury or illness, parts of body affected, and object/substance that directly injured or made person ill (e.g., Second degree burns on right forearm from acetylene torch)
___/___ month/day		

Classify the case

Using these four categories, check ONLY the most serious result for each case:

Death (G)	Days away from work (H)	Remained at work — Job transfer or restriction (I)	Remained at work — Other recordable cases (J)

Enter the number of days the injured or ill worker was:

On job transfer or restriction (K)	Away from work (L)
___ days	___ days

Check the "injury" column or choose one type of illness: (M)

Injury (1)	Skin disorder (2)	Respiratory condition (3)	Poisoning (4)	All other illnesses (5)

Page totals ▶

Be sure to transfer these totals to the Summary page (Form 300A) before you post it.

Injury (1)	Skin disorder (2)	Respiratory condition (3)	Poisoning (4)	All other illnesses (5)

Public reporting burden for this collection of information is estimated to average 14 minutes per response, including time to review the instructions, search and gather the data needed, and complete and review the collection of information. Persons are not required to respond to the collection of information unless it displays a currently valid OMB control number. If you have any comments about these estimates or any other aspects of this data collection, contact: US Department of Labor, OSHA Office of Statistics, Room N-3644, 200 Constitution Avenue, NW, Washington, DC 20210. Do not send the completed forms to this office.

Page ___ of ___

Figure 2.1 OSHA 300 Log

This form, which replaces the old Form 200 and can now be printed on legal size paper as opposed to its former large format. This is where employers are required to record work-related deaths, injuries and illnesses, and days away from work. Employers are also forbidden to enter employees' names on this form for some sensitive illness or injuries including sexual assaults, HIV infections, and mental illness.

Part 1904 of the Code of Federal Regulations (CFR) provides the basic requirements for the Log and Summary of Occupational Injuries and Illnesses. It requires the following:

> Each employer shall maintain in each establishment a log and summary of all recordable occupational injuries and illnesses for that establishment; and enter each recordable injury and illness on the log and summary as early as practicable but no later than seven working days after receiving information that a recordable injury or illness has occurred.

> OSHA Form No. 300 or an equivalent that is as readable and comprehensible to a person not familiar with, it shall be used. The log and summary shall be completed in the detail provided in the form and instructions on form OSHA No. 300. Any employer may maintain the log of occupational injuries and illnesses at a place other than the establishment or by means of data-processing equipment, or both, if:

> (1) There is available at the place where the log is maintained sufficient information to complete the log to date within seven working days after receiving information that a recordable case has occurred, and

> (2) At each of the employer's establishments, there is available a copy of the log which reflects separately the injury and illness experience of that establishment complete and current to a date within 45 calendar days.

The log consists of three parts: (1) A descriptive section which identifies the employee and briefly describes the injury or illness, (2) a section classifying the case and extent of the injuries recorded, and (3) a section on the type and extent of illnesses.

While most of the columns are self-explanatory, there are some important requirements to be considered when completing the log. The following discussion pertains to the descriptive sections of the log.

What Do We Need to Do if an Injury/Illness Occurs?

1. Within seven calendar days after you receive information about a case, decide if the case is recordable under the OSHA recordkeeping requirements.
2. Determine whether the incident is a new case or a recurrence of an existing one.
3. Establish whether the case was work-related.
4. If the case is recordable, decide which form you will fill out as the injury and illness incident report. You may use OSHA's 301: Injury and Illness Incident Report or an equivalent form. Some state workers' compensation, insurance, or other reports may

be acceptable substitutes, as long as they provide the same information as the OSHA 301.

How Should the OSHA 300 Form Be Filled Out?

Step One: Identify the employee

In Columns A-C enter a unique case number, the employee's name, and the job title.

Column A—Enter a number that is unique for each case. This is very important because each case must be identified and examined separately. The simplest method of numbering may be the best; i.e., 1, 2, 3. Employers may also number cases by month; for example, 7-15 would indicate the 15th case occurring during July.

Column B—Insert one of two entries: (1) First name, middle initial, and last name; or (2) first initial, middle initial, and last name.

Column C—Specify the injured or ill employee's regular job title even if the employee was working outside his or her regularly assigned occupation at the time of the injury or illness exposure.

Step Two: Describe the Case

In Columns D-F provide the date of the injury or illness. Describe where the event occurred and describe in more detail what parts of the body were affected.

Column D—Enter the date of the work accident that resulted in injury. For occupational illnesses, enter date of initial diagnosis of illness, or, if absence from work occurred before diagnosis, enter the first day of absence attributable to the illness that was later diagnosed or detected. Cases do not necessarily fall consecutively by date because injuries and illnesses are recorded as an employer learns that a case has occurred.

Column E—State the location where the event took place.

Column F—Briefly describe the nature of the injury or illness, part(s) of the body affected, and object/substance that directly injured or made person ill. For example, "amputation, finger" is not sufficiently detailed. A correct entry would be "amputation: second joint, forefinger, left hand." This tells which hand, which finger, and to what degree. The examples listed in the heading for column F on the log form are good indications of how entries should be made.

Step Three: Classify the Case

Using the four categories (G)-(J), check only the most serious result for each case. You must classify the case as either death, days away from work, job transfer or restriction, or other recordable cases. Then in column (K) or (L) you must enter the number of days the worker was injured or ill. In the last column, (M), you must then check the injury column or choose one type of illness.

Column G—Check death if an occupational injury results in a fatality. (In some cases, an employee may be injured but not die until several weeks or months later.) If the death occurs within five years following the end of the calendar year in which the injury occurred and the injury was work related, the entry on the log must be changed to reflect a fatality; the entries in columns 2 through 5 must be lined out.

Column H—If a case involves days away from work due to an injury, check this column. Days away from work should not include the day of the injury. The new rule makes employers count calendar days rather than workdays, capping their count at 180 days.

Column I—If the employee remained at work, then you must check either job transfer or restriction.

Column J—Check if the injury involves other recordable cases. These cases consist of relatively less serious injuries and illnesses which satisfy the criteria for recordability but which do not result in death or require the affected employee to have days away from work or days of restricted work activity beyond the date of injury or illness.

Column K—Enter the number of days the injured or ill worker was on job transfer/ restriction. Restricted work/job transfer activity occurs when, as the result of a work-related injury or illness, an employer or healthcare professional recommends keeping an employee from doing the routine functions of his or her job or from working the full workday that the employer would have been scheduled to work before the injury or illness occurred. Begin counting days from the day after the incident occurs. Stop counting restricted days once the total reaches 180 days.

Column L—Enter the number of calendar days the employee was away from work as a result of the recordable injury or illness. Do not count the days on which the injury or illness occurred in the number. Begin counting days from the day after the incident occurs. Stop counting days away from work once the total reaches 180 days.

Column M (1)—Injury—If the injury is recordable. A check must be entered to indicate that it was an injury. An injury is defined as any wound or damage to the body resulting from an event in the work environment.

Some examples would be a cut; puncture; laceration; abrasion; fracture; bruise; contusion; chipped tooth; amputation; insect bite; electrocution; or a thermal, chemical, electrical, or radiation burn. Sprain and strain injuries to muscles, joints, and connective tissues are classified as injuries when they result from a slip, trip, fall, or other similar accidents.

Column M (2)—Illness; Skin disorders—For occupational illnesses, an entry should be placed in one of the columns M2 through M5 if an illness is recordable. A check must be entered to indicate that it was an illness depending upon which column is applicable. It is important to keep in mind that illnesses must be evaluated individually. For example, heatstroke would require a check in column M5.

Skin diseases or disorders are illnesses involving the worker's skin that are caused by work exposure to chemicals, plants, or other substances.

Some examples include contact dermatitis, eczema, or rash caused by primary irritants and sensitizers or poisonous plants; oil acne; friction blisters; chrome ulcers; inflammation of the skin.

Column M (3)—Illness; Respiratory conditions—Respiratory conditions are illnesses associated with breathing hazardous biological agents, chrome ulcers; or inflammation of the skin.

Some examples would be: Silicosis, asbestosis, pneumonitis, pharyngitis, rhinitis, or acute congestion; farmer's lung, beryllium disease, tuberculosis, occupational asthma, reactive airways dysfunction syndrome (RADS), chronic obstructive pulmonary disease (COPD), hypersensitivity pneumonitis, toxic inhalation injury such as metal fume fever, chronic obstructive bronchitis, and other pneumoconiosis.

Column M (4)—Illness; Poisoning—Poisoning includes disorders evidenced by abnormal concentrations of toxic substances in blood, other tissues, other bodily fluids, or the breath that are caused by the ingestion or absorption of toxic substances into the body.

Some examples include poisoning by lead, mercury, cadmium, arsenic, or other metals; poisoning by carbon monoxide, hydrogen sulfide, or other gases; poisoning by benzene, bezol, carbon tetrachloride, or other organic solvents; poisoning by insecticide sprays such as parathion or lead arsenate; poisoning by other chemicals such as formaldehyde.

Column M (5)—All other Illnesses—This is a category of miscellaneous illnesses.

Some examples include heatstroke, sunstroke, heat exhaustion, heat stress and other effects of environmental heat; freezing, frostbite, and other effects of exposure to low temperatures; decompression sickness; effects of ionizing radiation (isotopes, x-rays, radium); effects of non-ionizing radiation (welding flash, ultra-violet rays, lasers); anthrax; bloodborne pathogenic diseases such as AIDS, HIV, hepatitis B, hepatitis C; brucellosis; malignant or benign tumors; histoplasmosis; coccidioidomycosis.

Step Four: Maintenance of the OSHA 300 Log

In addition to keeping the log on a calendar-year basis, employers are required to update this form to reflect changes that occur in recorded cases after the end of the calendar year. Maintenance of the log is different from the retention of records. Although all OSHA injury and illness records must be retained, only the log must be maintained by the employer.

If, during the five-year retention period, there is a change in the extent or outcome of an injury or illness which affects an entry on a previous year's log, then the first entry should be lined out and a corrected entry made on that log. Also, new entries should be made for previously unrecorded cases that are discovered or for cases that initially weren't recorded but were found to be recordable after the end of the year in which the case occurred.

The entire entry should be lined out for recorded cases that are later found non-recordable. Log totals should also be modified to reflect these changes. If a change in the ownership of an establishment occurs, the new owner must retain the previous owner's OSHA records for five years, but he or she is not responsible for maintaining the previous owner's log.

What Is the OSHA 300-A (Summary of Work-Related Injuries and Illnesses)?

The employer is responsible for preparing an annual summary of injuries and illnesses that occurred during the calendar year. The annual summary, OSHA form 300A, displays the totals from columns G through M of OSHA 300 Log. The summary also displays the calendar year covered, company name and address, annual average number of employees, and total hours worked by all employees covered by the OSHA 300 Log.

Form 300A is a separate form and does not display any personal information, as shown on OSHA 300 Log. Form 300A also makes it easier to calculate incident rates.

The annual summary must be

- Posted by February 1 and remain posted until April 30th
- Posted in areas where other notices are normally placed
- Certified (signed) by a company executive, stating that the information is correct and complete to the best of the employer's ability
- Retained for five years following the calendar year to which they relate

If no cases are recorded during a reporting period, a summary must still be posted. Zeros should be entered into all spaces provided on form 300A.

How Should the OSHA 300-A Summary Form Be Filled Out?

It should be filled out as follows:

Step One: Establishment Information

Your establishment name—Enter your company's name, street address, city, state, and zip code.

Industry description—Describe what type of product your company produces. For example, we manufacture automobile tires.

Standard Industrial Classification Code (SIC), if known—Enter your 4-digit SIC code. For example, general building contractors, single family houses would be SIC code 1521.

Employment Information—At the end of the year, OSHA requires you to enter the average number of employees and the total hours worked by your employees on the summary. See the worksheet (Figure 2.3) to help fill in the summary.

Sign Here—An executive of the company is required to certify to the best of their knowledge that the information on the summary is true, accurate, and complete. You must write the company executive's name, title, the phone number, and date.

OSHA's Form 300A

Summary of Work-Related Injuries and Illnesses

All establishments covered by Part 1904 must complete this Summary page, even if no work-related injuries or illnesses occurred during the year. Remember to review the Log to verify that the entries are complete and accurate before completing this summary.

Using the Log, count the individual entries you made for each category. Then write the totals below, making sure you've added the entries from every page of the Log. If you had no cases, write 0.

Employees, former employees, and their representatives have the right to review the OSHA Form 300 in its entirety. They also have limited access to the OSHA Form 301 or its equivalent. See 29 CFR Part 1904.35, in OSHA's recordkeeping rule, for further details on the access provisions for these forms.

Number of Cases

Total number of deaths	Total number of cases with days away from work	Total number of cases with job transfer or restriction	Total number of other recordable cases
____ (G)	____ (H)	____ (I)	____ (J)

Number of Days

Total number of days of job transfer or restriction	Total number of days away from work
____ (K)	____ (L)

Injury and Illness Types

Total number of . . .
(M)

(1) Injuries	____	(4) Poisonings	____
(2) Skin disorders	____	(5) All other illnesses	____
(3) Respiratory conditions	____		

Establishment information

Your establishment name _____

Street _____

City _____ State _____ ZIP _____

Industry description (e.g., Manufacture of motor truck trailers) _____

Standard Industrial Classification (SIC), if known (e.g., SIC 3715) __ __ __ __

Employment information *(If you don't have these figures, see the Worksheet on the back of this page to estimate.)*

Annual average number of employees _____

Total hours worked by all employees last year _____

Sign here

Knowingly falsifying this document may result in a fine.

I certify that I have examined this document and that to the best of my knowledge the entries are true, accurate, and complete.

Company executive _____ Title _____

(___) _____ _____
Phone   Date

Post this Summary page from February 1 to April 30 of the year following the year covered by the form.

Figure 2.2 OSHA 300A Form

Optional

Worksheet to Help You Fill Out the Summary

At the end of the year, OSHA requires you to enter the average number of employees and the total hours worked by your employees on the summary. If you don't have these figures, you can use the information on this page to estimate the numbers you will need to enter on the Summary page at the end of the year.

How to figure the average number of employees who worked for your establishment during the year:

❶ Add the total number of employees your establishment paid in all pay periods during the year. Include all employees: full-time, part-time, temporary, seasonal, salaried, and hourly.

The number of employees paid in all pay periods = ❶ _____

❷ Count the number of pay periods your establishment had during the year. Be sure to include any pay periods when you had no employees.

The number of pay periods during the year = ❷ _____

❸ Divide the number of employees by the number of pay periods.

$$\frac{❶ \underline{\qquad}}{❷} = ❸ \underline{\qquad}$$

❹ Round the answer to the next highest whole number. Write the rounded number in the blank marked *Annual average number of employees.*

The number rounded = ❹ _____

For example, Acme Construction figured its average employment this way:

For pay period...	Acme paid this number of employees...		
1	10		
2	0		
3	15		
4	30	Number of employees paid = 830	❶
5	40	Number of pay periods = 26	❷
...	...		
24	20	$\frac{830}{26} = 31.92$	❸
25	15		
26	+10	31.92 rounds to 32	❹
	830		

32 is the annual average number of employees

How to figure the total hours worked by all employees:

Include hours worked by salaried, hourly, part-time and seasonal workers, as well as hours worked by other workers subject to day to day supervision by your establishment (e.g., temporary help services workers).

Do not include vacation, sick leave, holidays, or any other non-work time, even if employees were paid for it. If your establishment keeps records of only the hours paid or if you have employees who are not paid by the hour, please estimate the hours that the employees actually worked.

If this number isn't available, you can use this optional worksheet to estimate it.

Optional Worksheet

Find the number of full-time employees in your establishment for the year.

_____

Multiply by the number of work hours for a full-time employee in a year.

X _____

This is the number of full-time hours worked.

_____

Add the number of any overtime hours as well as the hours worked by other employees (part-time, temporary, seasonal)

+ _____

Round the answer to the next highest whole number. Write the rounded number in the blank marked *Total hours worked by all employees last year.*

_____

Figure 2.3 OSHA worksheet

Step Two: Total Number of Cases

Enter the total number of deaths in line (G). This information can be transferred from the OSHA 300 Log to this sheet.

Enter the total number of cases with days away from work in line (H). This information can be transferred from the OSHA 300 Log to this sheet.

Enter the total number of cases with job transfer or restriction in line (I). This information can be transferred from the OSHA 300 Log to this sheet.

Enter the total number of other recordable cases in Line (J). This information can be transferred from the OSHA 300 Log to this sheet.

Step Three: Total Number of Days

Enter the total number of days of job transfer or restriction in Line (K). The information can be transferred from the OSHA 300 Log to this sheet.

Enter the total number of days away from work in Line (L).

Step Four: Total Number of Injury and Illness Types

Injuries—Enter the total number of injuries in line (M) (1). This information can be transferred from the OSHA 300 Log to this sheet. An injury is any wound or damage to the body resulting from an event in the work environment. Examples of injuries include abrasion; fracture; bruise; contusion; chipped tooth; amputation; insect bite; electrocution; or a thermal, chemical, electrical, or radiation burn. Sprain and strain injuries to muscles, joints, and connective tissues are classified as injuries when they result from a slip, trip, fall, or other similar accidents.

Skin disorders—Illness—Enter the total number of skin disorders in line (M) (2). This information can be transferred from the OSHA 300 Log to this sheet. Skin diseases or disorders are illnesses involving the worker's skin that are caused by work exposure to chemicals, plants, or other substances. Examples of skin disorders include: Contact dermatitis, eczema, or rash caused by primary irritants and sensitizers or poisonous plant; oil acne, friction blisters, chrome ulcers; inflammations of the skin.

Respiratory conditions—Illness—Enter the total number of respiratory conditions in line (M) (3). This information can be transferred from the OSHA 300 Log to this sheet. Respiratory conditions are illnesses associated with breathing hazardous biological agents, chemicals, dust, gases, vapors, or fumes at work. Examples of respiratory conditions include: pharyngitis, rhinitis or acute congestion; farmer's lung, beryllium disease, tuberculosis, occupational asthma, reactive airways dysfunction syndrome (RADS), chromic obstructive pulmonary disease (COPD), hypersensitivity pneumonitis, toxic inhalation injury, such as metal fume fever, chronic obstructive bronchitis, and other pneumonconioses.

Poisonings—Illness—Enter the total number of poisonings in line (M) (4). This information can be transferred from the OSHA 300 Log to this sheet. Poisoning includes disorders evidenced by abnormal concentration of toxic substances in blood, other tissues, other bodily fluids, or the breath that are caused by the ingestion or absorption of toxic substances into the body. Examples of poisoning include: poisoning by lead, mercury, cadmium, arsenic, or other metals; poisoning by carbon monoxide, hydrogen sulfide, or other gases; poisoning by benzene, benzol, carbon tetrachloride, or other organic solvents; poisoning by insecticide sprays, such as parathion or lead arsenate; poisoning by other chemicals, such as formaldehyde.

All other illnesses—Enter the total number of illnesses in line (M) (5). This information can be transferred from the OSHA 300 Log to this sheet. Examples of illnesses include: heatstroke, sunstroke, heat exhaustion, heat stress and other effects of environmental heat; freezing, frostbite, and other effects of exposure to low temperatures; decompression sickness; effects of ionizing radiation (isotopes, x-rays, radium); effects of non-ionizing radiation (welding flash, ultra-violet rays, lasers); anthrax; bloodborne pathongenic diseases, such as AIDS, HIV, hepatitis B or hepatitis C; brucellosis; malignant or benign tumors; histoplasmosis; coccidioidomycosis.

What Is the OSHA 301 (Injury and Illness Incident Report)?

For every injury or illness entered on the OSHA 300 Log, it is necessary to record additional information on the Injury and Illness Incident Report, OSHA No. 301. If an injury or illness is recordable, a supplementary form (e.g., OSHA 301) must be completed. This new form provides more information about the case and is printed on 8 ½" by 11" paper. The Incident Report describes the events leading up to the injury or illness, body parts affected, and object(s) or substance(s) involved, etc.

The OSHA 301 form is only suggested. A different form may be used if it contains the same information as the OSHA 301 form. Other suitable forms are state workers' compensation reports, insurance claim reports, or the employer's accident report form.

The following is a checklist to use for maintaining supplementary records:

- Record cases within seven calendar days of receiving information that a recordable case has occurred.
- Keep the OSHA 301 form current within 45 days at any given time.
- Each establishment must maintain an OSHA 301 form or similar form.
- Retain records for five years following the calendar year to which they relate.

How Should the OSHA 301 Form Be Filled Out?

The following information is general instructions for completing the OSHA 301 Form. The Injury and Illness Incident Report is one of the first forms you must fill out when a recordable work-related injury or illness has occurred. This form along with the OSHA 300 Form will help

OSHA's Form 301
Injury and Illness Incident Report

U.S. Department of Labor
Occupational Safety and Health Administration

Form approved OMB no. 1218-0176

Attention: This form contains information relating to employee health and must be used in a manner that protects the confidentiality of employees to the extent possible while the information is being used for occupational safety and health purposes.

This *Injury and Illness Incident Report* is one of the first forms you must fill out when a recordable work-related injury or illness has occurred. Together with the *Log of Work-Related Injuries and Illnesses* and the accompanying *Summary*, these forms help the employer and OSHA develop a picture of the extent and severity of work-related incidents.

Within 7 calendar days after you receive information that a recordable work-related injury or illness has occurred, you must fill out this form or an equivalent. Some state workers' compensation, insurance, or other reports may be acceptable substitutes. To be considered an equivalent form, any substitute must contain all the information asked for on this form.

According to Public Law 91-596 and 29 CFR 1904, OSHA's recordkeeping rule, you must keep this form on file for 5 years following the year to which it pertains.

If you need additional copies of this form, you may photocopy and use as many as you need.

Completed by _____

Title _____

Phone (_____) _____ - _____   Date ___/___/___

Information about the employee

1) Full name _____

2) Street _____

City _____ State _____ ZIP _____

3) Date of birth ___/___/___

4) Date hired ___/___/___

5) ☐ Male
   ☐ Female

Information about the physician or other health care professional

6) Name of physician or other health care professional _____

7) If treatment was given away from the worksite, where was it given?

Facility _____

Street _____

City _____ State _____ ZIP _____

8) Was employee treated in an emergency room?
☐ Yes
☐ No

9) Was employee hospitalized overnight as an in-patient?
☐ Yes
☐ No

Information about the case

10) Case number from the Log _____ *(Transfer the case number from the Log after you record the case.)*

11) Date of injury or illness ___/___/___

12) Time employee began work _____ AM / PM

13) Time of event _____ AM / PM ☐ Check if time cannot be determined

14) **What was the employee doing just before the incident occurred?** Describe the activity, as well as the tools, equipment, or material the employee was using. Be specific. *Examples:* "climbing a ladder while carrying roofing materials"; "spraying chlorine from hand sprayer"; "daily computer key-entry."

15) **What happened?** Tell us how the injury occurred. *Examples:* "When ladder slipped on wet floor, worker fell 20 feet"; "Worker was sprayed with chlorine when gasket broke during replacement"; "Worker developed soreness in wrist over time."

16) **What was the injury or illness?** Tell us the part of the body that was affected and how it was affected; be more specific than "hurt," "pain," or sore." *Examples:* "strained back"; "chemical burn, hand"; "carpal tunnel syndrome."

17) **What object or substance directly harmed the employee?** *Examples:* "concrete floor"; "chlorine"; "radial arm saw." *If this question does not apply to the incident, leave it blank.*

18) **If the employee died, when did death occur?** Date of death ___/___/___

Public reporting burden for this collection of information is estimated to average 22 minutes per response, including time for reviewing instructions, searching existing data sources, gathering and maintaining the data needed, and completing and reviewing the collection of information. Persons are not required to respond to the collection of information unless it displays a current valid OMB control number. If you have any comments about this estimate or any other aspects of this data collection, including suggestions for reducing this burden, contact: US Department of Labor, OSHA Office of Statistics, Room N-3644, 200 Constitution Avenue, NW, Washington, DC 20210. Do not send the completed forms to this office.

Figure 2.4 OSHA 301 Form

employers develop a picture of the extent and severity of work-related incidents in your workplace.

Within seven calendar days after you receive information that a recordable injury or illness has happened, fill out this form or an equivalent one. Most state workers' compensation reports may be acceptable substitutes. To be considered an equivalent form, any substitute must contain all the information asked on this form. The form should be filled out the following way.

Step One: Name of Person Completing the Form

>   Enter the name of the person who completed the form; title; phone number; and date of completion.

Step Two: Information about the Employee

1. Enter the employee's full name.
2. Enter the full street address, city, state, and zip code of the employee.
3. Enter the date of birth of the employee.
4. Enter the date the employee was hired.
5. Check whether the employee is male or female.

Step Three: Information about the Physician or other Healthcare Professional

6. Enter the name of the physician or other healthcare professional who rendered medical treatment.
7. If the employee received treatment away from the worksite, indicate the name of the healthcare facility, the address, city, state, and zip code.
8. If the employee was treated at a hospital, check whether he/she was admitted in an emergency room.
9. If the employee was treated at a hospital, check whether he/she was admitted as an inpatient.

Step Four: Information about the Case

10. Each OSHA Form 301 must contain the unique case number relating it to the corresponding case number on the OSHA Form 300 Log of Work-Related Injuries and Illnesses. You should transfer the case number from the OSHA 300 Log after you record the case.
11. Enter the date of injury or illness (month/day/year). For injuries, enter the date the employee was injured. For illnesses, enter the date the illness was recognized. If there is not a definite date of recognition, enter the date of the clinical diagnosis.
12. Enter the time the employee arrived at his/her initial worksite. Indicate A.M. or P.M.
13. Enter the time of the event which resulted in the injury or illness. Indicate A.M. or P.M. Check the box if time cannot be determined.
14. Describe the activity the employee was engaged in before the incident occurred. Describe the activity as well as the tools or materials the employee was using. For example, the employee was climbing a ladder while carrying roofing materials.

15. Describe how the injury occurred. For example, the employee was on the edge of the scaffold to inspect the work and he lost balance and fell six feet to the floor.
16. Describe how the injury or illness occurred. Describe the part of the body that was affected and how it was affected. For example, the employee was on the edge of the scaffold to inspect the work; he lost balance and fell six feet to the floor. The employee's right wrist was broken in the fall.
17. Describe what object or substances directly hampered the employee. For example, the employee stepped back to inspect the work and slipped on some scrap metal. As the employee lost balance he fell against the hot metal.
18. If the employee died, enter the date of death. Enter the month/day/year.

MAINTAINING YOUR RECORDS

Retention of Records

The regulations require that the log and summary, OSHA No. 300, and the supplementary record, OSHA No. 301, must be retained in each establishment for five calendar years following the end of the year to which they relate.

Establishments Required to Maintain Records

Injury and illness records must be kept for each establishment covered by the Occupational Safety and Health Act. The regulations require that records be located and maintained to assist government agencies in administering and enforcing the act, to increase employer-employee awareness, and to promote injury and illness prevention.

The establishment is the basic organizational unit in the private sector among different firms and operations in different industries. It is the focus of the requirements for records location and maintenance, since the regulations require that records be kept at the establishment level.

Therefore, it is imperative that everyone involved in the recordkeeping process have a clear understanding of what constitutes an establishment for recordkeeping purposes.

Section 1904.46 of the regulations provides a concise definition of the term "establishment." An establishment is defined as a single physical location where business is conducted or where services or industrial operations are performed. The examples provided include a factory, mill, store, hotel, restaurant, movie theater, farm, ranch, sales office, warehouse, or central administrative office.

The regulations specify that distinctly separate activities performed at single physical location (for example, where contract construction activities are operated from the same physical location as a lumber yard) shall each be treated as a separate establishment for recordkeeping purposes.

Location of Records

Employers must keep a separate OSHA 300 Log for each establishment that is expected to be in operation for one year or longer. The records for all of an employer's establishments can be kept at a headquarters or other central location if information on injuries and illnesses can be transferred from the establishment to the central location within seven calendar days and if records at the central location can be sent to establishments within acceptable time frames.

Employees Associated with Fixed Establishments

Records for these employees should be located as follows:

- Records for employees working at fixed locations, such as factories, stores, restaurants, warehouses, etc., should be kept at the work location.
- Records for employees who report to a fixed location but work elsewhere should be kept at the place to which the employees report each day. These employees are generally engaged in activities such as agriculture, construction, and transportation.
- Records for employees whose payroll or personnel records are maintained at a fixed location but who do not report or work at a single establishment should be maintained at the base from which they are paid or the base of their firm's personnel operations. This category includes generally unsupervised employees such as salespeople, technicians, or engineers.

Employees "Not" Associated with Fixed Establishments

Some employees are subject to common supervision but do not report or work at a fixed establishment on a regular basis. These employees are engaged in physically dispersed activities that occur in construction, installation, repair, or service operations. Records for these employees should be located as follows:

- Records may be kept at the field office or mobile base of operations.
- Records may also be kept at an established central location. If the records are kept centrally, the address and telephone number of the place where the records are kept must be available at the worksite and there must be someone available at the central location during normal business hours to provide information from the records.

Access to the Records

All OSHA records which are being kept for the five-year retention period must be available for inspection and copying by authorized federal and state government officials.

Employees, former employees, and their representatives must be provided access to the current log (OSHA 300 form) and all the logs being retained for the five-year period at the facility in which he/she currently works or previously worked.

INJURY AND ILLNESS INCIDENCE RATES

What Is an Incidence Rate?

An incidence rate is the number of recordable injuries and illnesses occurring among a given number of full-time workers (usually 100 full-time workers) over a given period of time (usually one year). To evaluate your firm's injury and illness experience over time or to compare your firm's experience with that of your industry as a whole, you need to compute your incidence rate. Because a specific number of workers and a specific period of time are involved, these rates can help you identify problems in your workplace and/or progress you may have made in preventing work-related injuries and illnesses.

How Do You Calculate an Incidence Rate?

You can compute an occupational injury and illness incidence rate for all recordable cases or for cases that involved days away from work for your firm quickly and easily. The formula requires that you follow instructions in paragraph (a) below for the total recordable cases or those in paragraph (b) for cases that involved days away from work, *and* for both rates the instructions in paragraph (c).

(a) *To find the total number of recordable injuries and illnesses that occurred during the year*

Count the number of line entries on your OSHA Form 300, or refer to the OSHA Form 300A and sum the entries for columns (G), (H), (I),and (J).

(b) *To find the number of injuries and illnesses that involved days away from work*

Count the number of line entries on your OSHA Form 300 that received a check mark in column (H), or refer to the entry for column (H) on the OSHA Form 300A.

(c) *To find the number of hours all employees actually worked during the year*

Refer to OSHA Form 300A and optional number worksheet to calculate this.

You can compute the incidence rate for all recordable cases of injuries and illnesses using the following formula:

Total number of injuries and illnesses ÷ Number of hours worked by all employees X 200,000 hours = Total recordable case rate.

The 200,000 figure in the formula represents the number of hours 100 employees working 40 hours per week, 50 weeks per year would work and provides the standard base for calculating incidence rates.

Table 2.4 Incidence Rates

2000 incidence rates for construction machinery manufacturing:	Total recordable cases of injuries and illnesses	Lost workday cases of injuries and illnesses
All workforce sizes	12.6	5.7
Firms with 50 to 249 employees	11.2	5.8
ABC Company (200 workers)	7.5	3.5

How Are Incidence Rates Used?

Incidence rates take on more meaning for an employer when the injury and illness experience of his or her firm is compared with that of other employers doing similar work with workforces of similar size. Information available from BLS permits detailed comparisons by industry and size of firm.

The following tables illustrate how detailed comparisons can help a firm evaluate its safety and health experience more precisely.

In this example, the injury and illness rates for ABC Company are below the industry wide and similar-size averages for construction machinery manufacturing.

Information available from BLS goes beyond giving the average incidence rate for a particular industry and employment-size class. Data show how individual establishment rates within an industry-size combination are distributed.

Points on these rate arrays, called the first quartile, median, and third quartile, help answer the following question: What proportion of comparable employers have rates that are lower than (or higher than) my firm's rates? The following table (Table 2.5) for construction machinery manufacturing firms employing 50 to 249 workers illustrates how these statistical measures work.

Table 2.5 Incidence Rates for Construction Machinery Manufacturing

2000 incidence rates for construction machinery manufacturing:	Total recordable cases of injuries and illnesses	Lost workday cases of injuries and illnesses
Average (mean) for all establishments	11.2	5.8
First quartile—One-fourth establishments had a rate lower than or equal to	6.5	4.6
Median—One-half of the establishments had a rate lower than or equal to	11.6	5.3
Third quartile—Three-fourths of the establishments had a rate lower than or equal to	14.6	7.0

When ABC Company extends its rate comparison to these measures, the company finds that its total recordable rate (7.5) falls between the corresponding first quartile and median rates for construction machinery manufacturers of similar size, and its lost workdays case rate (3.5) falls below the first first quartile rate. In other words, both of ABC Company's rates are lower than the rates for at least one-half of the medium-size construction machinery manufacturers. The total recordable case rate is higher than the rate for at least one-fourth of those construction machinery manufacturers of comparable size while the lost workday case rate puts the company within the first quartile of these establishments. This analysis reinforces earlier findings that ABC Company has a lower incidence rate of injury and illness in its workplace than does most other construction machinery manufacturers of its size.

Reports by the Bureau of Labor Statistics

1. How can I compare my firm's injury and illness experience to others?

The Bureau of Labor Statistics annually reports on the number of workplace injuries, illnesses, and fatalities. Such information is useful for identifying industries with high rates or large numbers of injuries, illnesses, and fatalities both nationwide and separately for those states participating in this program.

Incidence rates by industry, by establishment size, and for many different case types are available from the Bureau of Labor Statistics (BLS). Using incidence rates allows a firm to evaluate its injury and illness experience and compare its experience to other firms doing the same type of work and of the same employment size group.

2. What information does the BLS survey data provide about workers who are injured?

The age, sex, occupation, race, and length of service with employer are the attributes of the worker collected for days away from work cases. For the nation and for participating states, distributions of days away from work cases by the various categories comprising each worker characteristic can be developed. From those distributions, important worker groups can be identified and separate injury and illness profiles developed. For example, separate profiles for women, older workers, and nursing occupations can be developed.

One analytical approach to identifying relatively hazardous jobs will be to compare a job's share of total employment to its share of total days away from work cases. This employment-injury comparison also can be useful at the state level, although usually at a higher level of occupational aggregation. The Bureau's annual bulletin, *Geographic Profiles*, provides, for example, figures on women employed in farming, forestry, and fishery occupations, which can be compared to OSH state data for the same workers.

3. What information does the BLS survey data provide about the injuries that have occurred?

Physical condition (nature), part of the body affected, source, and event/exposure will be the principal case characteristics gleaned from employers' descriptions about the circumstances surrounding the incidents. The principal case characteristics and their categories can be presented in separate tabulations for the nation and for participating states.

Frequency distributions and incidence rates for most case characteristic categories can be generated. These incidence rates tell us, for example, how frequently disabling falls occur in the construction industry of various states. With this information, a state with a relatively high rate of such falls might devote more resources to the study of how employers and employees are dealing with this particular hazard and offer advice on working under adverse weather conditions or the use of safety gear.

4. What information does the BLS survey data provide about injury and illness severity?

The number of workdays lost and restricted is collected for individual cases involving days away from work. The severity of various worker and case characteristic categories, such as back injuries and wrist/hand disorders, can be compared. The survey is unique in collecting information on workdays lost and restricted for individual cases. This improves on the BLS annual survey's collection of this information for individual establishments (summary data). State workers' compensation reports do not contain this type of information on recuperation time.

Measures of recuperation time include the median number of lost workdays for various worker and case characteristics categories and the distribution of workdays lost and restricted for the aforementioned categories. The median helps to identify groups of workers and types of incidents associated with relatively long recuperation periods; the distribution helps focus in on those cases involving comparatively large numbers of workdays lost and restricted, such as 30 workdays or more. Both measures are useful to employers and others in setting their priorities for preventing injuries and illnesses.

5. Who uses the incident rates data?

Employers and employees, policymakers, safety standards writers, safety inspectors, health and safety consultants, and researchers are some of the most frequent users of this data.

Employers and employees need definitive statistics on what kinds of serious injuries and illnesses occur to others whose work and workforce size are similar to theirs. BLS safety and health data permit employers to learn about the circumstances surrounding those incidents so that they can disarm potential hazards where they work.

Policymakers need to know how the safety and health of workers in their state compares to workers in other states doing comparable work. The survey helps these managers determine the additional need for state safety and health programs.

Safety standards writers need to know the factors surrounding injuries and illnesses that the standards were meant to prevent. Do those standards need revision or just better enforcement? Are new standards needed for uncovered incidents? The survey supplemented by special studies can help answer important questions of this type.

Safety inspectors need to know how best to allocate their time among and within establishments. By targeting where injuries and illnesses most frequently occur and their characteristics, survey data help in selecting which firms to visit and what hazards to look for. These visits are also opportunities for inspectors and employers to consult on ways to eliminate work hazards.

Safety and health consultants need to understand job hazards fully to develop effective training packages and educational materials for employers and their employees. The survey collects information on work activity that will help consultants piece together what precipitated an accident or exposure. Special studies of work hazards can provide additional assistance.

Researchers need to direct their limited resources at widespread problems, such as the proper manual lifting techniques and the best designs for tools and safety gear. They find survey data useful in focusing on those work hazards.

DEFINITION OF TERMS

The following definitions will help you decide if the injury or illness should be recorded:

Death

An injury or illness that results in death must be reported by entering a check mark on the OSHA 300 Log in the space for fatal cases. Employers must also report work-related fatalities to OSHA within eight hours.

Days Away From Work

An employee is not considered to have a day away from work until the day following an illness or injury. In other words, the employer is not to count the day of the injury or illness as a day away, but to begin counting days on the following day.

Employers are required to record work-related injuries and illnesses that result in days away from work and count the total number of days away associated with a given case. The employer must record an injury or illness that involves one or more days away from work by placing a check mark on the OSHA 300 Log in the space reserved for days away from work in the column reserved for that purpose. The employer must also record the number of calendar days that the employee was unable to work as a result of the injury or illness, regardless of whether or not the employee was scheduled to work on those days. Weekend days, holidays, vacation days, or other days off are included in the total number of days recorded if the employee would not have been able to work on those days because of a work-related injury or illness. The employer may "cap" the total days away at 180 calendar days.

Restricted Work or Transfer to Another Job

Restricted work occurs when the employer keeps the employee from performing one or more of the routine functions of his or her normal job, or from working the full workday. Restricted work also occurs when a physician or other health care professional recommends that the employee not perform one or more of the routine functions of his or her job, or not works the full workday. The employer must record the number of calendar days that the employee was restricted or transferred.

The final rule allows the employer to stop counting days if the employee's job has been permanently modified in a way that eliminates the routing functions the employee has been restricted from performing. For example, a permanent modification would include reassigning an employee with a respiratory allergy to a job where such allergens are not present or adding a mechanical assist to a job that formerly required manual lifting.

Medical Treatment and First Aid

Distinguishing between medical treatment and first aid can cause most employers a headache that itself may necessitate "medical treatment." Generally speaking, one-time visits to a medical facility or subsequent visits requiring mere observation by medical personnel are not considered "medical treatment." However, the boundary between what constitutes first aid and what constitutes medical treatment has finally been simplified.

The new rule defines "first aid" as a specific list of treatments. Anything not on the list qualifies as medical treatment. This makes it much easier to decide how to record a injury or illness. Under the new rule, treatment that qualifies as first aid on a first visit to medical personnel continues to qualify as first aid.

Under the system adapted in the final rule, once the employer has decided that a particular response to a work-related illness or injury is, in fact, treatment, he or she can simply turn to the first aid list to determine without elaborate analysis whether the treatment is first aid and recordable.

The final rule has adopted a single list of 14 first aid treatments. Any treatment not included on this list is not considered first aid for OSHA recordkeeping purposes. OSHA considers the listed treatments to be first aid regardless of the professional qualifications of the person providing the treatment or exam. Even when these treatments are provided by a physician, nurse, or other healthcare professional, they are considered first aid from OSHA recordkeeping purposes.

Medical Versus First Aid Treatment

One of the most confusing aspects of recordkeeping is determining if an injury or illness is recordable based upon first aid or medical treatment. The revised standard sets new definitions of medical treatment and first aid to simplify recording decisions.

Medical Treatment is defined as

The management and care of a patient for the purpose of combating disease or disorder. For example:

- Administering immunizations, such as Hepatitis B or rabies (does not include tetanus)
- Using wound closing devices, such as sutures or staples
- Using rigid means of support to immobilize parts of the body
- Physical therapy or chiropractic treatment

Medical Treatment does not include

- Visits to a physician or other licensed health care professional solely for observation or counseling
- The conduct of diagnostic procedures, such as X-rays and blood tests, including the administration of prescription medications used solely for diagnostic purposes

First Aid is defined as

- Dispensing a nonprescription medication at nonprescription strength
- Providing tetanus immunizations
- Cleaning, flushing, or soaking wounds on the surface of the skin
- Applying wound coverings, such as bandages, Band-Aids ® , gauze pads, etc.
- Applying hot or cold therapy
- Applying any non-rigid means of support, such as elastic bandages, wraps, non-rigid back belts, etc.
- Applying temporary immobilization devices while transporting an accident victim (e.g. splints, slings, neck collars, back boards, etc.)
- Drilling of a fingernail or toenail to relieve pressure, or draining fluid from a blister
- Applying eye patches
- Removing foreign bodies from the eye using only irrigation or a cotton swab
- Removing splinters or foreign material from areas other than the eye by irrigation, tweezers, cotton swabs, or other simple means
- Applying finger guards
- Administering massage
- Providing oral fluids for relief of heat stress

This is a complete list of all treatments defined as first aid under the revised standard.

Loss of Consciousness

The term is defined as any time a worker becomes unconscious as a result of a workplace exposure to chemicals, heat, or an oxygen deficient environment; a blow to the head; or some other workplace hazard that causes loss of consciousness. The employer must record such cases.

Unconsciousness means a complete loss of consciousness and not a sense of disorientation, "feeling whoozy," or a diminished level of awareness. If the employee is unconscious for any length of time, no matter how brief, the case must be recorded on the OSHA 300 Log.

Significant Work-Related Injuries and Illnesses

Employers must also record work-related injuries and illnesses that are significant or meet any of the following additional criteria.

Significant Criteria

Significant Criteria is defined as any work-related injury or illness that is diagnosed by a physician or other licensed health care professional. Record any work-related case involving cancer, chronic irreversible disease, a fractured or cracked bone, or a punctured eardrum. The employer must record such cases within seven days of diagnosis.

Additional Criteria

Additional criteria include the following conditions when they are work-related:

- Any needlestick injury or cut from a sharp object that is contaminated with another person's blood or other potentially infectious material.
- Any case requiring an employee to be medically removed under the requirements of an OSHA health standard.
- Tuberculosis infection as evidenced by a positive skin test or diagnosis by a physician or other licensed healthcare professional after exposure to a known case of active tuberculosis.

Recording Criteria for Cases Involving Medical Removal

Cases involving medical removal require the employer to record the case on the OSHA 300 Log if an employee is medically removed under the medical surveillance requirements of an OSHA standard. Currently the medical surveillance requirements of the standards have medical removal requirements for the following:

1. **Benzene**—General industry standard (§1910.1028(i); Shipyard standard (§1915.1028); and Construction standard (§1926.1128)
2. **Cadmium**—General industry standard (§1910.1027(l)); Shipyard standard (§1915.1027); and Construction standard (§1926.1127)
3. **Formaldehyde**—General industry standard (§1910.1048(l); Shipyard standard (§1915.1048); and Construction standard (§1926.1148)
4. **Lead**—General industry standard (§1910.1025); Shipyard standard (§1915.1025); and Construction standard (§1926.62)
5. **Methylenedianiline**—General industry standard (§1910.1050(m)); Shipyard standard (§1915.1050); and Construction standard (§1926.60(n))
6. **Methylene Chloride**—General industry standard (§1910.1052(j)); Shipyard standard (§1915.1052); Construction standard (§1926.1152)
7. **Vinyl Chloride**—General industry standard (§1910.1017(k)); Shipyard standard (§1915.1517); and Construction standard (§1926.1117)

Privacy Concern Cases

The employer must protect the privacy of the injured or ill employee. The employer must not enter an employee's name on the OSHA 300 Log when recording a privacy case. The employer must keep a separate, confidential list of the case numbers and employee names, and provide it to the government upon request. If the work-related injury involves any of the following, it is to be treated as a privacy case:

1. An injury or illness to an intimate body part or the reproductive system
2. An injury or illness resulting from a sexual assault
3. A mental illness
4. HIV infection, Hepatitis, or Tuberculosis

5. Needlestick and sharps injuries that are contaminated with another person's blood or other potentially infectious material as defined by §1910.1030
6. Other illnesses—If the employee independently and voluntarily requests that his or her name not be entered on the OSHA 300 Log (This does not apply to injuries. See the definition of "Injury and Illness" in §1904.46.)

Physician or Other Licensed Healthcare Provider's Opinion

In cases where two or more physicians or other licensed healthcare providers make conflicting or differing recommendations, the employer must make a decision as to which recommendation is the most authoritative (best documented, best reasoned, or most persuasive), and record based on that recommendation.

DETERMINING IF A CASE IS RECORDABLE

Is This Injury or Illness Recordable?

As an employer, you are responsible for reporting all recordable injuries and illnesses. To help you determine if an injury or illness is recordable, refer to the flow chart below in Figure 2.5.

Determining if a case is OSHA recordable consists of a basic four-step decision-making process.

Step One: Determine whether a case has occurred. Has an employee suffered an injury, illness or death?

The first step in the decision-making process is the determination of whether an injury or illness occurred. Employers have nothing to record unless an employee has experienced a work-related injury or illness. In most instances, recognition of these injuries and illnesses is a fairly simple matter.

Employers must maintain injury and illness records for their own employees, but they are not responsible for maintaining records for employees of other companies or independent contractors working temporarily in their facility.

Employee status generally exists for recordkeeping purposes when the employer supervises not only the output, product, or result to be accomplished by the person's work but also the details, means, methods, and process by which the work is to be accomplished.

- Who pays the worker's wages?
- Who withholds the worker's Social Security taxes?
- Who hired the worker?
- Who has the authority to terminate the worker's employment?

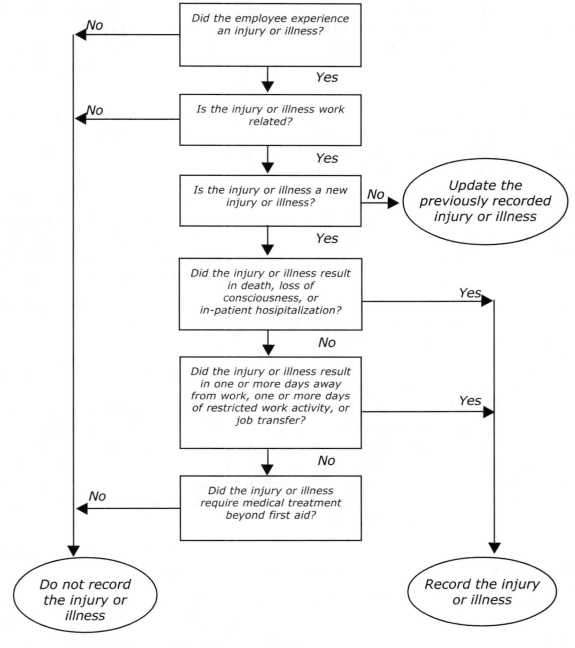

Figure 2.5 Do I Record the Injury or Not?

Step Two: Establish that the case was work-related. Did the injury, illness, or death result from an exposure in the work environment?

The work environment is primarily composed of the employer's premises and any other locations where employees are engaged in work-related activities or present as a condition of their employment. Company restrooms, hallways, and cafeterias are all considered to be a part of the employer's premises.

You must consider an injury or illness to be work-related if an event or exposure in the work environment either caused or contributed to the resulting condition or significantly aggravated a preexisting injury or illness. Work-relatedness is presumed for injuries and illnesses resulting from events or exposures occurring in the work environment unless an exception in § 1904.5(b)(2) specifically applies.

OSHA defines the work environment as "the establishment and other locations where one or more employees are working or are present as a condition of their employment. The work environment includes not only physical locations, but also the equipment or materials used by the employee during the course of his or her work."

For OSHA recordkeeping purposes, the work environment does not include employer-controlled athletic and recreational facilities and parking lots, provided the employee is not present as a condition of his or her employment. For example, an employee injured in the parking lot while arriving for or leaving work would not be a recordable case, but the same employee injured in the parking lot while performing maintenance would constitute a recordable case.

An employer can determine if a preexisting injury or illness has been significantly aggravated, for purposes of OSHA injury or illness recordkeeping, when an event or exposure in the work environment results in any of the following:

- Death, provided that the preexisting injury or illness would likely not have resulted in death but for the occupational event or exposure.
- Loss of consciousness, provided that the preexisting injury or illness would likely not have resulted in loss of consciousness but for the occupational event or exposure.
- One or more days away from work, or days of restricted work, or days of job transfer that otherwise would not have occurred but for the occupational event or exposure.
- Medical treatment, in a case where no medical treatment was needed for the injury or illness before the workplace event or exposure, or a change in medical treatment was necessitated by the workplace event or exposure.

An injury or illness is considered preexisting condition if it is resulted solely from a non-work-related event or exposure that occurred outside the work environment.

Injuries or illnesses that occur while an employee is on travel status are work related if, at the time of the injury or illness, the employee was engaged in work activities "in the interest of the employer." Examples of such activities include travel to and from customer contacts; conducting job tasks; and entertaining or being entertained to transact, discuss, or promote business (work-related entertainment includes only entertainment activities engaged in at the direction of the employer).

Step Three: Is the Injury or Illness a New Case?

You must consider whether the injury or illness to be a "new case" if

- The employee has not previously experienced a recorded injury or illness of the same type that affects the same part of the body

- The employee previously experienced a recorded injury or illness of the same type that affected the same part of the body but had recovered completely (all signs and symptoms had disappeared) from the previous injury or illness and an event or exposure in the work environment caused the signs or symptoms to reappear.

Where signs for occupational illnesses or symptoms may recur or continue in the absence of an exposure in the workplace, the case must only be recorded once. Examples may include cancer, asbestosis, byssinosis, and silicosis.

If the episode or recurrence was caused by an event or exposure in the workplace, the incident must be treated as a new case.

You are not required to seek the advice of a physician or other healthcare professional. However, if you do seek such advice, you must follow the physician's or other licensed healthcare professional's recommendation about whether the case is a new case or a recurrence. If you receive recommendations from two or more physicians or other licensed healthcare professionals, you must make a decision as to which recommendation is the most authoritative (best documented, best reasoned, or most authoritative), and record the case based upon that recommendation.

Table 2.6 Recordable Work-Related Injuries and Illnesses

A work-related injury or illness must be recorded if it results in one or more of the following:	
Death	See 29 CFR 1904.7(b)(2)
Days away from work	See 29 CFR 1904.7(b)(3)
Restricted work or transfer to another job	See 29 CFR 1904.7(b)(4)
Medical treatment beyond first aid	See 29 CFR 1904.7(b)(5)
Loss of consciousness	See 29 CFR 1904.7(b)(6)
A significant injury or illness diagnosed by a physician or other licensed healthcare professional	See 29 CFR 1904.7(b)(7)

Step Four: Is This an Injury or Illness Case?

You must consider an injury or illness to meet the general recording criteria, and therefore to be recordable, if it results in any of the following: death, days away from work, restricted work or transfer to another job, medical treatment beyond first aid, or loss of consciousness. You must also consider a case to meet the general recording criteria if it involves a significant injury or illness diagnosed by a physician or other licensed healthcare professional, even if it does not result in death, days away from work, restricted work or job transfer, medical treatment beyond first aid, or loss of consciousness.

REPORTING OCCUPATIONAL HEARING LOSS CASES[4]

The Occupational Safety and Health Administration (OSHA) is revising the hearing loss recording provisions of the Occupational Injury and Illness Recording and Reporting Requirements rule published January 19, 2001 and scheduled to take effect on January 1, 2003.

This final rule revises the criteria for recording hearing loss cases in several ways, including a requirement beginning in 2003 for the recording of Standard Threshold Shifts (10 dB shifts in hearing acuity) that have resulted in a total 25 dB level of hearing above audiometric zero, averaged over the frequencies at 2000, 3000, and 4000 Hz.

This final rule contains amended hearing loss recording criteria codified at 29 CFR 1904.10(a) and 1904.10(b)(1)-(7).

One of the major issues in releasing this revised recordkeeping rule was to determine the level of occupational hearing loss that constitutes a health condition serious enough to warrant recording. This was necessary because the final rule no longer requires recording of minor or insignificant health conditions that do not result in one or more of the general recording criteria such as medical treatment, restricted work, or days away from work. OSHA adopted the lower 10-dB threshold in the final rule based on an age-corrected STS as a large hearing change that can affect communicative competence.

OSHA requires employers to record audiometric results indicating a Standard Threshold Shift (STS) only when such STS cases also reflect a total hearing level of at least 25 dB from audiometric zero. The STS calculation uses audiometric results averaged over the frequencies 2000, 3000, and 4000 Hz using the original baseline and annual audiograms required by the OSHA noise standard § 1910.95.

The rule also allows the employer to adjust the employee's audiogram results used to determine an STS to subtract hearing loss caused by aging, allows the employer to retest the workers' hearing to make sure the hearing loss is persistent, and allows the employer to seek and follow the advice of a physician or licensed healthcare professional in determining whether the hearing loss was work-related.

The approach adopted in the final rule has several advantages. By using the STS definition from the OSHA noise standard § 1910.95, the § 1904.10 regulation uses a sensitive measure of hearing loss that has occurred while the employee is employed by his or her current employer. By requiring all STSs to exceed 25 dB from audiometric zero, the regulation ensures that all recorded hearing losses are significant illnesses. There is widespread agreement that a 25-dB shift from audiometric zero is a serious hearing loss.

The hearing loss recording level is also compatible with the final rule's definition of injury or illness, "an abnormal condition or disorder."

Is the Hearing Loss Requirement Similar to OSHA's Noise Standard?

The hearing loss recording requirements in § 1904.10 are different from the requirements of the OSHA noise standard (§ 1910.95) because under the noise standard the employer is required to take certain actions (employee notification, providing hearing protectors, or refitting of hearing protectors, etc.) for all 10-db standard threshold shifts while the § 1904 rule only requires the recording of STSs that also exceed the total 25-db level.

OSHA believes that this is an appropriate policy because 10-db shifts in hearing at higher levels (above 25 dB) are more significant.

When audiometric testing is done, test tones are presented at various sound levels, usually increasing or decreasing in 5-dB steps. The employee is asked to respond whenever a tone is heard, with the goal of finding the lowest level at which the employee can consistently hear. The standard measurement for measuring hearing level is decibels, a logarithmic scale. For the first increase in hearing level from 0 to 10 dB, the sound intensity increases tenfold. The next 10 dB is a 100-fold increase. By the time a person's hearing level changes from 0 to 30 dB hearing level, he or she needs 1,000 times more sound intensity to just barely hear.

Although the § 1904 recordkeeping regulation and the § 1910.95 noise standard treat the STS cases differently, this has no effect on the noise standard's requirements and does not have any effect on the requirement for employers to comply with § 1910.95. When employers detect work-related STS cases, they are required to take all of the follow-up actions required by the noise standard.

Employees may experience significant hearing loss in industries where audiometric testing is not required (construction, agriculture, oil and gas drilling and servicing, and shipbuilding industries) and is not provided voluntarily by the employer, and thus never be entered into the records. Likewise, an employee may experience gradual hearing loss while employed by several employers but never work for the same employer long enough to allow a recordable STS to be captured.

The new hearing loss recording rule will result in the recording of additional cases of hearing loss, not as a result of a change in the number of workers who experience hearing loss but simply because of the recordkeeping change.

The hearing loss recording requirements of § 1904.10 will not deprive employers and employees of information about noise hazards or diminish workers' protection against the hazards of noise in the workplace. The occupational noise exposure standard requires that employees in general industry be tested for hearing loss when noise exposure exceeds an eight-hour time-weighted average of 85dB, and that employees be informed, in writing, if a 10-dB shift has occurred. The audiometric test records must be retained for the duration of the affected employee's employment. (See 29 CFR 1910.95(g)(m).)

The noise standard also specifies the protective measures to be taken to prevent further hearing loss for employees who have experienced a 10-dB shift, including the use of hearing protectors and the referral for audiological evaluation where appropriate. (See 29 CFR 1910.95(g)(8).)

These requirements, which apply without regard to the recording criteria in the recordkeeping rule, will protect workers against the hazards of noise. The modified requirements of § 1904.10 will, therefore, not deprive employers and workers of the means to detect and prevent hearing loss.

Age Correction

The final rule allows the employers to age adjust an employee's annual audiogram when determining whether a 10-dB shift in hearing acuity has occurred.

While the final rule allows the employer to age correct the STS portion of the recording criteria, there is no allowance for age correction for determining a 25-dB hearing level. The method used in Appendix F of § 1910.95 is designed to age correct STS, not absolute hearing ability. The 25-dB criterion is used to ensure the existence of a serious illness and reflects the employee's overall health condition, regardless of causation. Age correcting the STS will provide adequate safeguards against recording age corrected hearing loss. Therefore, it would be inappropriate and unnecessary to age correct the 25-dB hearing level.

Record Consistently

One concern with the new hearing loss rule is how to verify that recorded hearing loss cases are recorded consistently. The OSHA Noise Standard addresses the issue of temporary hearing losses by allowing the employer to retest the employee's hearing within 30 days (1910.95(g)(7)(ii)). The 2001 rule adopted the same 30-day retest option at § 1904.10(b)(4) by allowing the employer to delay recording if a retest was going to be performed in the next 30 days.

OSHA agrees that the goal of the rule is to record only persistent hearing loss cases, and to help accomplish that goal, the Agency has carried forward the 30-day retest provision. However, OSHA has decided not to allow a longer retesting period. A longer retesting period would increase the likelihood that the employer would lose track of the case and therefore inadvertently fail to record the case. These errors would have a detrimental effect on the accuracy of the records and would run counter to OSHA's goal of improving the quality of the injury and illness data. The Agency also believes that using different time periods for retesting in the Part 1904 and § 1910.95 rules would result in increased confusion for employers.

In order to make it clear to employers that they may remove any cases that are found to be temporary, the final rule has adopted the removal option with three modifications. First, the final rule does not include the 15-month time limit. OSHA does not believe that a time limit is needed because any future audiogram that shows an improvement in hearing and refutes the recorded hearing loss would indicate a temporary hearing loss that should be removed from the records. Second, the regulatory text does not specify that the removal must be at the discretion of the reviewing professional. The OSHA noise standard, at § 1910.95(g)(3), requires that

> (3) Audiometric tests shall be performed by a licensed or certified audiologist, otolaryngologist, or other physician, or by a technician who is certified by the Council of

Accreditation in Occupational Hearing Conservation, or who has satisfactorily demonstrated competence in administering audiometric examinations, obtaining valid audiograms, and properly using, maintaining and checking calibration and proper functioning of the audiometers being used. A technician who operates microprocessor audiometers does not need to be certified. A technician who performs audiometric tests must be responsible to an audiologist, otolaryngologist or physician.

Third, the rule does not require the employer to maintain documentation concerning the removal of cases. Section 1910.95(m)(2) of the noise standard requires the employer to keep records of all audiometric tests that are performed, and those records will be available, should they be needed for future reference. As a result, there is no need to add a duplicative paperwork burden in the § 1904.10 rule.

If subsequent audiometric testing indicates that an STS is not persistent, you may erase or line out the recorded entry. OSHA has added this additional regulatory language to minimize the recording of temporary hearing loss cases while capturing complete data on the incidence of hearing loss disorders.

Baseline Reference and Revision of Baseline

The two-part test for recording that is adopted in the final rule uses the baseline audiogram as the reference point for determining whether the employee has had a change in hearing while employed by his or her current employer and then uses audiometric zero as the reference point for determining the overall hearing ability of the affected employee.

Using the baseline audiogram taken upon employment reduces the effect of any prior hearing loss the employee may have experienced, whether it is non-occupational hearing loss or occupational hearing loss caused by previous employment. Therefore, the final rule uses the employee's original baseline audiogram as the reference for the STS component of an initial hearing loss cases and uses the revised baseline audiogram from that initial case as the reference for future cases.

The 25-dB total hearing level component of an OSHA recordable hearing loss uses a reference of audiometric zero. This portion of the recording criteria is used to ensure that the employee's total hearing level is beyond the normal range of hearing, so it does not exclude hearing loss due to non-work causes, prior employment, or any other cause. The measurement simply reflects the employee's current hearing ability as reflected in the most recent audiogram. This comparison to audiometric zero is a simple matter, because audiometers are designed to provide results that are referenced to audiometric zero. The hearing level at each frequency is oftentimes printed by the equipment, so there is rarely a need to perform manual calculations.

Work Relationship

The final rule states that there are no special rules for determining work relationship and restates the rule's overall approach to determining work-relatedness: A case is work-related if one or more events or exposures in the work environment either caused or contributed to the hearing loss, or significantly aggravated a pre-existing hearing loss. It is possible for a worker who is exposed at or above the eight-hour 85-dBA action levels of the noise standard to experience a non-work-related hearing loss, and it is also possible for a worker to experience a work-related hearing loss and not be exposed above those levels.

The final rule allows the employer to seek the guidance of a physician or other licensed healthcare professional when determining the work-relatedness of hearing loss cases. The rule states that if a physician or other licensed healthcare professional determines that the hearing loss is not work related or has not been significantly aggravated by occupational noise exposure, the employer is not required to consider the case work related and therefore is not required to record it.

When evaluating the work-relatedness of a given hearing loss case, the employer should consider prior occupational and non-occupational noise exposure, evaluation of calibration records and the audiometric environment, investigation of related activities and personal medical conditions, and age correction before presuming that hearing loss is work related.

One important factor to consider is the effectiveness of the hearing protection program. When employees are exposed to high levels of noise in the workplace and do not wear appropriate hearing protection devices, a case of hearing loss is more likely to be work-related. If an employee's hearing protection devices are not appropriate for the noise conditions, if they do not fit properly, or if they are not used properly and consistently, they may not provide enough protection to prevent workplace noise from contributing to a hearing loss case.

Adding a Column to the 300 Log

OSHA has issued a separate *Federal Register* document proposing to delay the effective date of the hearing loss column requirement until January 1, 2004 and asking for comment on issues related to the hearing loss column. OSHA explained that it was adding a hearing loss column to the 300 Log so that BLS could produce more reliable statistics on occupational hearing loss cases.

OSHA has separated the requirement from § 1904.10(a) and placed it in a separate paragraph at § 1904.10(b)(7) which asks "How do I complete the 300 Log for a hearing loss case?" and answers "When you enter a recordable hearing loss case on the OSHA 300 Log, you must check the 300 Log column for hearing loss illnesses. OSHA has added a regulatory note to § 1904.10(b)(7) explaining that OSHA is delaying the applicability of § 1904.10(b)(7) until further notice while the Agency reconsiders the hearing loss column.

Frequently Asked Questions on Reporting Occupational Hearing Loss Cases

Recording cases involving Occupational Hearing Loss — § 1904.10

Q. What is the recording criterion for cases involving occupational hearing loss?

A. If an employee's hearing test (audiogram) reveals that the employee has experienced a work-related Standard Threshold Shift (STS) in hearing in one or both ears, and the employee's total hearing level is 25 decibels (dB) or more above audiometric zero (averaged at 2000, 3000, and 4000 Hz) in the same ear(s) as the STS, you must record the case on the OSHA 300 Log.

Q. What is a Standard Threshold Shift?

A. A Standard Threshold Shift, or STS, is defined in the Occupational Noise Exposure Standard at 29 CFR 1910.95(g)(10)(i) as a change in hearing threshold, relative to the baseline audiogram for that employee, of an average of 10 decibels (dB) or more at 2000, 3000, and 4000 hertz (Hz) in one or both ears.

Q. How do I evaluate the current audiogram to determine whether an employee has an STS level?

A. If the employee has never previously experienced a recordable hearing loss, you must compare the employee's current audiogram with that employee's baseline audiogram. If the employee has previously experienced a recordable hearing loss, you must compare the employee's current audiogram with the employee's revised baseline audiogram (the audiogram reflecting the employee's previous recordable hearing loss case).

Q. How do I evaluate the current audiogram to determine whether an employee has an 25-dB loss?

A. Audiometric test results reflect the employee's overall hearing ability in comparison to audiometric zero. Therefore, using the employee's current audiogram, you must use the average hearing level at 2000, 3000, and 4000 Hz to determine whether the employee's total hearing level is 25 dB or more.

Q. May I adjust the current audiogram to reflect the effects of aging on hearing?

A. Yes. When you are determining whether an STS has occurred, you may age adjust the employee's current audiogram results by using Tables F-1 or F-2, as appropriate, in Appendix F of 29 CFR 1910.95. You may not use an age adjustment when determining whether the employee's total hearing level is 25 dB or more above audiometric zero.

Q. Do I have to record the hearing loss if I am going to retest the employee's hearing?

A. No, if you retest the employee's hearing within 30 days of the first test and the retest does not confirm the recordable STS, you are not required to record the hearing loss case on the OSHA 300 Log. If the retest confirms the recordable STS, you must record the hearing loss illness within seven (7) calendar days of the retest. If subsequent audiometric testing performed under the testing requirements of the § 1910.95 Noise Standard indicates that an STS is not persistent, you may erase or line out the recorded entry.

Q. Are there any special rules for determining whether a hearing loss case is work-related?

A. No. You must use the rules in § 1904.5 to determine if the hearing loss is work related. If an event or exposure in the work environment either caused or contributed to the hearing loss, or significantly aggravated a pre-existing hearing loss, you must consider the case to be work related.

Q. If a physician or other licensed healthcare professional determines the hearing loss is not work related, do I still need to record the case?

A. If a physician or other licensed healthcare professional determines that the hearing loss is not work related or has not been significantly aggravated by occupational noise exposure, you are not required to consider the case work related or to record the case on the OSHA 300 Log.

Q. How do I complete the 300 Log for a hearing loss case?

A. When you enter a recordable hearing loss case on the OSHA 300 Log, you must check the 300 Log column for hearing loss.

Note to 1904.10(b)(7): The applicability of paragraph (b)(7) is delayed until further notice.

ENFORCEMENT OF THE RECORDKEEPING RULE[5]

Inspection Procedures

All OSHA inspections must review and record the establishment's injury and illness records for the three prior calendar years.

During the inspection the OSHA Inspector will compare this data with the OSHA 200 or OSHA 300 Logs for the three prior calendar years at the establishment.

Inspecting the OSHA 300 Log

The OSHA 300 Log serves as the primary source of data on occupational injuries and illnesses; it must be kept accurate and up-to-date. The Bureau of Labor Statistics, for example, relies upon the OSHA 300 Log to compile statistical data regarding occupational injuries and illnesses such as the Lost Workday Injury (LWDI) rate; OSHA in turn utilizes this data to allocate inspection resources and target areas of the workplace where violations, evidenced by high injury rates, may exist.

Employers, too, depend upon this data to pinpoint specific job functions that are impacting employee safety or identify certain areas of the workplace which are experiencing high rates of injury or illness.

When an OSHA inspector shows up at your workplace, the first request will usually be to see the OSHA 300 Log of employee injuries and illnesses (and perhaps other records). OSHA will want to look them over to see the kinds of injuries and illnesses listed, their frequency, and whether there is a pattern there (such as eye injuries, falls, etc., or a particular job function that is connected to each injury).

The OSHA inspector does that so he'll know where to concentrate his attention when he begins the physical inspection. The inspector may also decide to calculate your rate of injury/illness. If so, he'll ask for an estimate of the hours worked by employees over a year's time. He may also want to verify the accuracy of your injury logs. If so, he'll ask for other records to compare them with.

OSHA will also review your injury and illness records to see if they can determine any back or other musculoskeletal disorders (MSDs) that appear excessive from incidence calculations. To determine this, a particular trend must exist at least several years on the employee's OSHA 200 and 300 Log.

The information gained from the review of the OSHA Log is limited by each company's practices of recording injuries and illnesses. Some plants record everything and some record only those cases that are sent to see a physician. With back disorders, these cases are not always recognized and, therefore, are not recorded. The OSHA Compliance Officer will review what cases are recorded and when the cases were recorded.

Records Review

The employer must record cases on the OSHA 300 Log of Work-Related Injuries and Illnesses, and on the OSHA 301 Incident Report, (or equivalent form), as prescribed in Subpart C of §1904. Where no records are kept and there have been injuries or illnesses which meet the requirements for recordability, as determined by other records or by employee interviews, a citation for failure to keep records will normally be issued.

When the required records are kept but no entry is made for a specific injury or illness which meets the requirements for recordability, a citation for failure to record the case will normally be issued.

Where no records are kept and there have been no injuries or illnesses, as determined by employee interviews, no citation will be issued.

When the required records are kept but have not been completed with the detail required by the regulation, or the records contain minor inaccuracies, the records will be reviewed to determine if there are deficiencies that materially impair the understandability of the nature of hazards, injuries, and illnesses in the workplace. If the defects in the records materially impair the understandability of the nature of the hazards, injuries, and/or illnesses at the workplace, then an other-than-serious citation will normally be issued.

Incomplete Recorded Cases on the OSHA 300 or 301

If the deficiencies do not materially impair the understandability of the information, normally no citation will be issued. For example, an employer should not be cited solely for misclassifying an injury as an illness or vice versa. The employer will be provided information on keeping the records for the employer's analysis of workplace injury trends and on the means to keep the records accurately. The employer's promised actions to correct the deficiencies will be recorded and no citation will be issued.

Except for violation-by-violation citations, recordkeeping citations for improper recording of a case will be limited to a maximum of one citation item per form per year. This applies to both the OSHA 300 and the OSHA 301. Where the conditions for citation are met, an employer's failure to accurately complete the OSHA 300 Log for a given year would normally result in one citation item. Similarly, an employer's failure to accurately complete the OSHA 301, or equivalent, would normally result in one citation item. Multiple cases which are unrecorded or inaccurately recorded on the OSHA 300 or 301 during a particular year will normally be reflected as instances of the violation under that citation item.

A single citation item for an OSHA 301 violation would result where OSHA 301 Logs had not been completed, or where so little information had been put on the logs for multiple cases as to make the logs materially deficient.

Penalties

When a penalty is appropriate, there will be an unadjusted penalty of $1,000 for each year the OSHA 300 Log was not properly kept; an unadjusted penalty of $1,000 for each OSHA 301 Log that was not filled out at all (up to a maximum of $7,000); and an unadjusted penalty of $1,000 for each OSHA 301 Log that was not accurately completed (up to a maximum of $3,000).

When citations are issued, penalties will be proposed in the following cases:

1. When OSHA can document that the employer was previously informed of the requirements to keep records
2. When the employer's deliberate decision to deviate from the recordkeeping requirements, or the employer's plain indifference to the requirements, can be documented

Failure to Post Annual Summary

An other-than-serious citation will normally be issued, if an employer fails to post the OSHA 300A Summary by February as required by §1904.32(a)(1); and/or fails to certify the Summary as required by §1904.32(b)(3); and/or fails to keep it posted for three months, until May 1, as required by §1904.32(b)(6). The unadjusted penalty for this violation will be $1,000.

A citation will not be issued if the Summary which has not been posted or certified does not reflect injuries or illnesses and no injuries or illnesses actually occurred. The OSHA Inspector will verify that there were no recordable injuries or illnesses by interviews, or by review of workers' compensation or other records, including medical records.

Failure to Report Fatalities and Hospitalizations

An employer is required to report to OSHA within eight hours of the time the employer learns of the death of any employee or the inpatient hospitalization of three or more employees from a work-related incident. This includes fatalities at work caused by work-related heart attacks. There is an exception for certain work-related motor vehicle accidents or public transportation accidents.

The employer must orally report the fatality or multiple hospitalizations by telephone or in person to the OSHA Area Office (or State Plan office) that is nearest to the site of the incident. OSHA's toll-free telephone number may be used: 1-800-321-OSHA (1-800-321-6742).

An other-than-serious citation will normally be issued for failure to report such an occurrence. The unadjusted penalty will be $5,000.

If the Area Director determines that it is appropriate to achieve the necessary deterrent effect, the unadjusted penalty may be $7,000.

If the Area Director becomes aware of an incident required to be reported through some means other than an employer report prior to the elapse of the eight-hour reporting period and an inspection of the incident is made, a citation for failure to report will normally not be issued.

Access to Records for Employees

If the employer fails to provide upon request copies of records required to any employee, former employee, personal representative, or authorized employee representative by the end of the next business day, a citation will normally be issued. The unadjusted penalty will be $1,000 for each form not made available.

For example: If the OSHA 300 Log or the OSHA 300A Log for the current year and the three preceding years are not made available, the unadjusted penalty will be $4,000.

If the employer does not make available the OSHA 301 Logs, the unadjusted penalty will be $1,000 for each OSHA 301 Log not provided, up to a maximum of $7,000.

If the employer is to be cited for failure to keep records (OSHA 300, OSHA 300A, or OSHA 301), no citation for failure to give access will be issued.

Willful, Significant, and Egregious Cases

When an OSHA Inspector determines that there may be significant recordkeeping deficiencies, it may be appropriate to make a referral for a recordkeeping inspection, or to contact the Region's Recordkeeping Coordinator for guidance and assistance.

All willful recordkeeping cases and all significant cases with major recordkeeping violations will be initially reviewed by the Region's Recordkeeping Coordinator.

When willful violations are apparent, violation-by-violation penalties may be proposed in accordance with OSHA's egregious policy.

Enforcement Procedures for Occupational Exposure to Bloodborne Pathogens

The term "contaminated" under recording criteria for Needlestick and Sharps Injuries, incorporates the definition of "contaminated" from the Bloodborne Pathogens Standard at 29 CFR

1910.1030(b) ("Definitions"). Thus, contaminated means the presence or the reasonably anticipated presence of blood or other potentially infectious materials on an item or surface.

Employers may use the OSHA 300 and 301 forms to meet the sharps injury log requirement of §1910.1030(h) (5) if the employer enters the type and brand of the device causing the sharps injury on the Log and maintains the records in a way that segregates sharps injuries from other types of work-related injuries and illnesses, or allows sharps injuries to be easily separated.

Enforcement Procedures for Occupational Exposure to Tuberculosis

If any of your employees has been occupationally exposed to anyone with a known case of active tuberculosis (TB) and that employee subsequently develops a tuberculosis infection, as evidenced by a positive skin test or diagnosis by a physician or other licensed healthcare professional, you must record the case on the OSHA 300 Log by checking the Respiratory Condition column.

COMPLIANCE OFFICER CHECKLIST

The following provides guidance for a records evaluation for inspections.

Check Data for the Establishment

During the inspection, compare the establishment's data with the OSHA 200 or OSHA 300 Logs for the five prior years, or for as many years as there is data. For non-construction, you can use data from 1996 forward. For construction establishments, the first ODI will collect the 2001 injury and illness data, which will not be available until 2003.

Obtain Administrative Subpoena/Medical Access Orders

If it is anticipated after review of the history of the establishment that a subpoena or medical access orders will be needed, review the directives for guidance.

Verify the SIC Code

Verify the accuracy of the establishment's SIC code and enter the correct SIC code on the OSHA 1.

Ask for Information

1. Ask for the OSHA Logs, the total hours worked, the number of employees for each year, and a roster of current employees.
2. If you have questions regarding a specific case on the log, request the OSHA 301 Logs or equivalent form for that case.
3. Check if the establishment has an on-site medical facility; ask where the nearest emergency room is located where employees may be treated.

Ask Your Regional Recordkeeping Coordinator for Assistance

1. If significant recordkeeping deficiencies are suspected, you and your Area Director may request assistance from the Regional Recordkeeping Coordinator.
2. In some situations the Compliance Inspector may need to make a referral for a recordkeeping inspection.

Procedures for a Recordkeeping Inspection

1. For recorded cases, initially do a random review of the OSHA 301 Logs and medical records that pertain to the current employees.
2. Randomly select employees from the office roster, i.e., every tenth employee.
3. For those randomly selected employees, obtain the name and address of the medical provider(s).
4. If the random sample shows sufficient deficiencies, expand the review.
5. If the review will be expanded, contact the Regional Recordkeeping Coordinator soon for guidance and/or assistance. Early contact should be within the first month.

Endnotes:

[1] OSHA, *Recordkeeping Policies and Procedures Manual (RKM)*, January 1, 2002.

[2] *Ibid.*

[3] *Ibid.*

[4] Adapted from *Federal Register*, vol. 67, no. 126, 1 July 2002.

[5] Adapted from OSHA's *Recordkeeping Policies and Procedures Manual (RKM)*, 1 January 2002.

_____

Construction Industry Standards

INTRODUCTION

Construction safety standards were initially adopted by the U.S. Department of Labor in April 1971 to implement the Contract Work Hours and Safety Standards Act, 40 USC 333, also known as the Construction Safety Act. Shortly thereafter they were converted into OSHA standards under the authority that Congress delegated to the Secretary of Labor in section 6(a) of the OSH Act.

An employer must comply with the safety and health regulations in Part 1926 if its employees are "engaged in construction work" (29 CFR 1910.12(b)). In order to determine if such work, defined as "work for construction, alteration, and/or repair, including painting and decorating," is being performed, one must consult 29 CFR 1926.13 for a discussion which states the following:

> The terms 'construction'... or 'repair' mean all types of work done on a particular building or work at the site thereof... all work done in the construction or development of the project, including without limitation, altering, remodeling, installation (where appropriate) on the site of the work of items fabricated off-site, painting and decorating, the transporting of materials and supplies to or from the building or work... and the manufacturing or furnishing of materials, articles, supplies or equipment on the site of the building or work.... The 'site of the work' is limited to the physical place or places where the construction called for in the contract will remain when work on it has been completed....

The Part 1926 standards include 26 subparts (Subpart A through Subpart Z), but Subparts A and B apply only to determining the scope of section 107 of the Construction Safety Act, 40 USC 333. That Act applies only to employers who are engaged in construction under contract with the US government. OSHA does not base citations upon either Subpart A or B. Consequently, no further consideration will be given to them in this book.

The Remaining 24 Subparts Are Listed Below:

Subpart C—General Safety and Health Provisions

Subpart D—Occupational Health and Environmental Controls

Subpart E—Personal Protective and Life Saving Equipment

Subpart F—Fire Protection and Prevention

Subpart G—Signs, Signals, and Barricades

Subpart H—Materials Handling, Storage, Use, and Disposal

Subpart I—Tools (Hand and Power)

Subpart J—Welding and Cutting

Subpart K—Electrical

Subpart L—Scaffolding

Subpart M—Fall Protection

Subpart N—Cranes, Derricks, Hoists, Elevators, and Conveyors

Subpart O—Motor Vehicles, Mechanized Equipment, and Marine Operations

Subpart P—Excavations

Subpart Q—Concrete and Masonry Construction

Subpart R—Steel Erection

Subpart S—Underground Construction, Caissons, Cofferdams, and Compressed Air

Subpart T—Demolition

Subpart U—Blasting and the Use of Explosives

Subpart V—Power Transmission and Distribution

Subpart W—Rollover Protective Structures; Overhead Protection

Subpart X—Stairways and Ladders

Subpart Y—Commercial Diving Operations

Subpart Z—Toxic and Hazardous Substances

Employers engaged in the construction industry should identify and read over each of those subparts that could be applicable to their own operations. They should then read the discussion of the subparts that follows.

SUBPART C—GENERAL SAFETY AND HEALTH PROVISIONS

1926.20 General safety and health provisions

1926.21 Safety training and education

1926.23 First aid and medical attention

1926.24   Fire protection and prevention

1926.25   Housekeeping

1926.26   Illumination

1926.27   Sanitation

1926.28   Personal protective equipment

1926.29   Acceptable certifications

1926.30   Shipbuilding and ship repairing

1926.31   Incorporation by reference

1926.32   Definitions

1926.33   Access to employee exposure and medical records

1926.34   Means of egress

1926.35   Employee emergency action plans

General Safety and Health Provisions[1] — §1926.20

Contractor Requirements

No contractor or subcontractor for any part of contract work for construction, alteration, and/or repair, including painting and decorating, shall require any employee to work in surroundings or under working conditions which are unsanitary, hazardous, or dangerous to his/her health or safety.

Accident Prevention Responsibilities

It shall be the responsibility of the employer to provide for frequent and regular inspections of the job site, materials, and equipment by competent persons designated by the employer.

A competent person is defined in §1926.32(f) as "one who is capable of identifying existing and predictable hazards in the surroundings or working conditions which are unsanitary, hazardous, or dangerous to employees, and who has authorization to take prompt corrective measures to eliminate them."

The use of any machinery, tool, material, or equipment which is not in compliance with OSHA standards is prohibited and shall be identified as unsafe by tagging or locking the controls to render them inoperable or shall be physically removed from their place of operation.

The employer shall permit only those employees qualified by training or experience to operate equipment and machinery.

Inspections

Some OSHA inspectors insist that §1926.20(b) requires that each construction employer must adopt a written accident prevention program for each job site. The standard, however, does not include the word "written." Employers should pay close attention to §1926.20(b). It is one of the most cited construction safety standards.

When a construction inspection is performed, the following will be followed:

1. An evaluation of the safety and health program will be completed. (See the sample program below.) These guidelines will be modified, based on the OSHA Inspector's professional judgment, to account for size and type of construction. A key indicator of an effective program will be the degree of knowledge employees have of potential site-specific safety and health hazards. This knowledge requires training (site familiarization) of skilled as well as non-skilled crafts in hazard recognition based on the employee's specific work environment and job related hazards.

2. Program deficiencies such as lack of management policy or safety and health rules, inadequate assignment of responsibility, or poor employee awareness or participation shall be discussed with the employer.

3. Violations of the requirements for instruction, first aid, recordkeeping, and identification and control of hazards shall be cited as indicated in the appropriate section of 29 CFR 1926.20, 29 CFR 1926.21, 29 CFR 1926.23, or 29 CFR 1904.2.

4. Where the conditions warrant a citation for violation of 1926.20 or 1926.21, it may be issued even if additional 29 CFR 1926 alleged violations were not documented. Note that 1926.21(b) requires only safety and health instructions. Employers are required to implement a safety and health program in accordance with the above mentioned standards. However, employers should be encouraged to implement a formal safety and health training program with the sample shown below.

5. Violations for 29 CFR 1926.20(b) in a routine inspection may be cited as either other-than-serious or serious as circumstances warrant.

Sample Employer's Safety and Health Program

A. Management Commitment and Leadership

1. Policy statement: Goals established, issued, and communicated to employees.
2. Program revised annually.
3. Participation in safety meetings, inspections; agenda item in meetings.
4. Commitment of resources is adequate.
5. Safety rules and procedures incorporated into site operations.
6. Management observes safety rules.

B. Assignment of Responsibility

1. Safety designee on site, knowledgeable, and accountable.
2. Supervisors (including foremen) safety and health responsibilities understood.
3. Employees adhere to safety rules.

C. Identification and Control of Hazards

1. Periodic site safety inspection program involves supervisors.
2. Preventative controls in place (PPE, maintenance, engineering controls).
3. Action taken to address hazards.
4. Safety Committee, where appropriate.
5. Technical references available.
6. Enforcement procedures by management.

D. Training and Education

1. Supervisors receive basic training.
2. Specialized training taken when needed.
3. Employee training program exists, is ongoing, and is effective.

E. Recordkeeping and Hazard Analysis

1. Records of employee illnesses/injuries maintained and posted.
2. Supervisors perform accident investigations to determine causes and to propose corrective action.
3. Injuries, near misses, and illnesses are evaluated for trends, similar causes; corrective action initiated

F. First Aid and Medical Assistance

1. First aid supplies and medical service available.
2. Employees informed of medical results.
3. Emergency procedures and training, where necessary.

Section 1926.21 requires that employees be given instruction (not training) in the recognition and avoidance of unsafe conditions and the regulations applicable (to his or her work) regarding hazard control, as well as the necessary precautions and the necessary protective equipment for those employees who are required to enter "confined or enclosed spaces." Although §1926.21 is somewhat ambiguous, it is a favorite of OSHA inspectors. Innumerable citations have been issued under that standard.

The standards in §§1926.22 through 1926.28 apply to most construction jobs and are stated succinctly and briefly. Each employer should familiarize himself or herself with them.

Personal Protective Equipment— §1926.28

The employer is responsible to require the wearing of appropriate personal protective equipment in all operations where there is an exposure to hazardous conditions or where these regulations indicate the need for using such equipment to reduce the hazards to the employees.

Regulations governing the use, selection, and maintenance of personal protective and lifesaving equipment are described in Subpart E, *Personal Protective and Lifesaving Equipment.*

The "acceptable certifications" standard in §1926.29 applies only to boilers, pressure vessels, and similar equipment.

The provisions of §§1926.30 and 1926.31 do not impose any substantive requirements on employers.

The three remaining standards are General Industry standards that OSHA has incorporated into the Construction Standards. Section 1926.33 is the same as §1910.1020 (discussed earlier in this book under "OSHA Recordkeeping and Reporting Regulations"). Section 1926.34 requires that there shall be free egress from every building and structure. It is essentially the same as Part 1910.

Section 1926.35 is the same as §1910.38, a standard that is also part of Part 1910, Subpart E—Means of Egress.

Access to Employee Exposure and Medical Records[2]—§1926.33

Access

"Access," for the purpose of the standard, means the right and opportunity to examine and copy. Access to employee medical and exposure records must be provided in a reasonable manner and place. If access cannot be provided within 15 days after the employee's request, the employer must state the reason for the delay and the earliest date when the records will be made available. Responses to initial requests, and new information that has been added to an initial request, are to be provided without cost to the employee or representative. The employer may give employee copies of the requested records, give the employee the records and the use of mechanical copying facilities so the employee may copy the records, or lend employee their records for copying off the premises. In addition, medical and exposure records are to be made available, on request, to OSHA representatives to examine and copy.

Exposure Records

Upon request, the employer must provide the employee, or employee's designated representative, access to employee exposure records. If no records exist, the employer must provide records of other employees with job duties similar to those of the employee. Access to these records does not require the written consent of the other employees. In addition, these records must reasonably indicate the identity, amount, and nature of the toxic substances or harmful physical agents to which the employee has been exposed. Union representatives must indicate an occupational health need for requested records when seeking access to exposure records without the written authorization of the employee(s) involved.

Medical Records

The employer also must provide employees and their designated representatives access to medical records relevant to the employee. Access to the medical records of another employee may be provided only with the specific written consent of that employee. The standard provides a suitable sample authorization letter for this purpose. Prior to employee access to medical records, physicians, on behalf of employers, are encouraged to discuss with employees the contents of their medical records; physicians also may recommend ways of disclosing medical records other than by direct employee access.

Where appropriate, a physician representing the employer can elect to disclose information on specific diagnoses of terminal illness or psychiatric conditions only to an employee's designated representative, and not directly to the employee. In addition, a physician, nurse, or other responsible healthcare person who maintains medical records may delete from requested medical records the names of persons who provided confidential information concerning an employee's health status.

Analyses Using Exposure or Medical Records

The standard ensures that an employee (or designated representative), as well as OSHA, can have access to analyses that were developed using information from exposure or medical records about the employee's working conditions or workplaces. Personal identities, such as names, addresses, social security and payroll numbers, age, race, and sex, must be removed from the data analyses prior to access.

Trade Secrets

In providing access to records, an employer may withhold trade secret information but must provide information needed to protect employee health. Where it is necessary to protect employee health, the employer may be required to release trade secret information but may condition access on a written agreement not to abuse the trade secret or to disclose the chemical's identity.

An employer also may delete from records any trade secret that discloses the manufacturing process or the percentage of a chemical substance in a mixture. The employer must, however, state when such deletions are made. When the deletion impairs the evaluation of where or when exposure occurs, the employer must provide alternative information that is sufficient to permit the requester to make such evaluations.

The employer also may withhold a specific chemical identity when the employer can demonstrate it is a trade secret, the employer states this to the requester, and all other information on the properties and effects of the toxic substance is disclosed. The specific chemical identity, however, must be disclosed to an attending physician or nurse when that physician or nurse states that a medical emergency exists and the identity is necessary for treatment. When the emergency is over, the employer may require the physician or nurse to sign a confidentiality agreement.

The employer also must provide access to a specific chemical identity in non-emergency situations to an employee, an employee's designated representative, or a healthcare professional if it will be used for one or more of the following activities:

- Assess the hazards of the chemicals to which employees will be exposed
- Conduct or assess sampling of the workplace atmosphere to determine employee exposure levels
- Conduct pre-assignment or periodic medical surveillance of exposed employees
- Provide medical treatment to exposed employees
- Select or assess appropriate personal protective equipment for exposed employees
- Design or assess engineering controls or other protective measures for exposed employees
- Conduct studies to determine the health effects of exposure

In these instances, however, the employer may require the requester to submit a written statement of need and the reasons why alternative information will not suffice, and to sign a confidentiality agreement not to use the information for any purpose other than the stated health need or to release it under any circumstances, except to OSHA.

The standard further prescribes the steps employers must follow if they decide not to disclose the specific chemical identity requested by the health professional, employee, or designated representative. Briefly, these steps are as follows:

- Provide a written denial.
- Provide the denial within 30 days of the request.
- Provide evidence that the chemical identity is a trade secret.
- Explain why alternative information is adequate.
- Give specific reasons for the denial.

An employee, designated representative, or healthcare professional may refer such a denial to OSHA for review and comment.

Employee Information

At the time of initial employment and at least annually thereafter, employees must be told of the existence, location, and availability of their medical and exposure records. The employer also must inform each employee of his or her rights under the access standard and make copies of the standard available. Employees also must be told who is responsible for maintaining and providing access to records.

Transfer of Records

When an employer ceases to do business, he or she is required to provide the successive employer with all employee medical and exposure records. When there is no successor to receive the records for the prescribed period, the employer must inform the current affected employees of their access rights at least three months prior to the cessation of business and must notify the Director of the National Institute for Occupational Safety and Health (NIOSH) in writing at least three months prior to the disposal of records.

Retention of Records

Each employer must preserve and maintain accurate medical and exposure records for each employee. The access standard imposes no obligation to create records but does apply to any medical or exposure records created by the employer in compliance with other OSHA rules or at his or her own volition.

Exposure records and data analyses based on them are to be kept for 30 years. Medical records are to be kept for at least the duration of employment plus 30 years. Background data for exposure records such as laboratory reports and work sheets need be kept only for one year. Records of employees who have worked for less than one year need not be retained after employment, but the employer must provide these records to the employee upon termination of employment. First-aid records of one-time treatment need not be retained for any specified period.

OSHA does not mandate the form, manner, or process by which an employer preserves a record, except that chest X-ray films must be preserved in their original state.

Three months before disposing of records, employers must notify the Director of NIOSH.

Means of Egress[3] — §1926.34

In every building or structure exits shall be so arranged and maintained as to provide free and unobstructed egress from all parts of the building or structure at all times when it is occupied. No lock or fastening to prevent free escape from the inside of any building shall be installed except in mental, penal, or corrective institutions where supervisory personnel is continually on duty and effective provisions are made to remove occupants in case of fire or other emergency.

Exit Marking

Exits shall be marked by a readily visible sign. Access to exits shall be marked by readily visible signs in all cases where the exit or way to reach it is not immediately visible to the occupants.

Maintenance and Workmanship

Means of egress shall be continually maintained free of all obstructions or impediments to full instant use in the case of fire or other emergency.

Employee Emergency Action Plans[4] — §1926.35

This applies to all emergency action plans required by a particular OSHA standard. The emergency action plan shall be in writing (except for employers with ten or fewer employees) and shall cover those designated actions employers and employees must take to ensure employee safety from fire and other emergencies.

The following shall be included in the plan:

1. Emergency escape procedures and emergency escape route assignments
2. Procedures to be followed by employees who remain to operate critical plant operations before they evacuate
3. Procedures to account for all employees after emergency evacuation has been completed
4. Rescue and medical duties for those employees who are to perform them
5. The preferred means of reporting fires and other emergencies
6. Names or regular job titles of persons or departments who can be contacted for further information or explanation of duties under the plan

Alarm System

The employer shall establish an employee alarm system which complies with §1926.159.

If the employee alarm system is also used for alerting fire brigade members, or for other purposes, a distinctive signal for each purpose shall be used.

Evacuation

The employer shall establish in the emergency action plan the types of evacuation to be used in emergency circumstances.

Training

Before implementing the emergency action plan, the employer shall designate and train a sufficient number of persons to assist in the safe and orderly emergency evacuation of employees.

The employer shall review the plan with each employee covered by the plan at the following times:

- Initially when the plan is developed
- Whenever the employee's responsibilities or designated actions under the plan change
- Whenever the plan is changed

The employer shall review with each employee upon initial assignment those parts of the plan which the employee must know to protect the employee in the event of an emergency. The written plan shall be kept at the workplace and made available for employee review. For those employers with ten or fewer employees, the plan may be communicated orally to employees and the employer need not maintain a written plan.

SUBPART D—OCCUPATIONAL HEALTH AND ENVIRONMENTAL CONTROLS

1926.50   Medical services and first aid

1926.51   Sanitation

1926.52   Occupational noise exposure

1926.53   Ionizing radiation

1926.54   Non-ionizing radiation

1926.55   Gases, vapors, fumes, dusts, and mists

1926.56   Illumination

1926.57   Ventilation

1926.59   Hazard communication

1926.60   Methylenedianiline (MDA)

1926.61   Retention of DOT markings, placards, and labels

1926.62   Lead

1926.64   Process safety management of highly hazardous chemicals

1926.65   Hazardous waste operations and emergency response

1926.66   Criteria for design and construction for spray booths

There are seven Part 1910 General Industry Standards that apply to Subpart D.

§1910.141(a)(1)—Scope of General Industry sanitation standards

§1910.141(a)(2)(v)—"Potable water" defined

§1910.141(a)(5)—Vermin control

§1910.141(g)(2)—Eating and drinking areas

§1910.141(h)—Food handling

§1910.151(c)—Medical services and first aid

§1910.161(a)(2)—Carbon dioxide extinguishing systems, safety requirements

The medical services, sanitation, and illumination requirements in §§1926.50, 1926.51, and 1926.56 apply at all construction sites.

Section 1926.52 is the Noise Standard for the construction industry. It limits employee noise exposure to 90 dBA averaged over an eight-hour day but, unlike the General Industry Noise Standard, it does not require the same kind of hearing conservation program (monitoring, audiometric testing, etc.). If you have employees exposed to high noise levels, you should also observe §1926.102 (part of Subpart E) which requires ear protective devices. For all intents and purposes, compliance with §1926.102 will satisfy the construction industry hearing conservation requirement.

The §1926.53 standard applies to activities that involve ionizing radiation, radioactive materials, or x-rays. The §1926.54 standard applies when laser equipment (non-ionizing radiation) is used. The use of laser safety goggles is required by a Subpart E standard, §1926.102(b)(2).

Section 1926.55 is the construction equivalent of Part 1910, Subpart Z. As originally issued, it did not list the 400-odd regulated substances by name or the employee exposure limitations for those substances. It simply stated that American Conference of Governmental Industrial Hygienists (ACGIH) Threshold Limit Values of Airborne Contaminants for 1970 shall be avoided. If not, controls must first be implemented whenever feasible. When not feasible, personal protective equipment must be used. Those ACGIH limits are now listed in Appendix A of §1926.55. Construction employers should read the list of substances that appear in the standard's Appendix A. If the limits listed there are exceeded, the §1926.55 requirements must be observed.

Section 1926.56 sets forth specific illumination intensities for construction areas, ramps, runways, corridors, offices, shops, and storage areas.

If ventilation is used as an engineering control to limit exposure to airborne contaminants, the §1926.57 Ventilation Standard applies.

If your work involves exposure to Methelcredianiline (MDA), you must be familiar with §1926.60 and carefully observe its requirements. It includes innumerable details. This is also true for employers whose work involves exposure to lead. They must observe the numerous requirements included in the Lead Standard (§1926.62). Per §1926.61, each employer who receives a package of hazardous material that is required to be marked, labeled, or placarded in compliance with DOT's Hazardous Materials Regulations (49 CFR Parts 171 through 180) must keep those markings, labels, and placards on the package until the packaging is sufficiently cleaned of residue to remove any potential hazards. The markings, placards, and labels must be maintained in a manner that ensures that they are readily visible.

Process Safety Management of Highly Hazardous Chemicals (§1926.64) is the same as the Part 1910 General Industry Standard.

Hazardous Waste Operations and Emergency Response (§1926.65) is also the same as the Part 1910 General Industry Standard. See the discussion earlier in this book under the heading "Subpart H—Hazardous Materials."

Criteria for Design and Construction for Spray Booths (§1922.66) incorporates into the Construction Standards the General Industry Requirements (§1910.107) discussed earlier in this book under the heading "Subpart H—Hazardous Materials."

The Hazard Communication Standard is the single-most cited OSHA standard. Every employer must pay strict attention to §1926.59. It requires that chemical manufacturers and importers assess the hazards of all chemicals that they produce or import and furnish detailed material safety data sheets (MSDSs) to their customers for those chemicals determined to be hazardous. All employers must provide that information to their employees by means of a written hazard communication program, labels on containers, MSDS sheets, employee training, and access to written records of all of this.

The term "hazardous chemical" is defined very broadly, so virtually every employer is required to observe §1926.59.

OSHA Summary of Hazard Communication Standard[5]

Protection under OSHA's Hazard Communication Standard (HCS) includes all workers exposed to hazardous chemicals in all industrial sectors. This standard is based on a simple concept—that employees have both a need and a right to know the hazards and the identities of the chemicals they are exposed to when working. They also need to know what protective measures are available to prevent adverse effects from occurring.

Scope of Coverage

More than 30 million workers are potentially exposed to one or more chemical hazards. There are an estimated 650,000 existing hazardous chemical products, and hundreds of new ones are being introduced annually. This poses a serious problem for exposed workers and their employers.

Benefits

The HCS covers both physical hazards (such as flammability or the potential for explosions), and health hazards (including both acute and chronic effects). By making information available to employers and employees about these hazards, and recommending precautions for safe use, proper implementation of the HCS will result in a reduction of illnesses and injuries caused by chemicals. Employers will have the information they need to design an appropriate protective program. Employees will be better able to participate in these programs effectively when they understand the hazards involved, and better able to take steps to protect themselves. Together, these employer and employee actions will prevent the occurrence of adverse effects caused by the use of chemicals in the workplace.

Requirements

The HCS established uniform requirements to make sure that the hazards of all chemicals imported into, produced, or used in U.S. workplaces are evaluated and that this hazard information is transmitted to affected employers and exposed employees.

Chemical manufacturers and importers must convey the hazard information they learn from their evaluations to downstream employers by means of labels on containers and material safety data sheets (MSDSs). In addition, all covered employers must have a hazard communication program to get this information to their employees through labels on containers, MSDSs, and training.

This program ensures that all employers receive the information they need to inform and train their employees properly and to design and put in place employee protection programs. It also provides necessary hazard information to employees so they can participate in and support the protective measures in place at their workplaces.

All employers in addition to those in manufacturing and importing are responsible for informing and training workers about the hazards in their workplaces, retaining warning labels, and making MSDSs available with hazardous chemicals.

Some employees deal with chemicals in sealed containers under normal conditions of use (such as in the retail trades, warehousing, and truck and marine cargo handling). Employers of these employees must ensure that labels affixed to incoming containers of hazardous chemicals are kept in place. They must maintain and provide access to MSDSs received, or obtain MSDSs if requested by an employee. And they must train workers on what to do in the event of a spill or leak. However, written hazard communication programs will not be required for this type of operation.

All workplaces where employees are exposed to hazardous chemicals must have a written plan which describes how the standard will be implemented in that facility. The only work operations which do not have to comply with the written plan requirements are laboratories and work operations in which employees only handle chemicals in sealed containers.

The written program must reflect what employees are doing in a particular workplace. For example, the written plan must list the chemicals present at the site, indicate who is responsible for the various aspects of the program in that facility and indicate where written materials will be made available to employees.

The written program must describe how the requirements for labels and other forms of warning, material safety data sheets, and employee information and training are going to be met in the facility.

Effect on State Right-To-Know Laws

The HCS pre-empts all state (in states without OSHA-approved job safety and health programs) or local laws which relate to an issue covered by HCS without regard to whether the state law would conflict with, complement, or supplement the federal standard, and without regard to whether the state law appears to be "at least as effective as" the federal standard.

The only state worker right-to-know laws authorized would be those established in states and jurisdictions that have OSHA-approved state programs.

SUBPART E—PERSONAL PROTECTIVE AND LIFESAVING EQUIPMENT

1926.95   Criteria for personal protective equipment

1926.96   Occupational foot protection

1926.100   Head protection

1926.101   Hearing protection

1926.102   Eye and face protection

1926.103 Respiratory protection

1926.104 Safety belts, lifelines, and lanyards

1926.105 Safety nets

1926.106 Working over or near water

1926.107 Definitions applicable to this subpart

Application

Protective equipment, including personal protective equipment for eyes, face, head, and extremities; protective clothing; respiratory devices; and protective shields and barriers, shall be provided, used, and maintained in a sanitary and reliable condition wherever it is necessary by reason of hazards of processes or environment, chemical hazards, radiological hazards, or mechanical irritants encountered in a manner capable of causing injury or impairment in the function of any part of the body through absorption, inhalation, or physical contact.

Employee-Owned Equipment

Where employees provide their own protective equipment, the employer shall be responsible to ensure its adequacy, including proper maintenance and sanitation of such equipment.

What are the most frequently cited serious Personal Protective Equipment (PPE) violations?[6]

1. Failure to provide and use proper head protective equipment (1926.100(a))
2. Lack of proper eye or face protection (1926.102(a))
3. The lack of safety nets for fall hazards of more than 25 feet (1926.105(a))
4. Workers not using serviceable PPE when exposed hazards that could cause serious injuries (1926.95(a))
5. Not using personal floatation equipment when working over or near deep water (1926.106(a))
6. Not using properly approved respiratory protective equipment (1926.103(a)(1) and (2))

Effective Control Measures for These Violations[7]

Personal protective equipment (PPE) is one of the more common controls used, and the equipment must be maintained in serviceable condition and replaced when no longer useable.

Employers must establish a policy of what PPE is to be used by their employees for various jobs and must instruct workers on the proper care and use of PPE. Each worker must wear his or her hard hat when potentially exposed to falling objects, and eye protection must worn.

As part of the employer's safety and health program, a hazard survey of the work needs to be done to determine what control measures to use where hazards can not be eliminated. This survey can serve as a resource for which type of PPE needs to be used by workers to minimize injury or illness exposure.

The PPE used must be the proper type that provides the worker the necessary protection from the hazards found in the survey. Only NIOSH-approved respirator protective devices are acceptable to OSHA for respiratory hazards. All personal floatation devices must be approved by the U.S. Coast Guard to be acceptable to OSHA for those workers doing jobs over deep water or next to deep water hazards.

Where personal fall arrest equipment is not practical for all workers who are exposed to fall hazards over 25 feet, safety nets become the backup protection to protect all who work above the net. This is typically done in bridge construction work and high rise buildings.

SUBPART F—FIRE PROTECTION AND PREVENTION

1926.150   Fire protection

1926.151   Fire prevention

1926.152   Flammable and combustible liquids

1926.153   Liquefied petroleum gas (LP-Gas)

1926.154   Temporary heating devices

1926.155   Definitions applicable to this subpart

What are the serious most frequently cited fire hazard violations in descending order?[8]

1. Transporting or handling flammable liquids in non-approved containers (1926.152(a)(1))
2. Failure to have a class 2-A rated fire extinguisher within 100 feet (30.4 m) of an area where class A fire hazards exist within a building (1926.150(c)(1)(I)). Another frequent violation related to this one is not having at least one class 2-A rated fire extinguisher on each floor of a multi-story building located near the stairway (1926.150(c)(1)(iv))
3. Failure of the employer to develop and implement a fire protection program for all phases of work involving employees on the job site (1926.150(a)(1))
4. Failure to inspect and maintain portable fire extinguishers to keep them in serviceable condition (1926.150(c)(1)(iii))
5. Lack of "no smoking" signs where refueling operations are conducted (1926.152(g)(9)) and where operations which constitute a fire hazard, which commonly will include flammable liquids and flammable gases (1926.151(a)(3))

What are some effective control measures that can be used for these serious hazards?[9]

The importance of developing and implementing a fire protection program, which will be a part of the employer's safety and health program, will help to avoid being unprepared for fire emergencies. This is most important to those contractors that work with easily ignitable materials such as flammable liquids and gases and those that use equipment which depends on liquid or gaseous fuels. Material safety data sheets (MSDSs) from your supplier can be used to determine what materials are flammable and which ones are only combustible or will not burn. Also, the MSDS will tell you which fire extinguishing agent is effective on the specific material if it were to catch on fire.

As part of the program, preplanning for trash and rubbish removal is an important part of preventing fire hazards. Removing accumulations of Class A waste materials such as paper, cardboard, wood pallets and packing materials, and trash from the work area will help minimize unwanted fires. Also, minimize the spilling of flammable liquids by using approved safety cans or the DOT shipping containers. Approved containers will have a laboratory listing or label recognized by OSHA.

Where fire hazards cannot be removed from the work area, the types of fire extinguishing equipment to be used must match the type of materials being used and the job activities. For Class A hazards, OSHA would accept any approved 2-A rated fire extinguisher; a 55-gallon (208 l) open drum of water with two fire pails; 100 feet (30.4 m) of one-half inch (1.27cm) rubber or plastic garden type hose that can supply at least five gallons per minute (0.31 l/s) with a hose stream range of 30 feet (9.1 m) horizontally; or 100 feet (30.4 m) of an approved fire hose system that will deliver 25 gallons or more per minute (1.57 l/s).

Control of ignition sources is also an important component of the program. Bonding and grounding metal containers when transferring flammable liquids; posting "no smoking" signs near fire hazards areas; and using approved lighting for hazardous locations, such as in a confined space where flammable materials are being used, are some examples of good control of ignition sources.

SUBPART G—SIGNS, SIGNALS, AND BARRICADES

1926.200   Accident prevention signs and tags

1926.201   Signaling

1926.202   Barricades

1926.203   Definitions applicable to this subpart

Safety Standards for Signs, Signals, and Barricades[10]

The Occupational Safety and Health Administration (OSHA) has revised the construction industry safety standards to require that traffic control signs, signals, barricades, or devices protecting workers conform to Part VI of either the 1988 Edition of the Federal Highway Administration (FHWA) Manual on Uniform Traffic Control Devices (MUTCD) with 1993 revisions (Revision 3) or the Millennium Edition of the FHWA MUTCD (Millennium Edition), instead of the American National Standards Institute (ANSI) D6.1-1971, Manual on Uniform Traffic Control Devices for Streets and Highways (1971 MUTCD).

This final rule addresses the types of signs, signals, and barricades that must be used to protect construction employees from traffic hazards. The vast majority of road construction in the United States is funded through federal transportation grants. As a condition to receiving federal funding, the U.S. Department of Transportation's (DOT's) Federal Highway Administration requires compliance with its MUTCD.

It requires the posting of construction areas with legible traffic signs at points of hazard §1926.200(g), and the use of flagmen or other appropriate traffic controls when the necessary protection cannot be provided by signs, signals, and barricades (§1926.201). The other Subpart G standards simply provide the specifications and color coding for signs and tags if they are used or required by some other OSHA standard.

SUBPART H—MATERIALS, HANDLING, STORAGE, USE, AND DISPOSAL

1926.250   General requirements for storage

1926.251   Rigging equipment for material handling

1926.252   Disposal of waste materials

Materials Handling and Storing[11]

Handling and storing materials involves diverse operations such as hoisting tons of steel with a crane; driving a truck loaded with concrete blocks; manually carrying bags and material; and stacking drums, barrels, kegs, lumber, or loose bricks.

The efficient handling and storing of materials is vital to industry. These operations provide a continuous flow of raw materials, parts, and assemblies through the workplace and ensure that materials are available when needed. Yet, the improper handling and storing of materials can cause costly injuries.

Workers frequently cite the weight and bulkiness of objects being lifted as major contributing factors to their injuries. In 2000, back injuries resulted in 400,000 workplace accidents. The second factor frequently cited by workers as contributing to their injuries was body movement. Bending, followed by twisting and turning, were the more commonly cited movements that caused back injuries. Back injuries accounted for more than 20 percent of all occupational illnesses, according to data from the National Safety Council.

In addition, workers can be injured by falling objects, improperly stacked materials, or by various types of equipment. When manually moving materials, however, workers should be aware of potential injuries, including the following:

- Strains and sprains from improperly lifting loads, or from carrying loads that are either too large or too heavy
- Fractures and bruises caused by being struck by materials, or by being caught in pinch points

- Cuts and bruises caused by falling materials that have been improperly stored, or by incorrectly cutting ties or other securing devices.

Since numerous injuries can result from improperly handling and storing materials, it is important to be aware of accidents that may occur from unsafe or improperly handled equipment and improper work practices, and to recognize the methods for eliminating, or at least minimizing, the occurrence of those accidents. Consequently, employers and employees can and should examine their workplaces to detect any unsafe or unhealthful conditions, practices, or equipment and take the necessary steps to correct them.

SUBPART I—TOOLS (HAND AND POWER)

1926.300   General requirements

1926.301   Hand tools

1926.302   Power operated hand tools

1926.303   Abrasive wheels and tools

1926.304   Woodworking tools

1926.305   Jacks—lever and ratchet, screw and hydraulic

1926.306   Air Receivers

1926.307   Mechanical power-transmission apparatus

The Subpart I standards set forth the requirements to be observed whenever the identified kinds of tools and equipment are used in construction. The General Requirements Standard, §1926.300, requires maintenance "in a safe condition" of all hand tools, power tools, and similar equipment; the guarding of moving parts of equipment; and imposes requirements for use of personal protective equipment (when using hand and power tools) and on-off controls on power tools.

Section 1926.306 is the same as §1910.169. See the discussion earlier in this book under Subpart M—Compressed Gas and Compressed Air Equipment.

Section 1926.307 is the same as §1910.219—Machinery and Machine Guarding. OSHA has also identified parts of four General Industry Standards that apply to Subpart I:

- Machine point of operations guarding (§1910.212(a)(3) and §1910.212(a)(5))
- Anchoring fixed machinery (§1910.212(b))
- Abrasive blast cleaning nozzles (§1910.244(b))
- Jacks, operation, and maintenance (§1910.244(a)(2)(iii) through
- 1910.244(a)(2)(viii))

SUBPART J—WELDING AND CUTTING

1926.350 Gas welding and cutting

1926.351   Arc welding and cutting

1926.352   Fire prevention

1926.353   Ventilation and protection in welding, cutting, and heating

1926.354   Welding, cutting, and heating in way of preservative coatings

Subpart J regulates all aspects of welding, cutting, and heating when those operations are performed on a construction project.

What are the most frequently cited standards in Subpart I?[12]

1. Arc welding cables improperly spliced or in need of repair (electric shock hazard) (1926.351(b)(2))
2. Inadequate ventilation while welding within a confined space (explosion or respiratory hazard) (1926.353(b)(1))
3. Regulators and gauges on compressed gas cylinders not in proper working order (danger of uncontrolled release of compressed gases) (1926.350(h))

What are effective control measures that can be used to control these hazards?[13]

Welding cables need to be inspected daily, before use, and repaired or replaced if needed. All splicing of cables must maintain the insulated protection with no exposed metal parts. Cables in need of repair are not to be used until they are put into safe serviceable condition.

Confined spaces must be adequately ventilated while welding is underway to control hazardous welding fumes. See §§ 1926.353 and 1926.350 for additional specific criteria.

Regulators and gauges need to be inspected daily before welding or cutting begins. Cylinders with defective regulators or gauges are not to be used until the defective equipment is replaced or repaired. See § 1926.350 for specific criteria.

SUBPART K—ELECTRICAL

GENERAL

1926.400   Introduction

INSTALLATION SAFETY REQUIREMENTS

1926.402   Applicability

1926.403   General requirements

1926.404   Wiring design and protection

1926.405   Wiring methods, components, and equipment for general use

1926.406   Specific purpose equipment and installations

1926.407 Hazardous (classified) locations

1926.408 Special systems

SAFETY-RELATED WORK PRACTICES

1926.416 General requirements

1926.417 Lockout and tagging of circuits

SAFETY-RELATED MAINTENANCE AND ENVIRONMENTAL
CONSIDERATIONS

1926.431 Maintenance of equipment

1926.432 Environmental deterioration of equipment

SAFETY REQUIREMENTS FOR SPECIAL EQUIPMENT

1926.441 Battery locations and battery charging

Subpart K covers the electrical safety requirements for construction jobsites. They are separated into four major divisions as indicated above. The electrical standards are quite detailed, so the definitions in §1926.449 should be consulted when reading the standards.

Installation Safety Requirements[14]

Part I of the Standard is very comprehensive. Only some of the major topics and brief summaries of these requirements are included in this discussion.

Sections 29 CFR 1926.402 through 1926.408 contains installation safety requirements for electrical equipment and installations used to provide electric power and light at the jobsite. These sections apply to installations, both temporary and permanent, used on the jobsite, but they *do not* apply to existing permanent installations that were in place before the construction activity commenced.

Approval

The electrical conductors and equipment used by the employer must be approved.

Examination, Installation, and Use of Equipment

The employer must ensure that electrical equipment is free from recognized hazards that are likely to cause death or serious physical harm to employees. Safety of equipment must be determined by the following:

- Suitability for installation and use in conformity with the provisions of the standard. Suitability of equipment for an identified purpose may be evidenced by a listing, by labeling, or by certification for that identified purpose.

- Mechanical strength and durability. For parts designed to enclose and protect other equipment, this includes the adequacy of the protection thus provided.
- Electrical insulation.
- Heating effects under conditions of use.
- Arcing effects.
- Classification by type, size, voltage, current capacity, and specific use.
- Other factors that contribute to the practical safeguarding of employees who use or are likely to come in contact with the equipment.

Guarding

Live parts of electric equipment operating at 50 volts or more must be guarded against accidental contact. Guarding of live parts must be accomplished as follows:

- Location in a cabinet, room, vault, or similar enclosure accessible only to qualified persons
- Use of permanent, substantial partitions or screens to exclude unqualified persons
- Location on a suitable balcony, gallery, or platform elevated and arranged to exclude unqualified persons
- Elevation of eight feet or more above the floor

Entrance to rooms and other guarded locations containing exposed live parts must be marked with conspicuous warning signs forbidding unqualified persons to enter.

Electric installations that are over 600 volts and that are open to unqualified persons must be made with metal-enclosed equipment or enclosed in a vault or area controlled by a lock. In addition, equipment must be marked with appropriate caution signs.

Overcurrent Protection

The following requirements apply to overcurrent protection of circuits rated 600 volts, nominal, or less.

- Conductors and equipment must be protected from overcurrent in accordance with their ability to safely conduct current, and the conductors must have sufficient current-carrying capacity to carry the load.
- Overcurrent devices must not interrupt the continuity of the grounded conductor unless all conductors of the circuit are opened simultaneously, except for motor-running overload protection.
- Overcurrent devices must be readily accessible and not located where they could create an employee safety hazard by being exposed to physical damage or located in the vicinity of easily ignitable material.
- Fuses and circuit breakers must be located or shielded so that employees will not be burned or otherwise injured by their operation, e.g., arcing.

Grounding of Equipment Connected by Cord and Plug

Exposed non-current-carrying metal parts of cord- and plug-connected equipment that may become energized must be grounded in the following situations:

- When in a hazardous (classified) location
- When operated at over 150 volts to ground, except for guarded motors and metal frames of electrically heated appliances if the appliance frames are permanently and effectively insulated from ground
- With one of the types of equipment listed below. (But see Item 6 for exemption.)

    1. Hand held motor-operated tools
    2. Cord- and plug-connected equipment used in damp or wet locations or by employees standing on the ground or on metal floors or working inside metal tanks or boilers
    3. Portable and mobile X-ray and associated equipment
    4. Tools likely to be used in wet and/or conductive locations
    5. Portable hand lamps
    6. Tools likely to be used in wet and/or conductive locations need not be grounded if supplied through an isolating transformer with an ungrounded secondary of not over 50 volts. Listed or labeled portable tools and appliances protected by a system of double insulation, or its equivalent, need not be grounded. If such a system is employed, the equipment must be distinctively marked to indicate that the tool or appliance uses a system of double insulation.

Safety-Related Work Practices[15]

Protection of Employees

The employer must not permit an employee to work near any part of an electric power circuit that the employee could contact in the course of work, unless the employee is protected against shock by de-energizing the circuit and grounding it or by guarding it effectively by insulation or other means.

Where the exact location of underground electric power lines is unknown, employees using jack hammers or hand tools that may contact a line must be provided with insulated protective gloves.

Even before work is begun, the employer must determine by inquiry, observation, or instruments where any part of an exposed or concealed energized electric power circuit is located. This is necessary because a person, tool, or machine could come into physical or electrical contact with the electric power circuit.

The employer is required to advise employees of the location of such lines, the hazards involved, and protective measures to be taken as well as to post and maintain proper warning signs.

Passageways and Open Spaces

The employer must provide barriers or other means of guarding to ensure that workspace for electrical equipment will not be used as a passageway during the time when energized parts of electrical equipment are exposed. Walkways and similar working spaces must be kept clear of electric cords.

Other standards cover load ratings, fuses, cords, and cables.

Lockout and Tagging of Circuits

Tags must be placed on controls that are to be deactivated during the course of work on energized or de-energized equipment or circuits. Equipment or circuits that are de-energized must be rendered inoperative and have tags attached at all points where such equipment or circuits can be energized.

Safety-Related Maintenance and Environmental Considerations[16]

Maintenance of Equipment

The employer must ensure that all wiring components and utilization equipment in hazardous locations are maintained in a dust-tight, dust-ignition-proof, or explosion-proof condition without loose or missing screws, gaskets, threaded connections, seals, or other impairments to a tight condition.

Environmental Deterioration of Equipment

Unless identified for use in the operating environment, no conductors or equipment can be located

- In damp or wet locations
- Where exposed to gases, fumes, vapors, liquids, or other agents having a deteriorating effect on the conductors or equipment
- Where exposed to excessive temperatures

Control equipment, utilization equipment, and busways approved for use in dry locations only must be protected against damage from the weather during building construction.

For protection against corrosion, metal raceways, cable armor, boxes, cable sheathing, cabinets, elbows, couplings, fittings, supports, and support hardware must be of materials appropriate for the environment in which they are installed.

Safety Requirements for Special Equipment[17]

Batteries

Batteries of the unsealed type must be located in enclosures with outside vents or in well-ventilated rooms arranged to prevent the escape of fumes, gases, or electrolyte spray into other areas. Other provisions include the following:

- Ventilation—to ensure diffusion of the gases from the battery and to prevent the accumulation of an explosive mixture.
- Racks and trays—treated to make them resistant to the electrolyte.
- Floors—acid-resistant construction unless protected from acid accumulations.
- Face shields, aprons, and rubber gloves—for workers handling acids or batteries.
- Facilities for quick drenching of the eyes and body—within 25 feet (7.62 m) of battery handling areas.
- Facilities for flushing and neutralizing spilled electrolytes and for fire protection.

Battery Charging

Battery charging installations must be located in areas designated for that purpose. When batteries are being charged, vent caps must be maintained in functioning condition and kept in place to avoid electrolyte spray. Also, charging apparatus must be protected from damage by trucks.

What are the most frequently cited standards in Subpart K?[18]

1. Electric equipment must be approved by an OSHA accepted laboratory or agency. (1926.403(a))
2. Electric equipment must be free from recognized hazards and used in accordance with manufacturers and approval agency instructions. (1926.403(b))
3. Live parts of electric equipment operating at more than 50 volts must be guarded from contact by an approved enclosure or by other approved means. (See §1926.405(a)(2)(ii) (E), (a)(2)(iii), and (b) for specific guarding requirements for lamps on temporary wiring, for temporary wiring over 600 volts, and for electrical boxes, respectively.) (1926.403(I) (2) & (j))
4. Employers must provide either ground-fault circuit interrupters or assured equipment grounding conductor program. (1926.404(b) (1))
5. Electric equipment must be grounded under the conditions and in the manner given in Subpart K. (1926.404(f))
6. Extension cord sets used with portable tools and appliances must be of the three-wire, grounding type; and flexible cords must be designed for hard or extra-hard usage. (1926.405(a) (2)(ii)(J))
7. Flexible cords must be provided with strain relief. (1926.405(g) (2) (IV))

8. The employer must determine the location of electric circuits that might be contacted during the course of work and must implement measures to protect employees from accidental contact. (1926.416(a))
9. Worn or frayed electric cords and cables may not be used. (1926.416(e) (1))

What are some effective control measures that can be used to control electrical hazards listed in Subpart K?[19]

1. Make sure that electric equipment is listed or labeled by an Agency or testing laboratory acceptable to OSHA.
2. Use electric equipment in accordance with the manufacturer's instructions. Install electric equipment with exposed live parts inside approved enclosures, or install exposed live parts in such a manner that unqualified employees cannot readily gain access to them (for example, by installation at a height of eight feet or more above the floor).
3. Make sure that all non-double-insulated electric equipment is equipped with a grounding conductor. Maintain the equipment grounding conductor in good, operable condition.
4. Provide either ground-fault circuit interrupters or an assured equipment grounding conductor program (which includes the regular testing of all equipment grounding conductors) to protect employees from ground faults.
5. Use only three-wire extension cords sets, which provide an equipment grounding conductor. (This applies regardless of whether they are used "only" with double-insulated tools and equipment.) Use flexible cords and cables that are marked with one of the following types: S, SC, SCE, SCT, SE, SEO, SEOO, SJ, SJE, SJEO, SJEOO, SJO, SJT, SJTO, SJTOO, SO, SOO, ST, STO, STOO, G, PPE, or W.
6. Provide strain relief where flexible cords are connected to devices and fittings to prevent pull from being applied directly to joints or terminal screws.
7. Determine, by direct observation, by inquiry, or by the use of instruments, whether energized electric circuits are present in a work area where they might be contacted by employees. Deenergize and ground the circuits or guard them, or use insulation to protect employees. Inform employees of the presence and location of these circuits.
8. Inspect flexible cords for damage, and discard or repair any that are worn or frayed or that have damaged insulation.

What work practices are needed for protection against electrical hazards?[20]

1. Maintain a ten-foot minimum clearance from overhead power lines (1926.416, 1926.550(a)(15), and 1926.600(a)(6))
2. Use barriers or other forms of guarding when live parts of electric equipment and circuits are exposed (1926.416(b)(1))
3. Maintain electric equipment in good condition (1926.416(e) (1) and 1926.431)
4. Lockout and tag electric circuits are to be deenergized (1926.417)

SUBPART L—SCAFFOLDING

1926.450 Scope and application

1926.451 General Requirements

1926.452 Requirements for specific types of scaffolds

1926.453 Aerial lifts

1926.454 Training requirements

Employers who use scaffolds in construction work should be familiar with all the definitions included in §1926.450. Section 1926.451 contains all the general requirements for scaffolds. It also sets the minimum strength criteria for all scaffold components and connections. It requires that each scaffold component be capable of supporting, without failure, its own weight and at least four times the maximum intended load applied to it. Section1926.452 has specific requirements for 25 particular types of scaffolds listed below:

Specific Requirements for Scaffolds

Pole	§1926.452 (a)
Tube and coupler	§1926.452 (b)
Fabricated frame	§1926.452 (c)
Plasterers, decorators, and large area scaffolds	§1926.452 (d)
Bricklayers square	§1926.452 (e)
Horse	§1926.452 (f)
Form and carpenters	§1926.452 (g)
Roof bracket	§1926.452 (h)
Outrigger	§1926.452 (i)
Pump jack	§1926.452 (j)
Ladder jack	§1926.452 (k)
Window jack chairs	§1926.452 (l)
Crawling boards	§1926.452(m)
Step platform and trestle ladder	§1926.452 (n)
Single-point adjustable	§1926.452 (o)
Two-point adjustable suspension (Swing stages)	§1926.452 (p)

Multi-point suspension	§1926.452 (q)
Catenary	§1926.452 (r)
Float or ship	§1926.452 (s)
Interior hung	§1926.452 (t)
Needle beam	§1926.452 (u)
Multi-level	§1926.452 (v)
Mobile	§1926.452 (w)
Repair bracket	§1926.452 (x)
Stilts	§1926.452 (y)

Section 1926.453 sets some specific requirements for lift operations. It applies to equipment for vehicle mounted elevating and rotating work platforms.

Section 1926.454 adds a new section on training requirements. The standard distinguishes between the training needed by employees to erect and to dismantle scaffolds. The standard applies to all construction work. It requires employers to instruct each employee in the recognition and avoidance of unsafe conditions. It also sets certain criteria allowing employers to tailor training to fit the particular circumstances of each employer's workplace.

SUBPART M—FALL PROTECTION

1926.500 Scope, application, and definitions applicable to this subpart

1926.501 Duty to have fall protection

1926.502 Fall protection systems

1926.503 Training requirements

Subpart M contains the requirements for Fall Protection. It helps construction workplaces prevent employees from falling off, onto, or through levels, and to protect employees from being struck by falling objects.

Under the standard, employers will be able to select fall protection measures compatible with the type of work being performed. Fall protection generally can be provided through the use of guardrail systems, safety net systems, personal fall arrest systems, positioning device systems, and warning line systems, among others.

Introduction[21]

In the construction industry in the U.S., falls are the leading cause of worker fatalities. Each year, on average, between 150 and 200 workers are killed and more than 100,000 are injured as a result of falls at construction sites. OSHA recognizes that accidents involving falls are generally complex events frequently involving a variety of factors. Consequently the standard for fall protection deals with both the human and equipment-related issues in protecting workers from fall hazards. For example, employers and employees need to do the following:

- Where protection is required, select fall protection systems appropriate for given situations.
- Use proper construction and installation of safety systems.
- Supervise employees properly.
- Use safe work procedures.
- Train workers in the proper selection, use, and maintenance of all protection systems.

Duty to Have Fall Protection[22]

Section 1926.501 requires employers to assess the workplace to determine if the walking/working surfaces on which employees are to work have the strength and structural integrity to safely support workers. Employees are not permitted to work on those surfaces until it has been determined that the surfaces have the requisite strength and structural integrity to support the workers. Once employers have determined that the surface is safe for employees to work on, the employer must select one of the options listed for the work operation if a fall hazard is present.

For example, if an employee is exposed to falling six feet or more from an unprotected side or edge, the employer must select a guardrail system, safety net system, or personal fall arrest system to protect the worker.

Fall Protection Systems, Criteria, and Practices[23]

Under §1926.502, fall protection can be provided through the use of guardrail systems, safety net systems, or personal fall arrest systems. However, under certain conditions, employers that cannot use conventional fall protection equipment must develop and implement a written fall protection plan specifying alternative fall protection measures.

Guardrail Systems

If the employer chooses to use guardrail systems to protect workers from falls, the system must meet the following criteria: Guardrails require a top rail at around 42 inches, a mid-rail at around 21 inches, and toe boards when necessary to prevent objects from falling over the edge.

Safety Net Systems

If an employer chooses to use safety net systems, the system must be installed as close as practicable under the walking/working surface on which employees are working and never more than 30 feet (9.1 meters) below such levels. Defective nets cannot be used. Safety nets must be inspected at least once a week for wear, damage, and other deterioration. The maximum size of each safety net mesh opening is not to exceed 36 square inches (230 square centimeters) nor be longer than six inches (15 centimeters) on any side, and the openings, measured center-to-center of mesh ropes or webbing shall not exceed six inches (15 centimeters). All mesh crossings shall be secured to prevent enlargement of the mesh opening.

Safety nets must be capable of absorbing an impact force of a drop test consisting of a 400-pound (180 kilogram) bag of sand 30 inches (76 centimeters) in diameter dropped from the highest walking/working surface at which workers are exposed, but not from less than 42 inches (1.1 meters) above that level.

Personal Fall Arrest Systems

If an employer chooses to use a personal fall arrest system, it must consist of an anchorage, connectors, and a body harness and may include a deceleration device, lifeline, or suitable combinations. If a personal fall arrest system is used for fall protection, it must do the following:

- Limit maximum arresting force on an employee to 900 pounds
- Limit maximum arresting force on an employee to 1,800 pounds (eight kilonewtons) when used with a body harness
- Be rigged so that an employee can neither free fall more than six feet (1.8 meters) nor contact any lower level

Personal fall arrest systems must be inspected prior to each use for wear damage and other deterioration. Defective components must be removed from service. Dee-rings and snap hooks must have a minimum tensile strength of 5,000 pounds (22.2 kilonewtons) and must be proof-tested to a minimum tensile load of 3,600 pounds (16 kilonewtons) without cracking, breaking, or suffering permanent deformation.

Snap hooks must be sized to be compatible with the member to which they will be connected, or be of a locking configuration.

Horizontal lifelines must be installed and used under the supervision of a qualified person. Self-retracting lifelines and lanyards that automatically limit free fall distance to two feet (0.61 meters) or less must be capable of sustaining a minimum tensile load of 3,000 pounds (13.3 kilonewtons) applied to the device with the lifeline or lanyard in the fully extended position.

Self-retracting lifelines and lanyards that do not limit free fall distance to two feet (0.61 meters) or less, rip stitch lanyards, and tearing and deforming lanyards must be capable of sustaining a minimum tensile load of 5,000 pounds (22.2 kilonewtons) applied to the device with the lifeline or lanyard in the fully extended position.

Anchorages must be designed, installed, and used under the supervision of a qualified person as part of a complete personal fall arrest system that maintains a safety factor of at least two, i.e., capable of supporting at least twice the weight expected to be imposed upon it. Anchorages used to attach personal fall arrest systems must be independent of any anchorage being used to support or suspend platforms and must be capable of supporting at least 5,000 pounds (22.2 kilonewtons) per person attached.

Lanyards and vertical lifelines must have a minimum breaking strength of 5,000 pounds (22.2 kilonewtons).

Other acceptable fall protection systems under §1926.502 can be used, such as controlled access zones which are allowed as part of a fall protection plan when designed according to OSHA requirements. Safety monitoring systems will also be acceptable as part of a fall protection plan. Covers will be utilized to protect against fall hazards (like floor holes).

Written fall protection plans may be used when employers can demonstrate that conventional fall protection systems pose a greater hazard to employees than using alternative systems.

A warning line system can also be used. It can consist of ropes, wires, or chains. Warning lines must be erected around all sides of roof work areas.

When mechanical equipment is being used, the warning line must be erected not less than six feet (1.8 meters) from the roof edge parallel to the direction of mechanical equipment operation, and not less than ten feet (three meters) from the roof edge perpendicular to the direction of mechanical equipment operation.

Training Requirements

Section 1926.503 covers training requirements. Employers must provide a training program that teaches employees who might be exposed to fall hazards how to recognize such hazards and how to minimize them. Employees must be trained in the following areas:

- The nature of fall hazards in the work area.
- The correct procedures for erecting, maintaining, disassembling, and inspecting fall protection systems.
- The use and operation of controlled access zones and guardrail, personal fall arrest, safety net, warning line, and safety monitoring systems.
- The role of each employee in safety monitoring when the system is in use.
- The limitations on the use of mechanical equipment during the performance of roofing work on low-sloped roofs.
- The correct procedures for equipment and materials handling and storage and the erection of overhead protection.
- The employees' role in fall protection plans.
- The employer must prepare a written certification that identifies the employee trained and the date of the training. The employer or trainer must sign the certification record and retraining must be provided when necessary.

What are the most frequently cited violations of the Fall Protection Standard?[24]

1. Failure to protect workers from falls of six feet of more feet unprotected by sides, edges, floors, and roofs (1926.501(b)(1)(b)(10)and(b)(11))

2. Failure to protect workers from falling into or through holes and openings in floors and walls (1926.501(b)(4) and (b)(14))
3. Failure to provide guardrails on runways and ramps where workers are exposed to falls of six feet or more to a lower level (1926.501(b)(6))

What are some control measures that can be used for the serious hazards for which OSHA has most frequently cited employers?[25]

1. Determine if any of the work (even a small portion) can be performed at ground level or if a crane can be used to lift assembled portions (e.g., sections of roofing) into place, eliminating or reducing the number of workers exposed to falling.
2. Restrain the worker so he or she cannot reach the edge, thereby eliminating the fall hazard.
3. Consider the use of aerial lifts or elevated platforms to provide better working surfaces rather than walking top plates or beams.
4. Erect guardrail systems, warning lines, or control line systems to protect workers from falls off the edges of floors and roofs.
5. Place covers over holes as soon as they are created if no work is being done at the hole.
6. Use safety net systems or personal fall arrest systems (body harness).

What work practices can be used to provide protection from falls when conventional systems are not feasible?[26]

1. Designate one of the workers as a safety monitor to observe employees and to alert employees of hazards that could cause them to trip or fall.
2. Establish a designated area or control zone when conventional fall protection systems such as guardrails and personal fall arrest systems are not feasible or create a greater hazard.
3. Store materials in an area away from where workers are exposed to fall hazards.

SUBPART N—CRANES, DERRICKS, HOISTS, ELEVATORS, AND CONVEYORS

1926.550   Cranes and derricks

1926.551   Helicopter cranes

1926.552   Material hoists, personnel hoists, and elevators

1926.553   Base-mounted drum hoists

1926.554   Overhead hoists

1926.555   Conveyors

Requirements for Mobile Cranes[27]

Subpart N contains requirements for employers to lift materials. The employer shall comply with the manufacturer's specifications and limitations applicable to the operation of any and all cranes and derricks. Where manufacturer's specifications are not available, the limitations assigned to the equipment shall be based on the determinations of a qualified engineer competent in this field and such determinations will be appropriately documented and recorded. Attachments used with cranes shall not exceed the capacity, rating, or scope recommended by the manufacturer.

Rated load capacities, recommended operating speeds, and special hazard warnings or instruction shall be conspicuously posted on all equipment. Instructions or warnings shall be visible to operators while they are at their control stations.

Hand signals to crane and derrick operators shall be those prescribed by the applicable ANSI standard for the type of crane in use. An illustration of the signals shall be posted at the job site.

The employer shall designate a competent person who shall inspect all machinery and equipment prior to each use and during use to make sure it is in safe operating condition. Any deficiencies shall be repaired or defective parts replaced before continued use.

A thorough, annual inspection of the hoisting machinery shall be made by a competent person, or by a government or private agency recognized by the U.S. Department of Labor. The employer shall maintain a record of the dates and results of inspections for each hoisting machine and piece of equipment.

SUBPART O—MOTOR VEHICLES, MECHANIZED EQUIPMENT, AND MARINE OPERATIONS

1926.600   Equipment

1926.601   Motor vehicles

1926.602   Material handling equipment

1926.603   Pile driving equipment

1926.604   Site clearing

1926.605   Marine operations and equipment

1926.606   Definitions applicable to this subpart

The Subpart O standards apply to the equipment and operations listed in the titles of its six standards.

Section 1926.600 imposes various requirements for

1. Equipment that is left unattended at night
2. When employees inflate, mount, or dismount tires installed on split rims or rims equipped with locking rims or similar devices

3. Heavy equipment or machinery that is parked, suspended, or held aloft by slings, hoists or jacks
4. Care and charging batteries
5. When the equipment is being moved in the vicinity of power lines or energized transmitters
6. When rolling cars are on spur railroad tracks

This section also requires that cab glass must be safety glass (or equivalent) so that there will be no visible distortion to affect the safe operation of the machines that are regulated in Subpart O.

Section 1926.601 imposes various requirements for vehicles that operate within an off-highway jobsite that is not open to public traffic.

Section 1926.602 applies to earthmoving equipment, off-highway trucks, rollers, compactors, front-end loaders, bulldozers, tractors, lift-trucks, stackers, high-lift rider industrial trucks, and similar construction equipment. Pile driving and the equipment used in that operation is regulated by §1926.603.

Section 1926.604 requires rollover guards and canopy guards on equipment used in site clearing operations and provides that employees engaged in such operations be protected from irritant and toxic plants and instructed in the first aid treatment that is available.

Section 1926.605 imposes various requirements upon employers engaged in marine construction operations.

OSHA rules for other vehicles frequently used in construction are included in Subpart W. See the discussion of that subpart later in this chapter.

SUBPART P—EXCAVATIONS

1926.650   Scope, application, and definitions applicable to this subpart

1926.651   Specific excavation requirements

1926.652   Requirements for protective systems

The Subpart P standards apply to all open excavations (including trenches) made in the earth's surface. In some situations, it requires the use of written designs (approved by a registered professional engineer) for sloping, benching, and support systems.

There are five appendices to Subpart P that should be consulted by employers engaged in excavation work. They include illustrations, tables, and diagrams. They cover the following subjects:

Appendix A: Soil Classification

Appendix B: Sloping and Benching

Appendix C: Timber Shoring for Trenches

Appendix D: Aluminum Hydraulic shoring for Trenches

Appendix E: Alternatives to Timber shoring

Appendix F: Selection of Protective Systems

The Subpart P standards went into effect in 1990. They constitute a total revision of the OSHA trenching and excavation standards that had been in effect for the previous 18 years.

Section 1926.651 has specific requirements for excavations that need to be reviewed for further details. Daily inspections of excavations must be made by a competent person to make sure that cave-ins, failure of protective systems, or hazardous atmospheres are not found.

If the competent person discovers there are indications that could result in a possible cave-in, exposed employees must be removed from the hazardous area until the necessary precautions have been taken.

What are some control measures that can be used for hazards in Subpart P?[28]

1. The competent person should develop a checklist enumerating the items listed in P-1 and use the list to identify and correct unsafe or unhealthy conditions that exist on a particular worksite.
2. All excavations including trenches that are five feet (1.52 m) in depth or greater must be shored or sloped (1926.652). For excavations less than five feet (1.52 m) in depth, the competent person examines the excavation for potential cave-in hazards and makes a determination if protection is needed. (1926.652(a)(1))

SUBPART Q—CONCRETE AND MASONRY CONSTRUCTION

1926.700 Scope, application, and definitions applicable to this subpart

1926.701 General requirements

1926.702 Requirements for equipment and tools

1926.703 Requirements for cast-in-place concrete

1926.704 Requirements for pre-cast concrete

1926.705 Requirements for lift-slab construction operations

1926.706 Requirements for masonry construction

Subpart Q includes detailed requirements to be observed in concrete and masonry construction operations.

What are the most frequently cited violations of the Concrete and Masonry Construction Standard?[29]

1. Failure to protect employees from impalement—rebar not capped or covered (1926.701(b))
2. Failure to establish a limited access zone (LAZ) to limit the number of workers in the danger zone where a masonry wall is under construction (1926.706(a) (1))
3. Failure to brace unsupported section of masonry wall over eight feet in height (1926.706(b))
4. Failure to have drawings or plans at the jobsite to indicate jack layout and formwork placement (1926.703(a) (2))

What are some of the effective control measures that can be used to eliminate the hazard of being struck by flying brick and block in the event of a wall collapse? [30]

Until the wall has gained sufficient strength that overturning (collapse) is no longer a hazard, keep employees out of the area where the wall is being constructed unless they are actually engaged in constructing the wall. The most effective control measure is to follow the standard by marking off an area with tape, rope, or chain or any other material that will indicate to employees that they are not to enter the zone (area) that has been marked. The zone should be equal to the length of the wall under construction and extend out a distance equal to the height of the wall to be constructed plus four feet. So if the wall to be constructed is ten feet high and 30 feet long, the zone should be 14 feet by 30 feet.

What are some of the ways to brace a wall over eight feet to provide protection against the hazard of collapse?[31]

The project engineer or competent person should determine how best to brace the wall. According to the Magazine of Masonry Construction (November 1988) a typical masonry wall brace includes a vertical member, an inclined strut, stakes, and, if necessary, a strut-brace. Scaffold planks (two by tens) are typically used as the vertical member and the inclined strut, and two by fours for stakes and strut-brace. Two by fours and two by sixes are considered by most experts to be inadequate for vertical members or inclined struts. All lumber must be in serviceable condition.

The American National Standards Institute's Standard for Concrete and Masonry Work, ANSI A10.9-1983 standard, recommends that "[t]he support or bracing shall be designed by or under the supervision of a qualified person to withstand a minimum of 15 pounds per square foot. Local environmental conditions (e.g., strong winds) need to be considered in determining the bracing design. Braces or shores shall be secured in position."

Keep a copy of formwork drawings and plans at the jobsite for review by the employer, employees, and OSHA compliance personnel. Formwork must be installed as shown on the drawings or plans.

SUBPART R—STEEL ERECTION

1926.750 Scope

1926.751 Definitions

1926.752 Site layout, site-specific erection plan, and construction sequence

1926.753 Hoisting and rigging

1926.754 Structural steel assembly

1926.755 Column anchorage

1926.756 Beams and columns

1926.757 Open web steel joists

1926.758 Systems-engineered metal buildings

1926.759 Falling object protection

1926.760 Fall protection

1926.761 Training

The new steel erection standard enhances protections provided to iron workers by addressing the hazards that have been identified as the major causes of injuries and fatalities in the steel erection industry. These are hazards associated with working under loads; hoisting, landing and placing decking; column stability; double connections; landing and placing steel joints; and falling to lower levels.

The final rule protects all workers engaged in steel erection activities. It does not cover electric transmission towers, communications towers, broadcast towers, water towers or tanks.

Key provisions of the Steel Erection standard include the following:[32]

- Site Layout and Construction Sequence
  - ~ Requires certification of proper curing of concrete in footings, piers, etc., for steel columns
  - ~ Requires controlling contractor to provide erector with a safe site layout including pre-planning routes for hoisting loads
- Site-Specific Erection Plan
  - ~ Requires pre-planning of key erection elements, including coordination with controlling contractor before erection begins, in certain circumstances
- Hoisting and Rigging
  - ~ Provides additional crane safety for steel erection
  - ~ Minimizes employee exposure to overhead loads through pre-planning and work practice requirements
  - ~ Prescribes proper procedure for multiple lifts (Christmas-treeing)

- Structural Steel Assembly
    - ~ Provides safer walking/working surfaces by eliminating tripping hazards, and minimizes slips through new slip resistance requirements
    - ~ Provides specific work practices regarding safely landing deck bundles and promoting the prompt protection from fall hazards in interior openings
- Column Anchorage
    - ~ Requires four anchor bolts per column along with other column stability requirements
    - ~ Requires procedures for adequacy of anchor bolts that have been modified in the field
- Beams and Columns
    - ~ Eliminates extremely dangerous collapse hazards associated with making double connections at columns
- Open Web Steel Joists
    - ~ Requirements minimizing collapse of lightweight steel joists by addressing need for erection bridging and method of attachment
    - ~ Requirements for bridging terminus anchors with illustrations and drawings in a non-mandatory appendix (provided by SJI)
    - ~ New requirements to minimize collapse in placing loads on steel joists
- Systems-Engineered Metal Buildings
    - ~ Requirements to minimize collapse in the erection of these specialized structures which account for a major portion of steel erection in this country
- Fall Protection
    - ~ Controlled decking zone (CDZ) provisions to prevent decking fatalities
    - ~ Deckers in a CDZ and connectors must be protected at heights greater than two stories or 30 feet. Connectors between 15 and 30 feet must wear fall arrest or restraint equipment and be able to be tied off or be provided another means of fall protection.
    - ~ Requires fall protection for all others engaged in steel erection at heights greater than 15 feet
- Training
    - ~ Requires qualified person to train exposed workers in fall protection
    - ~ Requires qualified person to train exposed workers engaged in special, high-risk activities.

SUBPART S—UNDERGROUND CONSTRUCTION, CAISSONS, COFFERDAMS, AND COMPRESSED AIR

1926.800 Underground construction

1926.801 Caissons

1926.802 Cofferdams

1926.803 Compressed air

Subpart S applies only to underground construction work including tunnels, shafts, chambers and passageways, caisson work, work on cofferdams, and work conducted in a compressed air environment pay strict attention to them.

Introduction[33]

The final rule applies to the construction of underground tunnels, shafts, chambers, and passageways. It also applies to cut-and-cover excavations, both those physically connected to on-going underground construction tunnels and those that create conditions characteristic of underground construction. These hazards include reduced natural ventilation and light; difficult and limited access and egress; and exposure to air contaminants, fire, and explosion.

Workers in this industry traditionally have had a higher accident and injury rate than other workers in the heavy construction industry.

Many provisions have been revised to give the employer the flexibility to select from a variety of appropriate and effective methods of controlling workplace hazards in underground construction. The revisions in the final rule also give additional duties and responsibilities to a "competent person." In addition, an integral part of the standard is a safety program that focuses on instructing workers in topics relevant and appropriate to the specific jobsite.

Competent Person[34]

A "competent person" is one who is capable of identifying existing and predictable hazards in the workplace and is authorized to take corrective action to eliminate them [29 CFR 1926.32 (f)]. Under the standard, a competent person is responsible for determining whether air contaminants are present in sufficient quantities to be dangerous to life; for testing the atmosphere for flammable limits before restoring power and equipment and before returning to work after a ventilation system has been shut down due to hazardous levels of flammable gas or methane; for inspecting the work area for ground stability; for inspecting all drilling equipment prior to each use; and for inspecting hauling equipment before each shift and visually checking all hoisting machinery, equipment, anchorages, and rope at the beginning of each shift and during hoisting, as necessary.

Safety Instruction[35]

The standard requires that employees be taught to recognize and avoid hazards associated with underground construction. The instruction shall include the following topics, as appropriate for the jobsite:

- Air monitoring
- Ventilation and illumination
- Communications
- Flood control
- Mechanical and personal protective equipment

- Explosives; fire prevention and protection
- Emergency procedures—evacuation plans and check-in and check-out procedures.

Access and Egress[36]

Under this provision, the employer must provide safe access to and egress from all work stations and must prevent any unauthorized entry underground. Completed or unused sections of an underground work area must be barricaded. Unused openings must be covered, fenced off, or posted with warning signs indicating "Keep Out" or other similar language.

Check-In/Check-Out[37]

The employer is required to maintain a check-in/check-out procedure that ensures that aboveground personnel can have an accurate count of the number of persons underground in an emergency. At least one designated person is to be on duty aboveground whenever anyone is working underground. This person is also responsible for securing immediate aid for and keeping an accurate count of employees underground in case of an emergency.

A check-in/check-out procedure is not required, however, when the underground construction is sufficiently completed so that permanent environmental controls are effective and when remaining construction activity will not cause an environmental hazard or structural failure of the construction.

Hazardous Classifications[38]

The standard provides classification criteria for gassy or potentially gassy operations and identifies additional requirements for work in gassy operations.

Potentially Gassy Operations

Potentially gassy operations occur under either of the following circumstances:

- When air monitoring shows, for more than a 24-hour period, ten percent or more of the lower explosive limit (LEL) for methane or other flammable gases measured at 12 inches 1 0.25 inch from the roof, face, floor, or walls in any underground work area
- When the geological formation or history of the area shows that ten percent or more of the LEL for methane or other flammable gases is likely to be encountered in the underground operation

Gassy Operations

Gassy operations occur under the following conditions:

- When air monitoring shows, for three consecutive days, ten percent or more of the LEL for methane or other flammable gases measured at 12 inches 1 0.25 inch from the roof, face, floor, or walls in any underground work area

- When methane or other flammable gases mandating from the strata have ignited, indicating the presence of such gases
- When the underground operation is connected to a currently gassy underground work area and is subject to a continuous course of air containing a flammable gas concentration

When a gassy operation exists, additional safety precautions are required. These include using more stringent ventilation requirements, using diesel equipment only if it is approved for use in gassy operations, posting each entrance with warning signs, prohibiting smoking and personal sources of ignition, maintaining a fire watch when hot work is performed, and suspending all operations in the affected area until all special requirements are met or the operation is declassified. Additional air monitoring is also required during gassy conditions.

Air Monitoring[39]

Under the standard, the employer is required to assign a competent person to perform all air monitoring required to determine proper ventilation and quantitative measurements of potentially hazardous gases. In instances where monitoring of airborne contaminants is required by the standard to be conducted "as often as necessary," this individual is responsible for determining which substances to monitor and how frequently, taking into consideration factors such as jobsite location, geology, history, work practices, and conditions.

The atmosphere in all underground areas shall be tested quantitatively for carbon monoxide, nitrogen dioxide, hydrogen sulfide, and other toxic gases, dusts, vapors, mists, and fumes as often as necessary to ensure that prescribed limits (29 CFR 1926.55) are met. Quantitative tests for methane shall also be performed in order to determine whether an operation is gassy or potentially gassy and in order to comply with other sections of the standard ((j) (1) vii, viii, ix).

A record of all air quality tests (including location, date, time, substances, and amount monitored) is to be kept aboveground at the worksite and shall be made available to the Secretary of Labor upon request.

Oxygen[40]

Testing is to be performed as often as necessary to ensure that the atmosphere at normal atmospheric pressure contains at least 19.5 percent oxygen, but not more than 22 percent.

Hydrogen Sulfide[41]

When air monitoring indicates the presence of five parts per million (ppm) or more of hydrogen sulfide, testing is to be conducted in the affected area at the beginning and midpoint of each shift until the concentration of hydrogen sulfide has been less than 5 ppm for three consecutive days. Continuous monitoring shall be performed when hydrogen sulfide is present above 10 ppm. Employees must be notified when the concentration of hydrogen sulfide is above 10 ppm. At concentrations of 20 ppm, an alarm (visual and aural) must signal to indicate that additional measures might be required (e.g., respirators, increased ventilation, evacuation) to maintain the proper exposure levels.

Other Precautions[42]

When the competent person determines that there are contaminants present that are dangerous to life, the employer must post notices of the condition at all entrances to underground work areas and must ensure that the necessary precautions are taken.

In cases where five percent or more of the LEL for these gases is present, steps must be taken to increase ventilation air volume to reduce the concentration to less than five percent of the LEL (except when operating under gassy/potentially gassy requirements).

When ten percent or more of the LEL for methane or other flammable gases is detected where welding, cutting, or other 'hot' work is being performed, work shall be suspended until the concentration is reduced to less than ten percent of the LEL.

Where there is a concentration of 20 percent or more LEL, all employees shall be immediately withdrawn to a safe location aboveground, except those necessary to eliminate the hazard, and electrical power, except for acceptable pumping and ventilating equipment, shall be cut off to the endangered area until the concentration of the gas is less than 20 percent of the LEL.

Potentially gassy and gassy operations require additional air monitoring. These include testing for oxygen in the affected work areas; using flammable gas monitoring equipment (continuous automatic when using rapid excavation machines; manual as needed to monitor prescribed limits); performing local gas tests prior to doing, and continuously during, any hot work; testing continuously for flammable gas when employees are working underground using drill and blast methods and prior to reentry after blasting.

Ventilation[43]

There are a number of requirements for ventilation in underground construction activities. In general, fresh air must be supplied to all underground work areas in sufficient amounts to prevent any dangerous or harmful accumulation of dusts, fumes, mists, vapors, or gases. A minimum of 200 cubic feet of fresh air per minute is to be supplied for each employee underground. Mechanical ventilation, with reversible airflow, is to be provided in all of these work areas, except where natural ventilation is demonstrably sufficient. Where blasting or drilling is performed or other types of work operations that may cause harmful amounts of dust, fumes, vapors, etc., the velocity of airflow must be at least 30 feet per minute.

For gassy or potentially gassy operations, ventilation systems must meet additional requirements. Ventilation systems used during gassy operations also must have controls located aboveground for reversing airflow.

Illumination[44]

As in all construction operations, the standard requires that proper illumination be provided during tunneling operations, as specified in 29 CFR 1926.56. When explosives are handled, only acceptable portable lighting equipment shall be used within 50 feet of any underground heading.

Fire Prevention and Control[45]

In addition to the requirements of Subpart F, "Fire Protection and Prevention" (29 CFR 1926), open flames and fires are prohibited in all underground construction activities, except for hot work operations. Smoking is allowed only in areas free of fire and explosion hazards, and the employer is required to post signs prohibiting smoking and open flames where these hazards exist. Various work practices are also identified as preventive measures. For example, there are limitations on the piping of diesel fuel from the surface to an underground location. Also, the pipe or hose system used to transfer fuel from the surface to the storage tank must remain empty except when transferring the fuel. Gasoline is not to be used, stored, or carried underground. Gasses such as acetylene, liquefied petroleum, and methylacetylene propadiene (stabilized) may be used underground only for hot work operations. Leaks and spills of flammable or combustible fluids must be cleaned up immediately. The standard also requires fire prevention measures regarding fire-resistant barriers, fire-resistant hydraulic fluids, the location and storage of combustible materials near openings or access to underground operations, electrical installations underground, lighting fixtures, fire extinguishers, etc.

Hot Work[46]

During hot work operations such as welding, noncombustible barriers must be installed below work being performed in or over a shaft or raise. As mentioned earlier, during these operations, only the amount of fuel gas and oxygen cylinders necessary to perform welding, cutting, or other hot work over the next 24-hour period shall be kept underground. When work is completed, gas and oxygen cylinders shall be removed.

Cranes and Hoists[47]

The standard contains provisions applicable to hoisting that are unique to underground construction. The standard incorporates by reference § 1925.550 with respect to cranes; § 1926.550(g) with respect to crane-hoisting of personnel, except that the limitations in paragraph (g)(2) do not apply to the routine access of employees to the underground via a shaft; § 1926.552(a) and (b) with respect to requirements for material hoists; and § 1926.552(a), (c) and (d) with respect to requirements of personnel hoists and elevators. The provisions for underground construction include the following:

- Securing or stacking materials, tools, etc., being raised or lowered in a way to prevent the load from shifting, snagging, or falling into the shaft

- Using a flashing warning light for employees at the shaft bottom and subsurface shaft entrances whenever a load is above these locations or is being moved in the shaft
- Following procedures for the proper lowering of loads when a hoistway is not fully enclosed and employees are at the shaft bottom
- Informing and instructing employees of maintenance and repair work that is to commence in a shaft served by a cage, skip, or bucket
- Providing a warning sign at the shaft collar, at the operator's station, and at each underground landing for work being performed in the shaft
- Using connections between the hoisting rope and cage or skip that are compatible with the wire rope used for hoisting
- Using cage, skip, and load connections that will not disengage from the force of the hoist pull, vibration, misalignment, release of lift force, or impact
- Maintaining spin-type connections in a clean condition
- Assuring that wire rope wedge sockets, when used, are properly seated

Additional requirements for cranes include the use of limit switches, or anti-two-block devices. These operational aids are to be used only to limit travel of loads when operational controls malfunction and not as a substitute for other operational controls.

Emergencies[48]

At work sites where 25 or more employees work underground at one time, employers are required to provide rescue teams or rescue services that include at least two five-person teams (one on the jobsite or within one-half hour travel time and one within two hours travel time). Where there are fewer than 25 employees underground at one time, the employer shall provide or make available in advance one five-person rescue time on site or within one-half hour travel time.

Rescue team members have to be qualified in rescue procedures and in the use of firefighting equipment and breathing apparatus. Their qualifications must be reviewed annually.

The employer must ensure that rescue teams are familiar with the jobsite conditions. Rescue team members are required to practice donning and using self-contained breathing apparatus on a monthly basis for jobsites where flammable or noxious gases are encountered or anticipated in hazardous quantities.

As part of emergency procedures, the employer shall provide self rescuers (currently approved by NIOSH and MSHA) to be immediately available to all employees at underground work stations who might be trapped by smoke or gas. The selection, use, and care of respirators shall be in accordance with § 1926.103 (b) and (c).

A "designated," or authorized, person shall be responsible for securing immediate aid for workers and for keeping an accurate count of employees underground. Emergency lighting, a portable hand or cap lamp, shall be provided to all underground workers in their work areas to provide adequate light for escape.

SUBPART T—DEMOLITION

1926.850   Preparatory operations

1926.851   Stairs, passageways, and ladders

1926.852   Chutes

1926.853   Removal of materials through floor openings

1926.854   Removal of walls, masonry sections, and chimneys

1926.855   Manual removal of floors

1926.856   Removal of walls, floors, and material with equipment

1926.857   Storage

1926.858   Removal of steel construction

1926.859   Mechanical demolition

1926.860   Selective demolition by explosives

The Subpart T standards are restricted in their application to employers engaged in demolition operations. They are neither lengthy nor complicated. Employers engaged in those operations should have little difficulty understanding them.

*[handwritten: ↑20' - ENCLOSED LOOKING
Less than 20' DON'T NEED SHOOT]*

Preparatory Operations[49]

Before the start of every demolition job, the demolition contractor should take a number of steps to safeguard the health and safety of workers at the job site. These preparatory operations involve the overall planning of the demolition job, including the methods to be used to bring the structure down, the equipment necessary to do the job, and the measures to be taken to perform the work safely. Planning for a demolition job is as important as actually doing the work. Therefore all planning work should be performed by a competent person experienced in all phases of the demolition work to be performed.

The American National Standards Institute (ANSI) in its ANSI A10.6-1983—*Safety Requirements for Demolition Operations* states:

> No employee shall be permitted in any area that can be adversely affected when demolition operations are being performed. Only those employees necessary for the performance of the operations shall be permitted in these areas.

What are the most frequently cited Demolition Standards?[50]

1. Not performing a written engineering survey of the structure before commencing demolition work (1926.850(a))
2. Not providing fall protection for employees exposed to wall openings (1926.850(g))

3. Not providing shoring or bracing for walls to prevent premature collapse (1926.850(b))
4. Not properly inspecting and maintaining stairways and ladders in safe conditions for employee use (1926.851(b))
5. Not properly testing and removing hazardous materials from within the structure before commencing demolition work (1926.850(e)), and not providing sidewalk shed covers to protect employee entrances to structure (1926.850(h))

What are some control measures that can be used to protect employees from these hazards?[51]

An engineering survey needs to be done to determine areas where premature collapse may occur. While this survey is being done, the presence of hazardous materials needs to be noted including whether asbestos or lead is present. If hazardous materials are present, these hazards will need to be included in the work scheduling before the demolition work can begin. Knowledge of the structure's weak points and the presence of hazardous materials including the contents of equipment within the structure prior to beginning demolition work are essential for providing a safe work environment. See § 1926.850 preparatory operations, § 1926.62 lead, § 1926.1101 asbestos, and § 1926.150 fire protection for more specific criteria.

Fall protection measures are necessary for employees who will be using stairs, ladders, or working near wall openings. The measures required in Subpart M—Fall Protection and Subpart X—Stairways and Ladders will provide further clarification of what measures are necessary to ensure employees are protected from falling.

Safe entrance into the structure to be demolished is necessary to protect employees from objects falling onto them as they enter to begin work. This overhead shed protection needs to extend at least eight feet (2.46 m) out from the structure to prevent workers from being struck by falling objects.

SUBPART U—BLASTING AND USE OF EXPLOSIVES

1926.900 General provisions

1926.901 Blaster qualifications

1926.902 Surface transportation of explosives

1926.903 Underground transportation of explosives

1926.904 Storage of explosives and blasting agents

1926.905 Loading of explosives or blasting agents

1926.906 Initiation of explosive charges—electrical blasting

1926.907 Use of safety fuse

1926.908   Use of detonating cord

1926.909   Firing the blast

1926.910   Inspection after blasting

1926.911   Misfires

1926.912   Underwater blasting

1926.913   Blasting in excavation work under compressed air

1926.914   Definitions applicable to this subpart

Subpart U applies when employers use explosives or do blasting in their work. No employer should use explosives unless he or she is familiar with the Subpart U standards. They impose detailed restrictions on those working with explosives.

General Provisions[52]

- The employer shall permit only authorized and qualified persons to handle and use explosives.
- Smoking; firearms; matches; open flame lamps; and other fires, flame, or heat-producing devices and sparks shall be prohibited in or near explosive magazines or while explosives are being handled, transported, or used.
- No person shall be allowed to handle or use explosives while under the influence of intoxicating liquors, narcotics, or other dangerous drugs.
- All explosives shall be accounted for at all times:
- Explosives not being used shall be kept in a locked magazine, unavailable to persons not authorized to handle them.
- The employer shall maintain an inventory and use record of all explosives.
- Appropriate authorities shall be notified of any loss, theft, or unauthorized entry into a magazine.
- No explosives or blasting agents shall be abandoned.
- No fire shall be fought where the fire is in imminent danger of contact with explosives. All employees shall be removed to a safe area and the fire area guarded against intruders.
- Original containers, or Class II magazines, shall be used for taking detonators and other explosives from storage magazines to the blasting area.
- When blasting is done in congested areas or in proximity to a structure, railway, highway, or any other installation that may be damaged, the blaster shall take special precautions in the loading, delaying, initiation, and confinement of each blast with mats or other methods so as to control the throw of fragments, and thus prevent bodily injury to employees.
- Employees authorized to prepare explosive charges or conduct blasting operations shall use every reasonable precaution to ensure employee safety including, but not limited to, visual and audible warning signals, flags, or barricades.

- Insofar as possible, blasting operations above ground shall be conducted between sunup and sundown.
- Due precautions shall be taken to prevent accidental discharge of electric blasting caps from current induced by radar, radio transmitters, lightning, adjacent power lines, dust storms, or other sources of extraneous electricity. These precautions shall include:
- Short-circuited detonators in holes and which have been primed and shunted until wired into the blasting circuit.
- The suspension of all blasting operations and removal of persons from the blasting area during the approach and progress of an electric storm
- The prominent display of adequate signs, warning against the use of mobile radio transmitters, on all roads within 1,000 feet of blasting operations. Whenever adherence to the 1,000-foot distance would create a handicap, a competent person shall evaluate the situation and alternative provisions may be used to prevent premature firing of electric blasting caps. These alternative procedures shall be in writing, certified by the competent person, and maintained at the construction site during the duration of work.
- Specimens of signs which would meet the requirements of the above paragraph are the following:

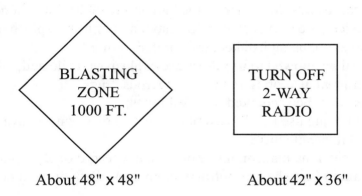

About 48" x 48"   About 42" x 36"

- Mobile radio transmitters which are less than 100 feet away from electric blasting caps, in other than original containers, shall be deenergized and effectively locked.
- Empty boxes and paper and fiber packing materials, which have previously contained high explosives, shall not be used again for any purpose, but shall be destroyed by burning at an approved location.
- Explosives, blasting agents, and blasting supplies that are obviously deteriorated or damaged shall not be used.
- Delivery and issue of explosives shall only be made by and to authorized persons and into authorized magazines or approved temporary storage or handling areas.
- Blasting operations in the proximity of overhead power lines, communication lines, utility services, or other services and structures shall not be carried on until the operators and/or owners have been notified and measures for safe control have been taken.
- The use of black powder shall be prohibited.

- All loading and firing shall be directed and supervised by competent persons thoroughly experienced in this field.
- All blasts shall be fired electrically with an electric blasting machine or properly designed electric power source, except as provided in §1926.906(a) and (r).

Buildings used for the mixing of blasting agents or water gels shall conform to the following requirements:

- Buildings shall be of noncombustible construction or sheet metal on wood studs.
- Floors in a mixing plant shall be of concrete or of other nonabsorbent materials.
- Where fuel oil is used, all fuel oil storage facilities shall be separated from the mixing plant and located in such a manner that in case of tank rupture, the oil will drain away from the mixing plant building.
- The building shall be well ventilated.
- Heating units which do not depend on combustion processes, when properly designed and located, may be used in the building. All direct sources of heat shall be provided exclusively from units located outside the mixing building.
- All internal-combustion engines used for electric power generation shall be located outside the mixing plant building, or shall be properly ventilated and isolated by a firewall. The exhaust systems on all such engines shall be located so any spark emission cannot be a hazard to any materials in or adjacent to the plant

SUBPART V — POWER TRANSMISSION AND DISTRIBUTION

1926.950 General requirements

1926.951 Tools and protective equipment

1926.952 Mechanical equipment

1926.953 Material handling

1926.954 Grounding for protection of employees

1926.955 Overhead lines

1926.956 Underground lines

1926.957 Construction in energized substations

1926.958 External load helicopters

1926.959 Lineman's body belts, safety straps, and lanyards

1926.960 Definitions applicable to this subpart

The Subpart V standards are limited in their application to construction, erection, alteration, conversion and improvement of electric transmission and distribution lines and equipment. Employers engaged in that type of work must familiarize themselves with those standards.

Employers who do not do work of that kind will have no reason to review the Subpart V standards.

Initial Inspections, Tests, or Determinations[53]

Existing conditions shall be determined before starting work, by an inspection or a test. Such conditions shall include, but are not be limited to, energized lines and equipment; conditions of poles; and the location of circuits and equipment, including power and communication lines, CATV, and fire alarm circuits.

Electric equipment and lines shall be considered energized until determined to be de-energized by tests or other appropriate methods or means.

Operating voltage of equipment and lines shall be determined before working on or near energized parts.

Clearances[54]

No employee shall be permitted to approach or take any conductive object without an approved insulating handle closer to exposed energized parts than shown in Table V-1, unless

- The employee is insulated or guarded from the energized part (gloves or gloves with sleeves rated for the voltage involved shall be considered insulation of the employee from the energized part)
- The energized part is insulated or guarded from the employee and any other conductive object at a different potential
- The employee is isolated, insulated, or guarded from any other conductive object(s), as during live-line bare-hand work

The minimum working distance and minimum clear hot stick distances stated in Table 3.1 shall not be violated. The minimum clear hot stick distance is that for the use of live-line tools held by linemen when performing live-line work.

TABLE 3.1 Alternating Current—Minimum Distances

Voltage Range (phase to phase) [Kilovolts]	Minimum Working and Clear Hot Stick Distance
2.1 to 15	2 ft. 0 in.
15.1 to 35	2 ft. 4 in.
35.1 to 46	2 ft. 6 in.
46.1 to 72.5	3 ft. 0 in.
72.6 to 121	3 ft. 4 in.
138 to 145	3 ft. 6 in.
161 to 169	3 ft. 8 in.
230 to 242	5 ft. 0 in.
345 to 362	7 ft. 0 in.*
500 to 552	11 ft. 0 in.*
700 to 765	15 ft. 0 in.*

* NOTE: From 345-362 kv., 500-552 kv., and 700-765 kv., the minimum working distance and the minimum clear hot stick distance may be reduced provided that such distances are not less than the shortest distance between the energized part and a grounded surface.

Conductor support tools, such as link sticks, strain carriers, and insulator cradles, may be used provided that the clear insulation is at least as long as the insulator string or the minimum distance specified in Table 3.1 for the operating voltage.

SUBPART W—ROLLOVER PROTECTIVE STRUCTURES; OVERHEAD PROTECTION

1926.1000 Rollover protective structures (ROPS) for material handling equipment

1926.1001 Minimum performance criteria for rollover protective structures for designated scrapers, loaders, dozers, graders, and crawler tractors

1926.1002 Protective frame (ROPS) test procedures and performance requirements for wheel-type agricultural and industrial tractors used in construction

1926.1003 Overhead protection for operators of agricultural and industrial tractors

Subpart W only applies to seven kinds of material handling equipment when used in construction work: 1) Rubber-tired, self-propelled scrapers; 2) rubber-tired front-end loaders; 3) Rubber-tired dozers; 4) Wheel-type agricultural and industrial tractors;(5) Crawler tractors; 6) Crawler-type loaders; and 7) Motor graders, with or without attachments.

Employers who use any of the seven kinds of equipment listed above should familiarize themselves with the applicable requirements. Those who don't use that equipment should have no reason to be concerned with Subpart W.

Remounting[55]

Rollover protective structures (ROPS) removed for any reason, shall be remounted with equal quality, or better, bolts or welding as required for the original mounting.

Labeling[56]

Each ROPS shall have the following information permanently affixed to the structure:

- Manufacturer or fabricator's name and address;
- ROPS model number, if any;
- Machine make, model, or series number that the structure is designed to fit.

Machines Meeting Governmental Requirements[57]

Any machine in use equipped with ROPS shall be deemed in compliance with this section if it meets the ROPS requirements of the State of California, the U.S. Army Corps of Engineers, or the Bureau of Reclamation of the U.S. Department of the Interior in effect on April 5, 1972. The requirements in effect are

- State of California: Construction Safety Orders, issued by the Department of Industrial Relations pursuant to Division 5, Labor Code, §6312, State of California
- U.S. Army Corps of Engineers: General Safety Requirements, EM-385-1-1 (March 1967)
- Bureau of Reclamation, U.S. Department of the Interior: Safety and Health Regulations for Construction. Part II (September 1971)

SUBPART X—STAIRWAYS AND LADDERS

1926.1050	Scope, application, and definitions applicable to this subpart
1926.1051	General requirements
1926.1052	Stairways
1926.1053	Ladders
1926.1060	Training requirements

The Subpart X standards went into effect in 1991. They set forth the conditions and circumstances under which ladders and stairways must be provided (and used). The standards apply to all stairways and ladders used in construction, alteration, repair, demolition, painting and repair work (however, some of the requirements do not apply to some ladders and stairways that were built prior to March 15, 1991).

Any employer who has one or more employees engaged in construction work who will use a ladder or stairway while working (and that covers the vast majority of construction employers) must become familiar with the Subpart X requirements.

Section 1926.1060 requires that a training program be provided for each employee using ladders and stairways so that he or she will be able to recognize the hazards related to ladders and stairways and know the procedures to be followed in order to minimize those hazards.

An appendix (Appendix A) has been added to Subpart X to assist employers in complying with the loading and strength requirements for ladders that are set forth in §1926.1053(a)(1).

What are the most frequently cited serious stairway and ladder violations?[58]

1 Not providing a handrail or stair rail system on stairs of four or more steps (1926.1052(c)(1))
2 Not securing a portable ladder or having it extended three feet (.9 m) above the upper landing before workers use it to reach an upper level (1926.1053(b)(1))
3 Not providing a safe means to gain access to a vertical rise in elevation of 19 inches (48 cm) or more (1926.1051(a))
4 Not providing a training program for workers on the proper construction, inspection, maintenance, care, use, and limitations of stairways and ladders (1926.1060(a))
5 Not marking or tagging a defective ladder so that it would not be used before it has been repaired (1926.1053(b) (16))

What are some effective control measures that can be used to protect workers from hazards in Subpart X?[59]

All stairways of four steps or more need to have a handrail. In addition, if there is a fall hazard of six feet (1.8 m) or more on an exposed side of the stairs then a stair rail system must be provided to prevent workers from falling off the side. Also, stairway landings six feet (1.8 m) or more above the surrounding area need to be provided with a guardrail system along the exposed perimeters of the landing.

When using a portable ladder, the top end must extend above the upper landing level by three feet (.9 m) or otherwise be tied off at the top to some secure point so that the ladder will not lose its position while a worker is using it. Portable ladders with structural defects are not to be used and are to be tagged or marked to indicate they are not to be used. Employees need to know what to look for to ensure the ladder is safe to use before they put it in use.

Training for workers who use ladders and stairways is to help them avoid overloading, to know the fall protection features that need to be present when using ladders, and to know the correct procedures for erecting and disassembling stairways or ladders.

SUBPART Y

Subpart Y is identical to the Part 1910 General Industry standards discussed earlier in this book under the heading Subpart T – Commercial Diving Operations.

SUBPART Z—TOXIC AND HAZARDOUS SUBSTANCES

1926.1101	Asbestos
1926.1102	Coal tar pitch volatiles: interpretation of term
1926.1103	13 carcinogens (4-Nitrobiophenyl, etc.)
1926.1104	alpha-Naphthylamine
1926.1106	Methyl chloromethyl ether
1926.1107	3.3'-Dichlorobenzidine (and its salts)
1926.1108	bis-Chloromethyl ether
1926.1109	beta-Naphthylamine
1926.1110	Benzidine
1926.1111	4-Aminodiphenyl
1926.1112	Ethyleneimine
1926.1113	beta-Propiolactone
1926.1114	2-Acetylaminofluorene
1926.1115	4-Dimethylaminoazobenzen
1926.1116	N-Nitrosodimethylamine
1926.1117	Vinyl chloride
1926.1118	Inorganic arsenic
1926.1127	Cadmium
1926.1128	Benzene
1926.1129	Coke oven emissions
1926.1144	1,2 dibromo-3-chloropropane
1926.1145	Acrylonitrile
1926.1147	Ethylene oxide

1926.1148   Formaldehyde

1926.1152   Methylene chloride

Subpart Z contains 24 substance-specific standards that are identical to the general industry standards that regulate those same substances (see the earlier discussion of Part 1910, Subpart Z—Toxic and Hazardous Substances). Those standards are designated in the construction industry standards.

Endnotes:

[1] 29 CFR 1926.20

[2] 29 CFR 1926.33

[3] 29 CFR 1926.34

[4] 29 CFR 1926.35

[5] OSHA Fact Sheet No. OSHA 93-26. Available at http://www.osha.gov/pls/oshaweb/owadisp.show_document?p_table=FACT_SHEETS&p_id=151

[6] http://www.osha-slc.gov/Publications/Const_Res_Man/1926sub-e-overview.html

[7] *Ibid.*

[8] OSHA, *Overview for Subpart F, Fire Protection and Prevention.* http://www.osha-slc.gov/Publications/Const_Res_Man/1926sub-f-overview.html

[9] *Ibid.*

[10] *Federal Register*, 67:57722-57736, *FR* Doc. 02-23142, 9/12/2002.

[11] U.S. Department of Labor, *Materials Handling and Storage*, OSHA 2236, 1998.

[12] OSHA, *Overview for Subpart J, Welding and Cutting.* Available at http://www.osha-slc.gov/Publications/Const_Res_Man/1926sub-j-overview.html

[13] *Ibid.*

[14] OSHA, *Electrical Standards for Construction, Construction Safety and Health Outreach Program.* Available at http://www.osha-slc.gov/doc/outreachtraining/htmlfiles/elecstd.html

[15] *Ibid.*

[16] *Ibid.*

[17] *Ibid.*

[18] OSHA, *Overview for Subpart K, Electrical.* Available at http://www.oshaslc.gov/Publications/Const_Res_Man/1926sub-k-overview.html

[19] *Ibid.*

[20] *Ibid.*

[21] OSHA, *Fall Protection, Construction Safety and Health Outreach Program.* Available at http://www.osha-slc.gov/doc/outreachtraining/htmlfiles/subpartm.html

[22] *Ibid.*

[23] *Ibid.*

[24] OSHA, *Overview for Subpart M, Fall Protection.* Available at http://www.osha.gov/Publications/Const_Res_Man/1926sub-m-overview.html

[25] *Ibid.*

[26] *Ibid.*

[27] OSHA, *Mobile Crane Inspection Guidelines for OSHA Compliance Officers*, 1994. Available at http://www.osha-slc.gov/SLTC/cranehoistsafety/mobilecrane/mobilecrane_3.html

[28] OSHA, *Overview for Subpart P, Excavations*, http://www.oshaslc.gov/Publications/Const_Res_Man/1926sub-p-overview.html

[29] OSHA, *Overview for Subpart Q, Concrete and Masonry*. Available at http://www.osha-slc.gov/Publications/Const_Res_Man/1926sub-q-overview.html

[30] *Ibid.*

[31] *Ibid.*

[32] OSHA, *Fact Sheet: Revised Steel Erection Standard*. Available at http://www.osha-slc.gov/steelerection/steelerectionfactsheet.html

[33] OSHA, *Underground Construction (Tunneling), Construction Safety and Health Outreach Program*. Available at http://www.osha-slc.gov/doc/outreachtraining/htmlfiles/tunnel.html

[34] *Ibid.*

[35] *Ibid.*

[36] *Ibid.*

[37] *Ibid.*

[38] *Ibid.*

[39] *Ibid.*

[40] *Ibid.*

[41] *Ibid.*

[42] *Ibid.*

[43] *Ibid.*

[44] *Ibid.*

[45] *Ibid.*

[46] *Ibid.*

[47] *Ibid.*

[48] *Ibid.*

[49] OSHA, *Demolition, Construction Safety and Health Outreach Program*. Available at http://www.osha-slc.gov/doc/outreachtraining/htmlfiles/demolit.html

[50] OSHA, *Overview for Subpart T, Demolition*. Available at http://www.osha-slc.gov/Publications/Const_Res_Man/1926sub-t-overview.html

[51] *Ibid.*

[52] OSHA, *Blasting and the Use of Explosives, Construction Safety and Health Outreach Program*. Available at http://www.osha-slc.gov/doc/outreachtraining/htmlfiles/blasting.html

[53] OSHA, *Power Transmission and Distribution, Construction Safety and Health Outreach Program*. Available at http://www.osha-slc.gov/doc/outreachtraining/htmlfiles/subpartv.html

[54] *Ibid.*

[55] OSHA, *Rollover Protective Structures (ROPS) for Material Handling Equipment, Construction Safety and Health Outreach Program*. Available at http://www.osha-slc.gov/doc/outreachtraining/htmlfiles/rollover.html

[56] *Ibid.*

[57] *Ibid.*

[58] OSHA, *Overview for Subpart X, Stairways and Ladders*. Available at http://www.osha-slc.gov/Publications/Const_Res_Man/1926sub-x-overview.html

[59] *Ibid.*

4 #3

Hazard Communication

Communicating the Hazards of Chemicals to Workers

INTRODUCTION

The Hazard Communication Standard is different from most OSHA standards. The principle purpose is to require employers to inform employees of the hazards associated with the use of chemicals in the workplace. Its requirements are intended to make employers communicate the hazards of chemicals to employees.

OSHA inspects for compliance with the requirement of the Hazard Communication Standard on every inspection it makes. The rate of non-compliance is very high in the construction industry. It is the number one most frequently cited standard against employers.

SCOPE

The Hazard Communication Standard covers all potential workplace exposures involving hazardous substances as defined by federal, state, and local regulations.

HAZARD DETERMINATION

Companies should not independently evaluate any of the hazardous substances purchased from suppliers and/or manufacturers. Rely upon the evaluation performed by the suppliers or by the manufacturers of the substances to satisfy the requirements for hazard determination.

CONTAINER LABELING

No container or hazardous substances should be released for use unless the container is correctly labeled and the label is legible.

All chemicals in bags, drums, barrels, bottles, boxes, cans, cylinders, reaction vessels, storage tanks, or the like should be checked by the receiving department to ensure the manufacturer's label is intact, is legible, and has not been damaged in any manner during shipment. Any containers found to have damaged labels should be quarantined until a new label has been installed.

123

The label must contain

- The chemical name of the contents
- The appropriate hazard warnings
- The name and address of the manufacturer and any other information required.

All secondary containers shall be labeled. The information must include details of all chemicals which are in the referenced container.

MATERIAL SAFETY DATA SHEETS (MSDS)

OSHA rules outline the content but not the exact form of every Material Safety Data Sheet. Here is what OSHA requires each data sheet to contain:

Identity—The data sheet must contain the name of the chemicals, which is found on the label. In addition, subject to deletion of legitimate trade secrets, it must give the chemical and common name of the substance. If the substance is a mixture and has not been tested as such, the data sheet must give the name of each hazardous constituent.

Characteristics—The data sheet must cite the physical and chemical characteristics of the chemical, such as vapor pressure and flash point.

Physical Hazards—Any potential for fire, explosion, or reaction must be included in the data sheet.

Health Hazards—Signs and symptoms of exposure must be entered, as must all medical conditions that are likely to be aggravated by exposure.

Routes of Entry—The data sheet must specify whether the chemical typically enters the system by ingestion, inhalation, dermal exposure, or some other route.

Exposure Limits—If OSHA has established an exposure limit for the chemical, or if a Threshold Limit Value has been established by the American Conference of Governmental Industrial Hygienists, these must be entered on the data sheet, as must any exposure limit used by the authority preparing the data sheet.

Carcinogens. The data sheet must indicate whether the chemical is listed as a carcinogen by the National Toxicology Program, by OSHA, or by the International Agency for Research in Cancer.

Use and Handling—The data sheet must cite any general applicable precautions for safe handling and use which are known to the firm preparing the data sheet, including hygiene practices, protective measures during repair and maintenance of contaminated equipment, and procedures for clean-up of spills and leaks. Industrial chemical consumers might often add site-specific procedures to the more general information offered by the chemical manufacturer.

Exposure Controls—The data sheet must include a description of special procedures to be employed in emergencies, as well as a description of appropriate first aid.

Dates—The sheet must bear the date of its preparation or of its latest revision.

Information Source—Finally, the sheet must cite the name, address, and telephone number of the person who prepared the data sheet or of some other person who can provide additional information relating to the chemical, such as citations to scientific literature or specialized emergency procedures.

Each location must maintain a master MSDS file as well as a department specific file. These Material Safety Data Sheets must be available to all employees, at all times, upon request.

The Safety Officer, Safety Committee, or a designee should review all incoming MSDSs for new and significant health/safety information. The company should ensure that any new information is passed on to the employees involved.   MATERIAL SAFETY DATA SHEET

If any MSDS is missing or obviously incomplete, a new MSDS will be requested from the manufacturer or distributor. OSHA is to be notified if the manufacturer or distributor will not supply the MSDS or if it is not received after 30 days from request. Any new information will be passed on to employees involved.

New materials should not be introduced into the work area until an MSDS has been received.

The purchasing department should make it an ongoing part of their function to obtain MSDSs for all new materials when they are first ordered.

The safety coordinator or his or her designee should coordinate with appropriate departments to make sure all MSDSs are obtained, distributed, and communicated.

LIST OF HAZARDOUS SUBSTANCES

Each facility must compile, annually review, and update as necessary a complete inventory of all substances present in that facility. The name of those materials determined to be hazardous are defined in applicable federal and state standards.

EMPLOYEE INFORMATION AND TRAINING

All employees must attend an orientation meeting for information and training on the following items prior to starting work with hazardous substances (training checklist is to be completed and kept on file):

- An overview of the requirements of the Hazard Communication Standard, including employee rights under this regulation.
- Information on where hazardous substances are present in work areas.
- Information regarding the use of hazardous substances in specific work areas.
- The location and availability of the written hazard communication program. A copy of the program will be given to all employees during the orientation meeting. Subsequent to this, the program will be available from managers and also from the office.
- The physical and health aspects of the substances in use.

- Methods and observation techniques used to determine the presence or release of hazardous substances in the work area.
- The controls, work practices, and personal protective equipment which are available for protection against possible exposure.
- Emergency and first aid procedures to follow if employees are exposed to hazardous substances.
- How to read labels and material safety data sheets to obtain the appropriate hazard information.

REFRESHER TRAINING SHALL BE CONDUCTED ANNUALLY

It is most important that all employees understand the information given in the orientation meetings. Questions regarding this information should be directed to the safety coordinator.

When new substances are introduced into the workplace, the department manager should review the above items with employees as they are related to the new materials.

The department manager should relay all the above information to new employees who will be working with hazardous substances, prior to their starting work.

An Acknowledgement Statement is to be completed by each employee receiving this information and training. These are to be kept on file in the human resources department.

NON-ROUTINE TASKS

Infrequently, employees may be required to perform non-routine tasks which involve the use of hazardous substances. Prior to starting work on such projects, each involved employee will be given information by his or her supervisor about hazards to which they may be exposed during such an activity.

This information will include

- The specific hazards
- Protective/safety measures which must be utilized
- The measures the company has taken to lessen the hazards, including special ventilation, respirators, the presence of another employee, air sample readings, and emergency procedures

INFORMING CONTRACTORS

To ensure that outside contractors work safely, and to ensure the safety of the contractor's employees, it will be the responsibility of management to provide contractors the following information:

- The hazardous substance to which they may be exposed while working in the plant.
- The precautions the contractor's employees must take to lessen the possibility of exposure (usage of appropriate measures).

- Rules and regulations regarding the protection of employee safety relevant to fire and ignition sources around flammable materials will be followed. The rules regarding smoking, welding, grinding will also be followed.

DATE OF REQUEST _____

DEPARTMENT_____

TO: _____

FROM: _____

I hereby request that I be given the Material Safety Data Sheets on the following hazardous substance(s):

_____
_____
_____
_____

Date Received _____

Requesting Employee_____

DEPARTMENT MANAGER_____

DATE _____

Figure 4.1 Sample Request for Material Safety Data Sheets

I have received information on the Hazard Communication Standard 29 CFR 1926.59 or the appropriate state standard and understand how to interpret and to use the labeling systems and Material Safety Data Sheets (MSDSs) that are in use and accessible to me in my work area. I agree to observe and follow the safe work practices as presented to me in the training sessions I attended on_____at_____.

_____   _____

Employee's Signature   Date

The above named employee has been informed and instructed by _____ on work practices, chemical hazards recognition, interpretation and use of chemical labels, MSDSs, the CFR 29, 1926.59 (e) or appropriate state standard, and the location at which these items are accessible to the employee.

_____   _____

Supervisor's Signature   Date

TRAINING ACKNOWLEDGEMENT (SPANISH)

Yo he recibido informacion sobre Communicacion de Peligros Estandarte 29 CFR 1926.59, y comprendo como usar y interpretar los systemas de marqueo junto con las Hojas de Datos sobre Seguridad (MSDS), que se encuentran en uso y acesible en mi area de trabajo. Yo estoy de acuerdo a observar y seguir las practicas seguras presentadas a mi en la secion de entrenamiento atendida este dia,_____ en_____

_____   _____

Firma de Empleado   Fecha

El empleado nombrado anterior, a sido informado y instruido por_____en las practicas seguras, a reconocer peligros de quimicas, la intrepetacion y uso de las etiquetas quimica, MSDS, la CFR 29. 1926.59, y la locacion en donde estos articulos son acesible a los empleados.

_____   _____

Firma de Supervisor o Mayordome   Fecha

Figure 4.2 *Sample Training Acknowledgment*

The MSDS is obtained from the hazardous substance manufacturer or supplier. You should become familiar with information on this sheet to avoid injury to yourself and fellow employees. Following is a description of the MSDSs principle sections. Not all sections are relevant to your safety, but brief descriptions will be provided.

Section I—Identification of Product

This identifies the chemical name, trade name or synonym, manufacturer's name, chemical formula, and emergency phone number for more detailed information.

Section II—Hazardous Ingredients

Hazardous ingredients are those substances which have been defined as hazardous due either to flammability characteristics or to their potential to have adverse health effects on the worker. The percentage of each hazardous ingredient in the product is provided, as well as the Threshold Limit Value.

Section III—Physical Data

This is primarily technical data providing little help to you unless you are a chemist. This data is used by chemists and industrial hygienists when doing calculations to determine the safe use parameters of the substance.

Section IV—Fire and Explosion Hazard Data

Not all hazardous substances are flammable or explosive. In this section, data is provided which describes the ability of the substance to burn or explode. The method for extinguishing a fire involving the substance is also provided. Pertinent data in this section is

1.   Flash Point—This is the lowest temperature at which the liquid gives off sufficient vapor to form an ignitable mixture with air and produce a flame when an ignition source is brought near the surface of the liquid.

2.   Extinguishing Media—The type of fire extinguishing material to be used when a particular substance is burning is provided here.

3.   Special Fire Fighting Procedures—These procedures describe the fire fighting equipment needed if the substance is involved in a fire. Some substances can give off toxic gases when burning; therefore, a special piece of personal protection equipment should be worn by persons fighting the fire. Talk to your supervisor regarding your actions in the event of a fire involving a hazardous substance.

4.   Unusual Fire and Explosion Hazards—This section provides information on substance incompatibility or its ability to react with other substances to create a flammable atmosphere.

Figure 4.3 Sample Material Safety Data Sheets

Section V—Health Hazard Data

Data included in this section is very important to you. This information will help you recognize the effects of overexposure to a particular hazardous substance and the emergency and first-aid procedures to take in the event of overexposure.

Here are some terms found in this section:

1. Threshold Limit Value—The value printed on the MSDS expresses the airborne concentration of material to which nearly all persons can be exposed day after day without adverse health effects. Threshold Limit Values (TLV) may be expressed in three ways: as Time Weighted (TWA), as Short Term Exposure Limit (STEL), and/or as Ceiling Exposure Limit (C). The TLV is used by engineers and industrial hygienists as a guide in the control of health hazards.

2. Effects of Overexposure—Describes what physical effects might be felt (dizziness, headaches, skin irritation, dermatitis, etc.).

3. Emergency and First-Aid Procedures—Explains the procedures to follow should it become necessary to provide first-aid treatment to a person who may be overcome by a hazardous substance. The procedures may address exposures that occur through inhalation of the substance, contact with skin, or ingestion (swallowing).

Section VI—Reactivity Data

This section presents information on reactive substances. Reactive substances are materials which, under certain environmental or induced conditions, enter into violent reaction with spontaneous generation of large quantities of heat, light, gases (flammable and non-flammable), or toxicants that can be destructive to life and property. Reactions occur often when incompatible materials are mixed together.

Some loosely categorized types of reactive chemicals are

1. Explosives—(example: nitroglycerin), reacts to friction, heat, or shock.

2. Acids—Don't mix with sensitives.

3. Oxidizers—Don't mix with reducers.

4. Water Sensitives— Should not be mixed with water.

5. Pyrophors—Those substances that generate sparks or heat when friction is applied, like a red-tip match head.

Figure 4.3 Sample Material Safety Data Sheets, continued

When reviewing a particular data sheet, note the conditions to avoid and incompatibility with certain materials. In general, isolate from other potentially reactive substances. Use appropriate personal protection gear that is recommended in Section VIII—Special Protection Information.

Section VII—Spill or Leak Procedures

This section directs persons to take certain actions in the event of a hazardous substance spill or leak. Do not attempt to contain a spill or leak by yourself! Get help from your supervisor!

Section VIII—Special Protection Information

This section specifies the proper personal protection devices for specific situations. Types of recommended equipment will include respirators, goggles, face shields, and safety glasses, gloves, protective aprons, footwear, etc.

Ventilation equipment will not necessarily be applicable. These requirements are based on amount used, container the substance is stored in, conditions use occurs in, etc.

Section IX—Special Precautions

Describes proper storage and handling procedures. This section is important and provides many of the dos and don'ts associated with the substance. It will also alert you to situations to avoid when handling or storing the substance.

Figure 4.3 Sample Material Safety Data Sheets, continued

	YES	NO

1. IDENTITY

Does it bear the name of the chemical as it appears on the label? ( ) ( )

Does it bear the chemical name and other names by which the substance is known? ( ) ( )

If it is a mixture, does it bear the name of each component that is hazardous? ( ) ( )

2. CHARACTERISTICS

Does it describe the chemical and physical properties of the substances? ( ) ( )

3. PHYSICAL HAZARDS

Does it describe conditions under which fire, explosion, or reaction may occur? ( ) ( )

Does it list other materials with which the substance may come into contact or be exposed to that would result in such hazards? ( ) ( )

4. EXPOSURE HAZARDS

Does it detail the signs and symptoms of various routes of entry (skin, eyes, ingestion, etc.) as well as various levels of exposure? ( ) ( )

Does it give medical conditions that are likely to be aggravated by such exposure? ( ) ( )

5. PRIMARY ROUTES OF EXPOSURE

Does it explain the circumstances under which hazardous exposure to the substance typically occurs through various routes? ( ) ( )

6. EXPOSURE LIMITS

Does it give the exposure limits to the substance set by OSHA or the American Conference of Governmental Industrial Hygienists? ( ) ( )

7. CANCER RISKS

Does it indicate whether or not the substance is listed as a cause of cancer by OSHA or by the International Agency of Research in Cancer? ( ) ( )

8. USE AND HANDLING

Does it specify:

Methods of preventing fire, explosion, or reaction? ( ) ( )

The proper protective equipment to be worn? ( ) ( )

Figure 4.4 Sample Material Safety Data Evaluation Checklist

132

Proper ventilation procedures? ( ) ( )

Special circumstances under which special protective measures are necessary? ( ) ( )

Procedures for guarding against spillage? ( ) ( )

Proper storage and labeling procedures? ( ) ( )

Actions that are prohibited for the purpose of reducing risks? ( ) ( )

9. CONTROL MEASURES

Does it specify:

Methods of diluting the substance? ( ) ( )

Methods of ventilating the area in which the substance is used? ( ) ( )

Personal protective gear to be worn? ( ) ( )

10. EMERGENCY AND FIRST AID PROCEDURES

Does it specify:

The proper responses to exposure through various routes, at various levels, and
according to the severity of signs and symptoms? ( ) ( )

Proper response to fire, explosion, and spillage? ( ) ( )

Proper procedures for disposal of spilled materials? ( ) ( )

11. DATE

Does it show the date it was prepared and/or the date of its latest revision?
( ) ( )

12. FURTHER INFORMATION

Does it include the name, address, and telephone number of someone who can be contacted for further information about the substance and for emergency procedures? ( ) ( )

If certain information is not given, does it state that the information is not available instead of having blank spaces? ( ) ( )

Is it written so that employees can be trained to read and understand it?
( ) ( )

13. GENERAL

If certain information is not given, does it state that the information is not available instead of having blank spaces? ( ) ( )

Is it written so that employees can be trained to read and understand it?
( ) ( )

Figure 4.4 Sample Material Safety Data Evaluation Checklist, continued

EXPLANATION OF TERMS USED ON MATERIAL SAFETY DATA SHEETS

Section I

Chemical Name and Synonyms: The product identification, the chemical or generic name of single elements and compounds, or for compounded products and mixtures.

Trade Names and Synonyms: The name under which the product is marketed and the common commercial name of the product.

Chemical Family: Refers to a grouping of chemicals which behave and react with other chemicals in a similar manner.

Formula: The chemical formula or single elements or compounds.

CAS Number: The Chemical Abstracts Service number, if applicable.

EPA: The code number assigned by the Environmental Protection Agency, if applicable.

DOT Classification: The appropriate classification as determined by the regulations of the Office of Hazardous Materials, Department of Transportation.

Section II

Hazardous Ingredients: The major components as well as any minor component(s) having potential for harm which are considered when evaluating the product.

TLV: Threshold Limit Value (TLV) indicates the permissible exposure concentration, a limit established by a governmental regulatory agency, or an estimate if none has been established.

Section III

Physical Data:

Boiling Point (degrees Fahrenheit): The temperature in degrees Fahrenheit at which the substances will boil.

Vapor Pressure: The pressure of saturated vapor above the liquid expressed in mm Hg at 20 ° Celcius.

Vapor Density: The relative density or weight of a vapor or gas (with no air present) compared with an equal volume of air at ambient temperature.

Solubility in Water: The solubility of a material by weight in water at room temperature. The terms are negligible, less than 0.1%; slight 0.1 to 1%; moderate 1 to 10%; appreciable 10% or greater.

Appearance and Odor: The general characterization of the material, i.e., powder, colorless liquid, aromatic odor, etc.

Specific Gravity (H$_2$O = 1): The ratio of the density of a volume of the material to the density of an equal volume of water.

Percent, Volatile by Volume (percent): The percent by volume of the material that is considered volatile (the tendency or ability of a liquid to vaporize.)

Evaporation Rate: The ratios of the time required to evaporate a measured volume of a liquid to the time required to evaporate the same volume of a reference liquid (ethyl ether) under ideal test conditions. The higher the ratio, the slower the evaporation rate.

Section IV

Flash Point (Method Used): The temperature in degrees Fahrenheit at which a liquid will give off enough flammable vapor to ignite in the presence of a source of ignition.

Conditions to Avoid: The substance present could become unstable if these conditions exist.

Incompatibility (Materials to Avoid): Materials which will react with the substance.

Hazardous Decomposition Products: Refers to reactions which take place at a rate which releases large amounts of energy. Indicates whether or not it may occur and under what storage conditions.

Section V

Health Hazard Data: Possible health hazards as derived from human observation, animal studies, or from the results of studies with similar products.

Threshold Limit Value (TLV): The values for airborne toxic material which are to be used as guides in the control of health hazards and represent concentrations to which nearly all workers may be exposed eight hours per day over extended periods of time without adverse effects.

Effects of Overexposure: The effects on or to an individual who has been exposed beyond the specified limits.

Emergency and First Aid Procedures: First aid and emergency procedures in case of eye and/or skin contact, ingestion, and inhalation.

Section VI

Stability: Whether the substance is stable or unstable. An unstable substance is one that will vigorously polymerize, decompose, condense, or will become self-reactive under conditions of shock, pressure, or temperature.

Section VII

Spill or Leak Procedures: Steps to be taken in case material is released or spilled, including methods and materials used to clean up or contain the spill or leak.

Waste Disposal Method: Method and type of disposal site to use.

Section VIII

Special Protection Information:

Respiratory Protection: Specific type should be specified (examples include dust mask, NIOSH-approved cartridge respirator with organic-vapor cartridge).

Ventilation: Type of ventilation recommended, i.e., local exhaust, mechanical, etc.

Protective Gloves: Refers to the gloves that should be worn when handling the product (examples include cotton or rubber).

Eye Protection: Refers to the type of eye protection that is to be worn when handling the product or in proximity to the product.

Flammable Limits: The range of gas or vapor concentration (percent by volume in air) which will burn or explode if an ignition source is present. (Lel) means the lower explosive limits and (Uel) the upper explosive limits given in percent.

Extinguishing Media: Specifies the fire fighting agent(s) that should be used to extinguish fires involving the substance.

Special Fire Fighting Procedures/Unusual Fire and Explosion Hazards: Refer to special procedures required if unusual fire or explosion hazards are involved.

THE MOST FREQUENTLY ASKED QUESTIONS ON HAZARD COMMUNICATION[1]

Q. What are my responsibilities as an end user of chemicals?

A. Employers who use chemicals in the workplace must obtain information concerning these chemicals and provide it to employees. Employers handling only sealed containers of chemicals need only obtain information appropriate for emergency response because their employees would not normally be exposed to chemical hazards. Laboratories have limited responsibilities under this standard since they are more fully covered under another OSHA standard.

General employers who use chemicals and whose employees might be exposed to chemical hazards have the most comprehensive requirements.

Employers who use chemicals must do the following:

- Analyze the hazards of the chemicals they use or obtain such information from the manufacturer or importer. This will result in a compilation of material safety data sheets.
- Develop a written Hazard Communication Program.
- Train employees about the hazards of the chemicals in the workplace and how to read and interpret MSDS information.
- Ensure that containers of chemicals are labeled as to their content and potential hazard.

Q. What are the requirements of the Hazard Communication Standard?

A. The Hazard Communication Standard requires providing information about the hazardous chemicals that employees are exposed to, product labels and other forms of warning, material safety data sheets, appropriate training, and a written hazard communication program.

Employees who work in areas where hazardous chemicals are stored, handled, or used are responsible for creating and maintaining an inventory of all hazardous chemicals; ensuring proper labeling of all hazardous chemicals; acquiring and maintaining material safety data sheets for all hazardous chemicals located in the work area; informing employees of any operations in their work area where hazards chemicals are present; informing employees of the location and availability of the written hazard communication program, the chemical inventory, and material safety data sheets; and training employees about hazardous chemicals used in the work area.

Q. What are the training requirements for the Hazard Communication Standard?

A. The standard sets forth the minimum requirements for the required training. It is the employer's responsibility to match the nature and the content of the training to the job, the employee, and the particular chemicals involved.

The four basic minimums upon which training must be given are

1. The methods and observations that may be used to detect the presence or release of a hazardous chemical in the work area, such as monitoring devices, odor, and visual appearance.
2. Both the physical and health hazards of the particular chemicals to which such employee may be exposed.
3. The measures the employee can take to protect himself from these hazards, like evacuation or other emergency procedures, protective clothing, and equipment.
4. The details of the written Hazard Communication Program which each employer is required to adopt. This includes an explanation of container labels, MSDSs and how employees can obtain and use the appropriate hazard information. In addition, employees must be given information about the operations in their work area where hazardous chemicals are present, the location and availability of the written hazard communication program, and the minimal OSHA training requirements mentioned above.

Q. How regularly must training be provided?

A. Employees are to be trained at the time they are assigned to work with a hazardous chemical. The intent of this provision is to have information prior to exposure to prevent the occurrence of adverse health effects. This purpose cannot be met if training is delayed until a later date. Additional training is to be done whenever a new hazard is introduced into the work area; if a new solvent is brought into the workplace and it has hazards similar to existing chemicals for which training has already been conducted, no new training is required.

The Material Safety Data Sheet must also be available, and the product must be properly labeled. If the newly purchased chemical is a suspect carcinogen and there has never been a carcinogenic hazard in the workplace before, then new training for carcinogen hazards must be conducted in the work areas where employees will be exposed to it. Complete retraining of an employee does not automatically have to be conducted when an employer hires a new employee if the employee has received prior training by a past employer, an employee union, or any other entity.

If it is determined that an employee has not received training, the current employer will be held responsible. An employer, therefore, has a responsibility to evaluate an employee's level of knowledge with regard to the training and information requirements of the standard.

Q. What are the requirements for refresher training?

A. Additional training is to be done whenever a new physical or health hazard is introduced into the work area, not a new chemical. For example, if a new solvent is brought into the workplace, and it has hazards similar to existing chemicals for which training has already been conducted, then no new training is required. As with initial training, and in keeping with the intent of the standard, the employer must make employees specifically aware which hazard category (i.e., corrosive, irritant, etc.) the solvent falls within. The substance-specific data sheet must still be available, and the product must be properly labeled. If the newly introduced solvent is a suspect carcinogen and there has never been a carcinogenic hazard in the workplace before, then new training for carcinogenic hazards must be conducted for employees in those work areas where employees will be exposed.

Q. What are the requirements for retraining a new employee?

A. It is not necessary that the employer retrain each new employee if that employee has received prior training by a past employer, an employee union, or any other entity. The employer, however, maintains the responsibility to ensure that their employees are adequately trained and are equipped with the knowledge and information necessary to conduct their jobs safely.

It is likely that additional training will be needed since employees must know the specifics of their new employer's programs such as where the MSDSs are located, details of the employer's in-plant labeling system, and the hazards of new chemicals to which they will be exposed. For example, the standard requires that employees be trained on the measures they can take to protect themselves from hazards, including specific procedures the employer has implemented such as work practices, emergency procedures, and personal protective equipment to be used. An employer, therefore, has a responsibility to evaluate an employee's level of knowledge with regard to the hazards in the workplace, their familiarity with the requirements of the standard, and the employer's hazard communication program.

Q. What is the purpose of the Hazard Communication Standard?

A. The purpose of the OSHA Hazard Communication Standard (HCS) is to alert workers to the existence of potentially dangerous substances in the workplace and the proper means and methods to protect themselves against them. Numerous states and some local governments have enacted substantially similar laws.

HCS is completely different from ordinary OSHA standards. OSHA standards obligate an employer to prevent (or minimize) workplace hazards through such means and methods as mandatory limitations upon noise levels and airborne contaminants, or guarding requirements for machinery and elevated workstations. The HCS does not impose either mandatory limitations or requirements to abate hazardous conditions. It requires, rather, that information be developed, obtained, and provided. That is also true of state and local Right-to-Know laws.

Q. Who does the Hazard Communication Standard apply to?

A. Any chemical manufacturer who produces or imports chemicals, any distributor of chemicals, and any employer who uses chemicals in the workplace are subject to the hazard communication standard. The only exemptions are for chemical products that are regulated by other laws and regulations, such as hazardous wastes, drugs, cosmetics, and alcohol and tobacco products.

Q. What are the Hazard Communication requirements for temporary agency employers?

A. In meeting the requirements of OSHA's Hazard Communication Standard, the temporary agency employer is expected to provide generic hazard training and information concerning categories of chemicals employees may potentially encounter. The host employers would then be responsible for providing site-specific hazard training.

Q. Does an employer have to inform a outside contractor and employees about chemical hazards on site and the appropriate protective equipment?

A. Yes. An employer who hires an outside contractor must provide that contractor with MSDSs, label information, and precautionary measures for the hazardous chemicals workers may encounter in the facility. Appropriate protective measures are also to be discussed. It is the responsibility of the contractor to inform his or her employees.

Q. Is there a requirement that each employer have a written Hazard Communication Program?

A. Yes, each employer must develop and implement a Hazard Communication Program. It must be in writing and it must be readily accessible to employees and the OSHA inspector. This means you must put in writing precisely how you plan to meet your MSDS, training, and labeling obligations.

Three things must be included in your written program:

1. A list of the hazardous chemicals known to be present. It may be compiled for the workplace as a whole or broken up by work areas, production lines, etc.

2. The methods to be used to inform your employees of the non-routine tasks like cleaning of reactor vessels and the means you will use to apprise employees of the hazards of the chemicals in pipes and piping systems, including protective measures they can take in the event of exposure situations.
3. The methods to be used to advise contractors whose employees are working on your premises of the hazardous chemicals to which they may be exposed while they are there and any suggestions for appropriate protective measures.

Q. Do employees have to be informed with written materials?

A. No. Written materials may be helpful training tools, but the employer is only required to develop in writing a Hazard Communication Program.

Q. Does each job site have to have a copy of the written program?

A. Yes. The written program is a training information tool which has to be made available to employees on request. The written program contains information employees need access to, such as the description of the labeling system.

Q. What are the container labeling requirements?

A. Under HCS, the manufacturer, importer, or distributor is required to label each container of hazardous chemicals. If the hazardous chemicals are transferred into unmarked containers, these containers must be labeled with the required information, unless the container into which the chemical is transferred is intended for the immediate use of the employee who performed the transfer.

Q. Some of my employees don't speak English. Do I have to make any special provisions for these workers?

A. Yes. The OSHA Hazard Communication Standard 29 CFR 1910.1200 requires that your employees are adequately trained in:

- The hazards of all chemicals they use
- Labeling
- Signage

MSDSs are part of the hazard communication process and training requirements; if your employees do not understand English and you do not provide training in a language they comprehend, then you are not in compliance.

Q. Are there any exceptions to the Hazard Communication labeling requirements?

A. Yes, there are three exceptions to the container labeling provision which states the following:

Portable containers into which hazardous chemicals are transferred from labeled containers don't have to be labeled if this is done only for the immediate use of the employee who performs the transfer which means that it will be under his control, and used only by him, during the same work shift in which it is placed in the portable container.

For stationary, permanent, or fixed containers: signs, placards, process sheets, batch tickets, operating procedures, or other such written material may be used in lieu of a label so long as they identify the applicable container and include the two things necessary to all labels: Identity of the hazardous chemicals and appropriate hazard warnings.

Containers used in laboratories that are part of manufacturing facilities do not have to be labeled. However, when a container comes into the laboratory already bearing a label, it must be left there. It cannot be removed or defaced.

Q. How does the Hazard Communication Standard apply to pharmaceutical drugs?

A. The HCS only applies to pharmaceuticals that the drug manufacturer has determined to be hazardous and that are known to be present in the workplace in such a manner that employees are exposed under normal conditions of use or in a foreseeable emergency. The pharmaceutical manufacturer and the importer have the primary duty for the evaluation of chemical hazards. The employer may rely upon the hazard determination performed by the pharmaceutical manufacturer or importer.

Q. What are the requirements of the Hazard Communication Standard to an office environment?

A. Office workers who encounter hazardous chemicals only in isolated instances are not covered by the rule. OSHA considers most office products (such as pens, pencils, or adhesive tape) to be exempt under the provisions of the rule, either as articles or as consumer products. For example, OSHA has previously stated that intermittent or occasional use of a copying machine does not result in coverage under the rule. However, if an employee handles the chemicals to service the machine, or operates it for long periods of time, then the program would have to be applied.

Q. What is considered a "hazardous chemical"?

A. The term "hazardous chemical" is defined very broadly (for example table salt is included), so virtually every employer is required to observe the Hazard Communication Standard (§ 1910.1200). It is also the most cited OSHA standard. If you don't think you use hazardous chemicals in your business or don't qualify as a chemical manufacturer, you should be cautioned that it is OSHA's contention that a "hazardous chemical" includes ordinary household detergents and that a chemical manufacturer is any employer who supplies his customer with a product that, in normal use or a foreseeable emergency, will release or otherwise result in exposure to a hazardous chemical. If either of these apply to your facility, so does HCS.

Q. What is a Material Safety Data Sheet (MSDS)?

A. An MSDS sheet is essentially a technical bulletin, usually two to four pages in length, that contains information about a hazardous chemical or a product containing one or more hazardous chemicals, such as its composition, its chemical and physical characteristics, its health and safety hazards, and the precautions for safe handling and use. For example, the manufacturer of your favorite household detergent probably has a MSDS available. The MSDS for that product will probably provide the detailed information regularly included on an MSDS required under the HCS.

The product manufacturer is usually happy to supply the MSDS to anyone who requests it. The MSDS is the centerpiece of the Hazard Communication Standard. Labels are keyed to it and the employee training and information requirements are based upon it. The MSDS serves as the primary vehicle for transmitting detailed hazard information to both employers and employees.

Q. What OSHA regulation requires employers to keep MSDSs sheets?

A. OSHA has a Hazard Communication Standard 29CFR 1910.1200. The purpose of this standard is to ensure that the hazards of all chemicals produced or imported are evaluated, and that information concerning their hazards is transmitted to employers and employees. This transmittal of information is to be accomplished by means of a comprehensive Hazard Communication Program, which includes container labeling and other forms of warning, material safety data sheets, and employee training.

Q. What formats are required on an MSDS sheet?

A. OSHA has suggested a format for MSDSs on Form 174. While this format is not mandatory, it is a frequently utilized format. An MSDS can contain more information than required by OSHA but not less.

Form 174 for Material Safety Data Sheets has the following sections:

1. Section 1: Manufacturer's Name and Contact Information
2. Section 2: Hazardous Ingredients/Identity Information
3. Section 3: Physical/Chemical Characteristics
4. Section 4: Fire and Explosion Hazard Data
5. Section 5: Reactivity Data
6. Section 6: Health Hazard Data
7. Section 7: Precautions for Safe Handling and Use
8. Section 8: Control Measures

Q. Is it required that an employer have an MSDS sheet for each hazardous chemical?

A. Yes, every employer must have an MSDS for each hazardous chemical. It does not matter where such a product came from or how it got on the premises. Perhaps the employer's most difficult job will not be obtaining MSDS copies from the chemical manufacturers or even making them readily accessible to the employees concerned, it will be finding out what hazardous chemicals are on the premises, so the employer can meet the MSDS responsibilities for them.

Q. What are the requirements and limits to using generic MSDSs?

A. Regarding the suitability of a generic material safety data sheet, MSDSs must be developed for hazardous chemicals used in the workplace and must list the hazardous chemicals that are found in a product in quantities of one percent or greater, or 0.1 percent or greater if the chemical is a carcinogen. The MSDS does not have to list the amount that the hazardous chemical occurs in the product.

A single MSDS can be developed for the various combinations of [chemicals], as long as the hazards of the various mixtures are the same. This 'generic' MSDS must meet all of the minimum requirements found in 29 CFR 1910.1200(g), including the name, address, and telephone number of the responsible party preparing or distributing the MSDS who can provide additional information.

Q. How come I don't always get MSDSs when I order chemicals?

A. MSDS sheets are required. But they are only required to give you one copy under the following conditions:

1. The first time you order a particular chemical.
2. When there is a change in the product's chemical composition.
3. In the event of a new significant information related to the chemical (e.g., new health data or a new OSHA standard that addresses a component of the substance).

The second and third items only occur if you purchase the chemical before the sheet changed, few firms would even attempt to let you know. But the next time you buy that chemical, they have to send you the updated sheet.

4. Some chemical suppliers ship MSDSs with every shipment; others hold to the minimum standard. The latter practice can be a real pain at a large organization where it may be the first time you ordered that chemical, but it's not the first time your employer (i.e., the purchaser) did. Fortunately, most manufacturers are happy to provide additional copies of an MSDS if you simply contact their customer service department and ask. In fact, many of them have realized the benefit of making all their MSDSs freely available on the Internet.

Q. If I have the same chemical from different manufacturers, do I need to keep all their MSDSs?

A. You probably should keep one from each. However, you can get away with one sheet for the chemical. The key points to remember are

1. The identity on the label must cross-reference to the MSDS
2. All employees must be trained that you are using on MSDS as representative of all vendors (so there isn't confusion during an emergency)
3. The MSDS must be complete and accurate
4. The manufacturer listed on the MSDS is willing to act as the responsible party in the event of an emergency.

Q. Is an MSDS required for a non-hazardous chemical?

A. No. MSDSs that represent non-hazardous chemicals are not covered by the HCS. The standard requires that the employer shall maintain in the workplace copies of the required MSDSs for each hazardous chemical and ensure that they are readily accessible during each work shift to employees when they are in their work area(s). OSHA does not require employers to maintain MSDSs for non-hazardous chemicals.

Q. Do we need to maintain MSDSs for common household chemicals used by our employees?

A. OSHA's general interpretation appears to be that if a chemical such as floor wax, toilet bowl cleaner, or window cleaner, is used in the same occasional fashion as a typical household consumer then no, you don't.

However, if your employee washes windows, floors, toilets, etc., several hours a day, then they are occupationally exposed to the material and an MSDS must be provided. If your employees use these materials, you can get the MSDS from the company that supplies the cleaning supplies to you. If you purchased these at a retail store ask, for them at their customer service desk.

Q. Do you need to keep MSDSs for commercial products such as Windex® and Wite•out®?

A. No. OSHA does not require that MSDSs be provided to purchasers of household consumer products when the products are used in the workplace in the same manner that a consumer would use them, i.e.; where the duration and frequency of use (and therefore exposure) is not greater than what the typical consumer would experience.

This exemption in OSHA's regulation is based, however, not upon the chemical manufacturer's intended use of his product, but upon how it actually is used in the workplace. Employees who are required to work with hazardous chemicals in a manner that result in a duration and frequency of exposure greater than what a normal consumer would experience have a right to know about the properties of those hazardous chemicals.

Q. Where can I find an MSDS sheet for an old chemical?

A. Under the OSHA Hazard Communication standard, manufacturing employers are not legally required to obtain, maintain, and make available upon request copies of MSDSs for chemicals obtained prior to May 26, 1986. OSHA would require you to make a 'good faith' effort to locate the manufacturer. If all else fails, see if you can find an MSDS for an identical formulation from another manufacturer. As long as you've made efforts such as these and you don't have any willful violations for any type you are probably looking at a *de minimus* violation, i.e., one that OSHA might note during an inspection but not assess a penalty.

Q. Does OSHA determine what information is required in the MSDS sheet under health hazard information or can employers use their own data?

A. The quality of a hazard communication program is largely dependent upon the adequacy and accuracy of the hazard determination. The hazard determination requirement of this standard is performance-oriented. Chemical manufacturers, importers, and employers evaluating chemicals are not required to follow any specific methods for determining hazards, but they must be able to demonstrate that they have adequately ascertained the hazards of the chemicals produced or imported in accordance with the criteria set forth in the standard.

Also note that the OSHA standard says: "'Health hazard' means a chemical for which there is statistically significant evidence based on at least one study conducted in accordance with established scientific principles that acute or chronic health effects may occur in exposed employees. The term 'health hazard' includes chemicals which are carcinogens; toxic or highly toxic agents; reproductive toxins; irritants; corrosives; sensitizes; hepatotoxoins; nephrontoxins; agents which act on the menatopoietic system; and agents which damage the lungs, skin, eyes, or mucous membranes."

Q. Can I throw away old and outdated MSDS sheets?

A. The question is not whether you can but whether you would dare to do so. A harmless chemical may later be found to cause cancer or other disease sometime in the future (asbestos is one good example). It is certainly to your company's legal benefit to be able to produce documentation showing that you had supplied all the protective equipment and procedures necessary according the MSDSs you had at the time.

Q. Could OSHA demand to see my MSDS sheets?

A. Yes. OSHA permits 'paperless compliance' because the Hazard Communication standard is performance-based. This means that while you don't have to have a paper copy on hand in the event of an emergency, you should have ready access to a collection of MSDSs in electronic format.

The most popular types of electronic format include CD-ROM subscriptions (renewable on quarterly or annual basis), fax on demand services (that will fax you a copy as soon as you request one), internet-based suppliers, and 'in-house' solutions in which corporations scan their MSDS sheets into a database rather than use a paper filing system.

Q. Does an employer have to keep every MSDS sheet they receive?

A. It really depends on how MSDSs are handled at your organization. If your employer has a CD-ROM, Internet, or fax-on-demand service for MSDSs then maybe not (check with your supervisor; the answer depends on what state and federal agencies have jurisdiction. When you get the same chemical with the same formulation from several different manufacturers or if you receive duplicate (i.e., identical) copies for MSDSs that you already have, then you only need to keep one of them.

Q. When does an MSDS sheet need to be revised or replaced?

A. As required by OSHA, MSDS are to be revised when any of the following happen:

- Any significant change has been made to the chemical compound
- Research has revealed a health or physical hazard different from what was originally stated
- The product has been listed as carcinogenic by a recognized government agency.

A new MSDS must be issued within three months of any of these events. The old MSDS sheets should be retained to potentially limit future liability. Recognize that this update requirement does not require the manufacturer to send a new MSDS to you if you purchased a chemical prior to the MSDS update. In other words, your MSDSs will go out of date from time to time. It is a good idea to review your MSDS collection and to update your sheets. You should consider setting a corporate policy on this matter. You can also subscribe to an MSDS service so you will always have up-to-date sheets.

Q. What's the best solution for handling my MSDS collection?

A. Every case is different and we really don't feel comfortable (from a legal and moral stand-point) trying to address such a complex issue. The only recommendation we will make is that you maintain your paper copies.

Q. Does having on-site MSDS sheets satisfy OSHA requirements?

A. Yes, the Hazard Communication standard is a performance-based standard which means that OSHA does not concern itself with how you comply, just that you manage to do it. While your method of managing your MSDS collection is flexible, there is criteria you need to meet with your MSDS collection as follows:

1. MSDSs must be maintained on site (including electronic access methods).
2. They must be readily accessible during each work shift to employees when they are in their work area(s). OSHA has said: "Employees should not have to ask for an MSDS, as this could be perceived by employees as a barrier to access. For example, if an employee must go through a supervisor to receive an MSDS, the employee may feel that this singles him or her out. This could very well dampen the employee's resolve to seek out the necessary hazard information."

Q. Does OSHA recommend how employers should be in compliance with filing their MSDS sheets?

A. No. The OSHA Hazard Communication standard is a performance-based standard. That means OSHA doesn't care how you organize your MSDS files, only that you ensure that they are readily accessible during each work shift to employees when they are in their work area(s). This can be by hardcopy, computer, fax, etc., as long as the employee is able to use your system to find any sheet he or she needs. But for most employers who simply file paper copies, how to organize a filing cabinet or three-ring binder is a more difficult question. OSHA doesn't care how you choose to do it, but employees have to be able to find the information when they need it. For this reason, many people prefer to file their MSDSs alphabetically by name.

Q. We have a large site. Does having one site-wide MSDS repository satisfy OSHA?

A. It depends. The Hazard Communication Standard is a performance-based standard which means that OSHA does not concern itself with how you comply, just that you manage to do it. While your method of managing your MSDS collection is flexible, specific criteria you need to meet are

1. MSDSs must be maintained on site (including electronic access methods)
2. MSDSs must be readily accessible during each work shift to employees when they are in their work area(s).

Q. How does OSHA want me to organize my MSDS filing system?

A. The OSHA Hazard Communication standard is a performance-based standard. That means OSHA is not concerned with *how* you organize your MSDS files, only that you "ensure that they are readily accessible during each work shift to employees when they are in their work area(s)." This can be by hardcopy, computer, fax, etc., as long as the employee is able to use your system to find any sheet when he or she needs it.

Q. Where can MSDSs be stored for mobile employees?

A. Where employees must travel between workplaces at more than one geographical location within a work shift, the MSDSs may be kept at a central workplace facility (the employee must still be able to obtain the MSDS in a reasonable amount of time). The employer must ensure that employees can immediately obtain the MSDS information in an emergency. This provision requires MSDSs to be electronically accessible and maintained on site. The key to compliance with this provision is that employees have no barriers to access to the information and that the MSDSs be available during the work shift.

When direct and immediate access to paper or hard-copy MSDSs does not exist, OSHA inspectors will evaluate the performance of the employer's system by requesting a specific MSDS. Requested information orally via telephone is not acceptable. Employees must have access to the MSDS and be able to get the information when they need it in order for an employer to be in compliance with the rule.

Q. Can Material Safety Data Sheets (MSDSs) be stored on a computer?

A. Yes, if the employee's work area includes the area where the MSDSs can be obtained, then maintaining them on a computer would be in compliance. If the MSDS sheets can only be accessed out of the employee's work area(s), then the employer would not be in compliance.

Q. Does our company have to supply MSDS sheets to employees who do not speak English?

A. Yes, MSDSs are a part of the Hazard Communication Standard. If any of your employees can't read English and you don't provide a translated copy then you are not in compliance.

Q. What are the penalties for non-compliance with MSDS requirements?

A. Full compliance means that every employee who uses hazardous chemicals in the workplace (or who could be expected to be exposed in a "foreseeable" emergency) has ready access to an MSDS. If you are not compliant, you could face OSHA fines up to $70,000 or more depending how they count your violation(s) and whether they were willful violations. In reality, most fines for MSDS violations are substantially lower. You might simply get a warning if you have only a few minor infractions.

Improper compliance with the Hazard Communication Standard (which includes MSDSs) is the most frequently cited violation in the manufacturing, transport, wholesale, retail, and services industries, and it is in the top five in all other categories.

Q. Does OSHA have enforcement responsibilities over retailers?

A. When retailers distribute hazardous materials (including consumer products falling under the definition of a hazardous chemical), the hazard communication rules require MSDS sheets to be provided to purchasing employers. Retailers who distribute primarily to consumers (that is, at least 75 percent of their customers are consumers and not employers) do not need to provide MSDSs with each hazardous material sold.

However, these consumer-oriented retailers must assist in obtaining MSDSs from the manufacturer of the hazardous materials when requested by purchasing employers. Retailers who distribute primarily to employers must ensure that MSDSs and labeling information are provided to these purchasing employers.

Q. Is our company protected from liability if someone is injured because our MSDS sheet is wrong but we had no way of knowing about the error?

A. Employers are not to be held responsible for inaccurate information on the MSDS sheets or labels which they did not prepare and they have accepted in good faith from the chemical manufacturer, importer, or distributor. Your company will not receive an OSHA citation, but that doesn't mean you won't get hit with a civil lawsuit.

Endnote:

[1] Adapted from http://www.osha.gov/html/faq-hazcom.html

5

Training Requirements in OSHA

Construction Standards Guidelines

INTRODUCTION[1]

Many standards promulgated by the Occupational Safety and Health Administration (OSHA) explicitly require the employer to train employees in the safety and health aspects of their jobs. Other OSHA standards make it the employer's responsibility to limit certain job assignments to employees who are "certified," "competent," or "qualified"—meaning that they have had special previous training, in or out of the workplace. The term "designated" personnel means selected or assigned by the employer or the employer's representative as being qualified to perform specific duties. These requirements reflect OSHA's belief that training is an essential part of every employer's safety and health program for protecting workers from injuries and illnesses. Many researchers conclude that those who are new on the job have a higher rate of accidents and injuries than more experienced workers. If ignorance of specific job hazards and of proper work practices is even partly to blame for this higher injury rate, then training will help to provide a solution.

As an example of the trend in OSHA safety and health training requirements, the *Process Safety Management of Highly Hazardous Chemicals Standard (Title 29 Code of Federal Regulations Part 1926.64)* contains several training requirements. This standard was promulgated under the requirements of the Clean Air Act Amendments of 1990. The Process Safety Management Standard requires the employer to evaluate or verify that employees comprehend the training given to them. This means that the training must have established goals and objectives regarding what is to be accomplished. Subsequent to the training, an evaluation would be conducted to verify that the employees understood the subjects presented or acquired the desired skills. If the established goals and objectives of the training program were not achieved as expected, the employer then would revise the training program to make it more effective, or conduct more frequent refresher training, or some combination of these.

The length and complexity of OSHA standards may make it difficult to find all the references to training. So, to help employers, safety and health professionals, training directors, and others with a need to know, OSHA's training-related requirements have been excerpted and collected in this discussion. Requirements for posting information, warning signs, labels, and the like are excluded, as are most references to the qualifications of people assigned to test workplace conditions or equipment.

It is usually a good idea for the employer to keep a record of all safety and health training. Records can provide evidence of the employer's good faith and compliance with OSHA standards. Documentation can also supply an answer to one of the first questions an accident investigator will ask: "Was the injured employee trained to do the job?"

Training in the proper performance of a job is time and money well spent, and the employer might regard it as an investment rather than an expense. An effective program of safety and health training for workers can result in fewer injuries and illnesses, better morale, and lower insurance premiums, among other benefits.

VOLUNTARY TRAINING GUIDELINES

The Occupational Safety and Health Act of 1970 does not address specifically the responsibility of employers to provide health and safety information and instruction to employees, although Section 5(a)(2) does require that each employer " . . . shall comply with occupational safety and health standards promulgated under this Act." However, more than 100 of the Act's current standards do contain training requirements.

Therefore, the Occupational Safety and Health Administration has developed voluntary training guidelines to assist employers in providing the safety and health information and instruction needed for their employees to work at minimal risk to themselves, to fellow employees, and to the public.

The guidelines are designed to help employers (1) determine whether a worksite problem can be solved by training; (2) determine what training, if any, is needed; (3) identify goals and objectives for the training; (4) design learning activities; (5) conduct training; (6) determine the effectiveness of the training; and (7) revise the training program based on feedback from employees, supervisors, and others.

The development of the guidelines is part of an agency-wide objective to encourage cooperative, voluntary safety and health activities among OSHA, the business community, and workers. These voluntary programs include training and education, consultation, voluntary protection programs, and abatement assistance.

TRAINING MODEL

The guidelines provide employers with a model for designing, conducting, evaluating, and revising training programs. The training model can be used to develop training programs for a variety of occupational safety and health hazards identified at the workplace. Additionally, it can assist employers in their efforts to meet the training requirements in current or future occupational safety and health standards.

A training program designed in accordance with these guidelines can be used to supplement and enhance the employer's other education and training activities. The guidelines afford employers significant flexibility in the selection of content and training program design. OSHA encourages a personalized approach to the informational and instructional programs at individual worksites, thereby enabling employers to provide the training that is most needed and applicable to local working conditions.

Assistance with training programs or the identification of resources for training is available through such organizations as OSHA full-service Area Offices, state agencies which have their own OSHA-approved occupational safety and health programs, OSHA-funded state on-site consultation programs for employers, local safety councils, the OSHA Office of Training and Education, and OSHA-funded New Directions grantees.

Review Commission Implications

OSHA does not intend to make the guidelines mandatory. And they should not be used by employers as a total or complete guide in training and education matters which can result in enforcement proceedings before the Occupational Safety and Health Review Commission. However, employee training programs are always an issue in Review Commission cases which involve alleged violations of training requirements contained in OSHA standards.

The adequacy of employee training may also become an issue in contested cases where the affirmative defense of unpreventable employee misconduct is raised. Under case law well-established in the Commission and the courts, an employer may successfully defend against an otherwise valid citation by demonstrating that all feasible steps were taken to avoid the occurrence of the hazard, and that actions of the employee involved in the violation were a departure from a uniformly and effectively enforced work rule of which the employee had either actual or constructive knowledge.

In either type of case, the adequacy of the training given to employees in connection with a specific hazard is a factual matter which can be decided only by considering all the facts and circumstances surrounding the alleged violation. The general guidelines in this publication are not intended, and cannot be used, as evidence of the appropriate level of training in litigation involving either the training requirements of OSHA standards or affirmative defenses based . upon employer training programs.

TRAINING GUIDELINES

OSHA's training guidelines follow a model that consists of

- Determining if training is needed
- Identifying training needs
- Identifying goals and objectives
- Developing learning activities
- Conducting the training
- Evaluating program effectiveness
- Improving the program

The model is designed to be one that even the owner of a business with very few employees can use without having to hire a professional trainer or purchase expensive training materials. Using this model, employers or supervisors can develop and administer safety and health training programs that address problems specific to their own business, fulfill the learning needs of their own employees, and strengthen the overall safety and health program of the workplace.

Determining If Training is Needed

The first step in the training process is a basic one to determine whether a problem can be solved by training. Whenever employees are not performing their jobs properly, it is often assumed that training will bring them up to standard. However, it is possible that other actions (such as hazard abatement or the implementation of engineering controls) would enable employees to perform their jobs properly.

Ideally, safety and health training should be provided before problems or accidents occur. This training would cover both general safety and health rules and work procedures, and would be repeated if an accident or near-miss incident occurred.

Problems that can be addressed effectively by training include those that arise from lack of knowledge of a work process, unfamiliarity with equipment, or incorrect execution of a task. Training is less effective (but still can be used) for problems arising from an employee's lack of motivation or lack of attention to the job. Whatever its purpose, training is most effective when designed in relation to the goals of the employer's total safety and health program.

Identifying Training Needs

If the problem is one that can be solved, in whole or in part, by training, then the next step is to determine what training is needed. For this, it is necessary to identify what the employee is expected to do and in what ways, if any, the employee's performance is deficient. This information can be obtained by conducting a job analysis which pinpoints what an employee needs to know in order to perform a job.

When designing a new training program, or preparing to instruct an employee in an unfamiliar procedure or system, a job analysis can be developed by examining engineering data on new equipment or the safety data sheets on unfamiliar substances. The content of the specific federal or state OSHA standards applicable to a business can also provide direction in developing training content. Another option is to conduct a Job Hazard Analysis. This is a procedure for studying and recording each step of a job, identifying existing or potential hazards, and determining the best way to perform the job in order to reduce or eliminate the risks. Information obtained from a Job Hazard Analysis can be used as the content for the training activity.

If an employer's learning needs can be met by revising an existing training program rather than developing a new one, or if the employer already has some knowledge of the process or system to be used, appropriate training content can be developed through such means as

1. Using company accident and injury records to identify how accidents occur and what can be done to prevent them from recurring.

2. Requesting employees to provide, in writing and in their own words, descriptions of their jobs. These should include the tasks performed and the tools, materials and equipment used.

3. Observing employees at the worksite as they perform tasks, asking about the work, and recording their answers.

4. Examining similar training programs offered by other companies in the same industry, or obtaining suggestions from such organizations as the National Safety Council (which can provide information on Job Hazard Analysis), the Bureau of Labor Statistics, OSHA-approved state programs, OSHA full-service Area Offices, OSHA-funded state consultation programs, or the OSHA Office of Training and Education.

The employees themselves can provide valuable information on the training they need. Safety and health hazards can be identified through the employees' responses to such questions as whether anything about their jobs frightens them, if they have had any near-miss incidents, if they feel they are taking risks, or if they believe that their jobs involve hazardous operations or substances.

Once the kind of training that is needed has been determined, it is equally important to determine what kind of training is *not* needed. Employees should be made aware of all the steps involved in a task or procedure, but training should focus on those steps on which improved performance is needed. This avoids unnecessary training and tailors the training to meet the needs of the employees.

Identifying Goals and Objectives

Once the employees' training needs have been identified, employers can then prepare objectives for the training. Instructional objectives, if clearly stated, will tell employers what they want their employees to do, to do better, or to stop doing.

Learning objectives do not necessarily have to be written, but in order for the training to be as successful as possible, clear and measurable objectives should be thought out before the training begins. For an objective to be effective it should identify as precisely as possible what the individuals will do to demonstrate that they have learned or that they have reached the objective. The objectives should also describe the important conditions under which the individual will demonstrate competence and define what constitutes acceptable performance.

Using specific, action-oriented language, the instructional objectives should describe the preferred practice or skill and its observable behavior. For example, rather than using the statement: "The employee will understand how to use a respirator" as an instructional objective, it would be better to say: "The employee will be able to describe how a respirator works and when it should be used." Objectives are most effective when worded in sufficient detail that other qualified persons can recognize when the desired behavior is exhibited.

Developing Learning Activities

Once employers have stated precisely what the objectives for the training program are, then learning activities can be identified and described. Learning activities enable employees to demonstrate that they have acquired the desired skills and knowledge. To ensure that employees transfer the skills or knowledge from the learning activity to the job, the learning situation should simulate the actual job as closely as possible. Thus, employers may want to arrange the objectives and activities in a sequence which corresponds to the order in which the tasks are to be performed on the job, if a specific process is to be learned. For instance, if an employee must learn the beginning processes of using a machine, the sequence might be (1) to check that the power source is connected, (2) to ensure that the safety devices are in place and are operative, (3) to know when and how to throw the switch, and so on.

A few factors will help to determine the type of learning activity to be incorporated into the training. One aspect is the training resources available to the employer. Can a group training program that uses an outside trainer and film be organized, or should the employer personally train the employees on a one-to-one basis? Another factor is the kind of skills or knowledge to be learned. Is the learning oriented toward physical skills (such as the use of special tools) or toward mental processes and attitudes? Such factors will influence the type of learning activity designed by employers. The training activity can be group-oriented, with lectures, role play, and demonstrations; or designed for the individual as with self-paced instruction.

The determination of methods and materials for the learning activity can be as varied as the employer's imagination and available resources will allow. The employer may want to use charts, diagrams, manuals, slides, films, viewgraphs (overhead transparencies), videotapes, audiotapes, blackboard and chalk, or any combination of these and other instructional aids. Whatever the method of instruction, the learning activities should be developed in such a way that the employees can clearly demonstrate that they have acquired the desired skills or knowledge.

Conducting the Training

With the completion of the steps outlined above, the employer is ready to begin conducting the training. To the extent possible, the training should be presented so that its organization and meaning are clear to the employees. To do so, employers or supervisors should (1) provide overviews of the material to be learned; (2) relate, wherever possible, the new information or skills to the employee's goals, interests, or experience; and (3) reinforce what the employees learned by summarizing the program's objectives and the key points of information covered. These steps will assist employers in presenting the training in a clear, unambiguous manner.

In addition to organizing the content, employers must also develop the structure and format of the training. The content developed for the program, the nature of the workplace or other training site, and the resources available for training will help employers determine for themselves the frequency of training activities, the length of the sessions, the instructional techniques, and the individual(s) best qualified to present the information.

In order to be motivated to pay attention and learn the material that the employer or supervisor is presenting, employees must be convinced of the importance and relevance of the material. Among the ways of developing motivation are (1) explaining the goals and objectives of instruction; (2) relating the training to the interests, skills, and experiences of the employees; (3) outlining the main points to be presented during the training session(s); and (4) pointing out the benefits of training (e.g., the employee will be better informed, more skilled, and thus more valuable both on the job and on the labor market; or the employee will, if he or she applies the skills and knowledge learned, be able to work at reduced risk).

An effective training program allows employees to participate in the training process and to practice their skills or knowledge. This will help to ensure that they are learning the required knowledge or skills and permit correction if necessary. Employees can become involved in the training process by participating in discussions, asking questions, contributing their knowledge and expertise, learning through hands-on experiences, and through role-playing exercises.

Evaluating Program Effectiveness

To make sure that the training program is accomplishing its goals, an evaluation of the training can be valuable. Training should have, as one of its critical components, a method of measuring the effectiveness of the training. A plan for evaluating the training session(s), either written or thought out by the employer, should be developed when the course objectives and content are developed. It should not be delayed until the training has been completed. Evaluation will help employers or supervisors determine the amount of learning achieved and whether an employee's performance has improved on the job. Supervisors are in good positions to observe an employee's performance both before and after the training and note improvements or changes and workplace improvements. The ultimate success of a training program may be changes throughout the workplace that result in reduced injury or accident rates.

However it is conducted, an evaluation of training can give employers the information necessary to decide whether the employees achieved the desired results, and whether the training session should be offered again at some future date.

Improving the Program

If, after evaluation, it is clear that the training did not give the employees the level of knowledge and skill that was expected, then it may be necessary to revise the training program or provide periodic retraining. At this point, asking questions of employees and of those who conducted the training may be of some help. Among the questions that could be asked are

- Were parts of the content already known and, therefore, unnecessary?
- What material was confusing or distracting?
- Was anything missing from the program?
- What did the employees learn, and what did they fail to learn?

It may be necessary to repeat steps in the training process, that is, to return to the first steps and retrace one's way through the training process. As the program is evaluated, the employer should ask

- If a job analysis was conducted, was it accurate?
- Was any critical feature of the job overlooked?
- Were the important gaps in knowledge and skill included?
- Was material already known by the employees intentionally omitted?
- Were the instructional objectives presented clearly and concretely?
- Did the objectives state the level of acceptable performance that was expected of employees?
- Did the learning activity simulate the actual job?
- Was the learning activity appropriate for the kinds of knowledge and skills required on the job?
- When the training was presented, was the organization of the material and its meaning made clear?
- Were the employees motivated to learn?
- Were the employees allowed to participate actively in the training process?
- Was the employer's evaluation of the program thorough?

A critical examination of the steps in the training process will help employers to determine where course revision is necessary.

MATCHING TRAINING TO EMPLOYEES

While all employees are entitled to know as much as possible about the safety and health hazards to which they are exposed, and while employers should attempt to provide all relevant information and instruction to all employees, the resources for such an effort frequently are not, or are not believed to be, available. Thus, employers are often faced with the problem of deciding who is in the greatest need of information and instruction.

One way to differentiate between employees who have priority needs for training and those who do not is to identify employee populations which are at higher levels of risk. The nature of the work will provide an indication that such groups should receive priority for information on occupational safety and health risks.

Identifying Employees at Risk

One method of identifying employee populations at high levels of occupational risk (and thus in greater need of safety and health training) is to pinpoint hazardous occupations. Even within industries which are hazardous in general, there are some employees who operate at greater risk than others. In other cases the hazards of an occupation are influenced by the conditions under which it is performed, such as noise, heat or cold, or safety or health hazards in the surrounding area. In these situations, employees should be trained not only on how to perform their job safely but also on how to operate within a hazardous environment.

A second method of identifying employee populations at high levels of risk is to examine the incidence of accidents and injuries, both within the company and within the industry. If employees in certain occupational categories are experiencing higher accident and injury rates than other employees, training may be one way to reduce that rate. In addition, thorough accident investigations can identify not only specific employees who could benefit from training but could also identify company-wide training needs.

Research has identified the following variables as being related to a disproportionate share of injuries and illnesses at the worksite on the part of employees:

1. The age of the employee (younger employees have higher incidence rates)
2. The length of time on the job (new employees have higher incidence rates)
3. The size of the firm (in general terms, medium-size firms have higher incidence rates than smaller or larger firms)
4. The type of work performed (incidence and severity rates vary significantly by SIC Code)
5. The use of hazardous substances (by SIC Code)

These variables should be considered when identifying employee groups for training in occupational safety and health. Information is readily available to help employers identify which employees should receive safety and health information, and education and training, and who should receive it before others. Employers can request assistance in obtaining information by contacting such organizations as OSHA Area Offices, the Bureau of Labor Statistics, OSHA-approved state programs, state on-site consultation programs, the OSHA Office of Training and Education, or local safety councils.

Training Employees at Risk

Determining the content of training for employee populations at higher levels of risk is similar to determining what any employee needs to know, but more emphasis is placed on the requirements of the job and the possibility of injury. One useful tool for determining training content from job requirements is the Job Hazard Analysis described earlier. This procedure examines each step of a job, identifies existing or potential hazards, and determines the best way to perform the job in order to reduce or eliminate the hazards. Its key elements are (1) job description; (2) job location; (3) key steps (preferably in the order in which they are performed); (4) tools, machines and materials used; (5) actual and potential safety and health hazards associated with these key job steps; and (6) safe and healthful practices, apparel, and equipment required for each job step.

Material Safety Data Sheets (MSDS) can also provide information for training employees in the safe use of materials. These data sheets, developed by chemical manufacturers and importers, are supplied with manufacturing or construction materials and describe the ingredients of a product, its hazards, protective equipment to be used, safe handling procedures, and emergency first-aid responses. The information contained in these sheets can help employers identify employees in need of training (i.e., workers handling substances described in the sheets) and train employees in safe use of the substances. Material Safety Data Sheets are generally available from suppliers, manufacturers of the substance, large employers who use the substance on a regular basis, or they can be developed by employers or trade associations. MSDSs are particularly useful for those employers who are developing training on chemical use as required by OSHA's Hazard Communication Standard.

Conclusion

In an attempt to assist employers with their occupational health and safety training activities, OSHA has developed a set of training guidelines in the form of a model. This model is designed to help employers develop instructional programs as part of their total education and training effort. The model addresses the questions of who should be trained, on what topics, and for what purposes. It also helps employers determine how effective the program has been and enables them to identify employees who are in greatest need of education and training. The model is general enough to be used in any area of occupational safety and health training, and allows employers to determine for themselves the content and format of training. Use of this model in training activities is just one of many ways that employers can comply with the OSHA standards that relate to training and enhance the safety and health of their employees.

CONSTRUCTION TRAINING REQUIREMENTS

The following training requirements have been excerpted from Title 29, *Code of Federal Regulations* Part 1926. Note that in addition to these requirements, Part 1910, relating to general industry, also contains applicable training standards.

General Safety and Health Provisions

1926.20(b) (2) and (4)

Such programs [as may be necessary to comply with this part] shall provide for frequent and regular inspections of the job sites, materials, and equipment to be made by competent persons [capable of identifying existing and predictable hazards in the surroundings or working conditions which are unsanitary, hazardous, or dangerous to employees, and who have authorization to take prompt corrective measures to eliminate them designated by the employers].

The employer shall permit only those employees qualified [one who, by possession of a recognized degree, certificate, or professional standing, or who by extensive knowledge, training, and experience, has successfully demonstrated his ability to solve or resolve problems relating to the subject matter, the work, or the project] by training or experience to operate equipment and machinery.

Safety Training and Education

1926.21(a)

General requirements—The Secretary shall, pursuant to section 107(f) of the Act, establish and supervise programs for the education and training of employers and employees in the recognition, avoidance, and prevention of unsafe conditions in employments covered by the Act.

1926.21(b) (1) through (6) (i) and (ii)

The employer should avail himself of the safety and health training programs the Secretary provides.

The employer shall instruct each employee in the recognition and avoidance of unsafe conditions and the regulations applicable to his work environment to control or eliminate any hazards or other exposure to illness or injury.

Employees required to handle or use poisons, caustics, and other harmful substances shall be instructed regarding their safe handling and use, and be made aware of the potential hazards, personal hygiene, and personal protective measures required.

In job site areas where harmful plants or animals are present, employees who may be exposed shall be instructed regarding the potential hazards and how to avoid injury, and the first aid procedures to be used in the event of injury.

Employees required handling or using flammable liquids, gases, or toxic materials shall be instructed in the safe handling and use of these materials and made aware of the specific requirements contained in Subparts D, F, and other applicable subparts of this part.

All employees required to enter into confined or enclosed spaces shall be instructed as to the nature of the hazards involved, the necessary precautions to be taken, and in the use of protective and emergency equipment required. The employer shall comply with any specific regulations that apply to work in dangerous or potentially dangerous areas.

"Confined or enclosed space" means any space having a limited means of egress which is subject to the accumulation of toxic or flammable contaminants or has an oxygen deficient atmosphere. Confined or enclosed spaces include, but are not limited to, storage tanks, process vessels, bins, boilers, ventilation or exhaust ducts, sewers, underground utility vaults, tunnels, pipelines, and open top spaces more than four feet in depth such as pits, tubs, vaults, and vessels.

Medical Services and First Aid

1926.50(c)

In the absence of an infirmary, clinic, hospital, or physician that is reasonably accessible in terms of time and distance to the worksite which is available for the treatment of injured employees, a person who has a valid certificate in first-aid training from the U.S. Bureau of Mines, the American Red Cross, or equivalent training that can be verified by documentary evidence, shall be available at the worksite to render first aid.

Ionizing Radiation

1926.53(b)

Any activity which involves the use of radioactive materials or X-rays, whether or not under license from the Atomic Energy Commission [Nuclear Regulatory Commission] shall be performed by competent persons specially trained in the proper and safe operation of such equipment. In the case of materials used under Commission license, only persons actually licensed, or competent persons under the direction and supervision of the licensee, shall perform such work.

Non-ionizing Radiation

1926.54(a) and (b)

Only qualified and trained employees shall be assigned to install, adjust, and operate laser equipment.

Proof of qualification of the laser equipment operator shall be available and in possession of the operator at all times.

Gases, Vapors, Fumes, Dusts, and Mists

1926.55(b)

To achieve compliance with this section, administrative or engineering controls must first be implemented whenever feasible. When such controls are not feasible to achieve full compliance, protective equipment or other protective measures shall be used to keep the exposure of employees to air contaminants within the limits prescribed in this section. Any equipment and technical measures used for this purpose must first be approved for each particular use by a competent industrial hygienist or other technically qualified person. Whenever respirators are used, their use shall comply with § 1926.103.

Asbestos

1926.58(k) (3) (i) through (iii) (A) through (E) and (4) (i) and (ii)

The employer shall institute a training program for all employees exposed to airborne concentrations of asbestos in excess of the action level and/or excursion limit and shall ensure their participation in the program.

Training shall be provided prior to or at the time of initial assignment (unless the employee has received equivalent training within the previous 12 months) and at least annually thereafter.

The training program shall be conducted in a manner that the employee is able to understand. The employer shall ensure that each such employee is informed of the following:

Methods of recognizing asbestos:

- The health effects associated with asbestos exposure
- The relationship between smoking and asbestos in producing lung cancer
- The nature of operations that could result in exposure to asbestos
- The importance of necessary protective controls to minimize exposure including, as applicable, engineering controls, work practices, respirators, housekeeping procedures, hygiene facilities, protective clothing, decontamination procedures, emergency procedures, and waste disposal procedures, and any necessary instruction in the use of these controls
- The purpose, proper use, fitting instructions and limitations of respirators as required by 29 CFR 1910.134.

Access to training materials—The employer shall make readily available to all affected employees without cost all written materials relating to the employee training program, including a copy of this regulation.

The employer shall provide to the Assistant Secretary and the Director, upon request, all information and training materials relating to the employee information and training program.

Hazard Communication, Construction

1926.59(h) (2) (i) through (iv)

Employee training should include at least

Methods and observations that may be used to detect the presence or release of a hazardous chemical in the work area (such as monitoring conducted by the employer, continuous monitoring devices, appearance or odor of hazardous chemicals when being released, etc.)

The physical and health hazards of the chemicals in the work area

The measures employees can take to protect themselves from these hazards, including specific procedures the employer has implemented to protect employees from exposure to hazardous chemicals, such as appropriate work practices, emergency procedures, and personal protective equipment to be used

The details of the hazard communication program to be developed by the employer, including an explanation of the labeling system and the material safety data sheet, and how employees can obtain and use the appropriate hazard information

Lead in Construction

1926.62(l) (1) (i) through (iv); (2)(i) through (viii) and (3)(i) and (ii)

The employer shall communicate information concerning lead hazards according to the requirements of OSHA's Hazard Communication Standard for the construction industry, 29 CFR 1926.59, including but not limited to the requirements concerning warning signs and labels, material safety data sheets (MSDS), and employee information and training. In addition, for all employees who are subject to exposure to lead at or above the action level on any day or who are subject to exposure to lead compounds which may cause skin or eye irritation (e.g., lead arsenate, lead azide), employers shall comply with the following requirements

The employer shall provide the program in accordance with paragraph (l) (2) of this section and ensure employee participation.

The employer shall provide the training program as initial training prior to the time of job assignment or prior to the start up date for this requirement, whichever comes last.

The employer shall also provide the training program at least annually for each employee who is subject to lead exposure at or above the action level on any day.

Training program—The employer shall ensure that each employee is trained in the following

The content of this standard and its appendices

The specific nature of the operations which could result in exposure to lead above the action level

The purpose, proper selection, fitting, use, and limitations of respirators

The purpose and a description of the medical surveillance program, and the medical removal protection program including information concerning the adverse health effects associated with excessive exposure to lead (with particular attention to the adverse reproductive effects on both males and females and hazards to the fetus and additional precautions for employees who are pregnant)

The engineering controls and work practices associated with the employee's job assignment including training of employees to follow relevant good work practices described in Appendix B of this section

The contents of any compliance plan in effect

Instructions to employees that chelating agents should not routinely be used to remove lead from their bodies and should not be used at all except under the direction of a licensed physician

The employee's right of access to records under 29 CFR 1910.20

Access to information and training materials—The employer shall make readily available to all affected employees a copy of this standard and its appendices.

The employer shall provide, upon request, all materials relating to the employee information and training program to affected employees and their designated representative, and to the Assistant Secretary and the Director.

Process Safety Management of Highly Hazardous Chemicals

1926.64(g) (1) (i) and (ii)

Each employee presently involved in operating a process, and each employee before being involved in operating a newly assigned process, shall be trained in an overview of the process and in the operating procedures as specified in paragraph (f) of this section. The training shall include emphasis on the specific safety and health hazards, emergency operations including shutdown, and safe work practices applicable to the employee's job tasks.

In lieu of initial training for those employees already involved in operating a process on May 26, 1992, an employer may certify in writing that the employee has the required knowledge, skills, and abilities to safely carry out the duties and responsibilities as specified in the operating procedures.

1926.64(g) (2)

Refresher training shall be provided at least every three years, and more often if necessary, to each employee involved in operating a process to ensure that the employee understands and adheres to the current operating procedures of the process. The employer, in consultation with the employees involved in operating the process, shall determine the appropriate frequency of refresher training.

1926.64(g) (3)

Training documentation—The employer shall ascertain that each employee involved in operating a process has received and understood the training required by this paragraph. The employer shall prepare a record which contains the identity of the employee, the date of training, and the means used to verify that the employee understood the training.

Contract Employer Responsibilities 1926.64(h) (3) (i) through (iv)

The contract employer shall ensure that each contract employee is trained in the work practices necessary to safely perform his/her job.

The contract employer shall ensure that each contract employee is instructed in the known potential fire, explosion, or toxic release hazards related to his or her job and the process, and the applicable provisions of the emergency action plan.

The contract employer shall document that each contract employee has received and understood the training required by this paragraph. The contract employer shall prepare a record which contains the identity of the contract employee, the date of training, and the means used to verify that the employee understood the training.

The contract employer shall ensure that each contract employee follows the safety rules of the facility including the safe work practices required by paragraph (f) (4) of this section.

Mechanical Integrity 1926.64(j) (3)

Training for process maintenance activities—The employer shall train each employee involved in maintaining the ongoing integrity of process equipment in an overview of that process and its hazards and in the procedures applicable to the employee's job tasks to ensure that the employee can perform the job tasks in a safe manner.

Hearing Protection

1926.101(b)

Ear protective devices inserted in the ear shall be fitted or determined individually by competent persons.

Respiratory Protection

1926.103(c) (1)

Employees required to use respiratory protective equipment approved for use in atmospheres immediately dangerous to life shall be thoroughly trained in its use. Employees required to use other types of respiratory protective equipment shall be instructed in the use and limitations of such equipment.

Fire Protection

1926.150(a) (5)

As warranted by the project, the employer shall provide a trained and equipped firefighting organization (Fire Brigade) to ensure adequate protection to life. "Fire Brigade" means an organized group of employees that are knowledgeable, trained, and skilled in the safe evacuation of employees during emergency situations and in assisting in firefighting operations.

1926.150(c) (1) (viii)

Portable fire extinguishers shall be inspected periodically and maintained in accordance with *Maintenance and Use of Portable Fire Extinguishers*, NFPA No. 10A-1970.

ANSI Standard 10A-1970:

> The owner or occupant of a property in which fire extinguishers are located has an obligation for the care and use of these extinguishers at all times. By doing so, he is contributing to the protection of life and property. The nameplate(s) and instruction manual should be read and thoroughly understood by all persons who may be expected to use extinguishers.

To discharge this obligation he should give proper attention to the inspection, maintenance, and recharging of this fire protective equipment. He should also train his personnel in the correct use of fire extinguishers on the different types of fires which may occur on his property.

Persons responsible for performing maintenance operations come from three major groups:

- Trained industrial safety or maintenance personnel
- Extinguisher service agencies
- Individual owners (e.g., self-employed)

Signaling

RAMSETS –shoot + certification

1926.201(a) (2)

Signaling directions by flagmen shall conform to American National Standards Institute D6.1-1971, *Manual on Uniform Traffic Control Devices for Streets and Highways*.

Powder-Operated Hand Tools

1926.302(e) (1) and (12)

Only employees who have been trained in the operation of the particular tool in use shall be allowed to operate a powder-actuated tool.

Powder-actuated tools used by employees shall meet all other applicable requirements of American National Standards Institute, A10.3-1970, *Safety Requirements for Explosive-Actuated Fastening Tools*.

Woodworking Tools

1926.304(f)

All woodworking tools and machinery shall meet other applicable requirements of American National Standards Institute, 01.1-1961, *Safety Code for Woodworking Machinery*.

ANSI Standard 01.1-1961: Selection and Training of Operators.

> Before a worker is permitted to operate any woodworking machine, he shall receive instructions in the hazards of the machine and the safe method of its operation.

Learn the machine's applications and limitations, as well as the specific potential hazards peculiar to this machine. Follow available operating instructions and safety rules carefully.

Keep working area clean and be sure adequate lighting is available.

Do not wear loose clothing, gloves, bracelets, necklaces, or ornaments. Wear face, eye, ear, respiratory, and body protection devices, as indicated for the operation or environment.

Do not use cutting tools larger or heavier than the machine is designed to accommodate. Never operate a cutting tool at greater speed than recommended.

Keep hands well away from saw blades and other cutting tools. Use a push stock or push block to hold or guide the work when working close to cutting tool.

Whenever possible, use properly locked clamps, jig, or vise to hold the work.

Combs (feather boards) shall be provided for use when an applicable guard cannot be used.

Never stand directly in line with a horizontally rotating cutting tool. This is particularly true when first starting a new tool, or a new tool is initially installed on the arbor.

Be sure the power is disconnected from the machine before tools are serviced.

Never leave the machine with the power on.

Be positive that hold-downs and anti kickback devices are positioned properly, and that the work piece is being fed through the cutting tool in the right direction.

Do not use a dull, gummy, bent, or cracked cutting tool.

Be sure that keys and adjusting wrenches have been removed before turning power on.

Use only accessories designed for the machine.

Adjust the machine for minimum exposure of cutting tool necessary to perform the operation.

Gas Welding and Cutting

1926.350(d) (1) through (6)

Use of fuel gas—The employer shall thoroughly instruct employees in the safe use of fuel gas as follows:

Before a regulator to a cylinder valve is connected, the valve shall be opened slightly and closed immediately. (This action is generally termed "cracking" and is intended to clear the valve of dust or dirt that might otherwise enter the regulator.) The person cracking the valve shall stand to one side of the outlet, not in front of it. The valve of a fuel gas cylinder shall not be cracked where the gas would reach welding work, sparks, flame, or other possible sources of ignition.

The cylinder valve shall always be opened slowly to prevent damage to the regulator. For quick closing, valves on fuel gas cylinders shall not be opened more than one and one-half turns. When a special wrench is required, it shall be left in position on the stem of the valve while the cylinder is in use so that the fuel gas flow can be shut off quickly in case of an emergency. In the case of manifold or coupled cylinders, at least one such wrench shall always be available for immediate use. Nothing shall be placed on top of a fuel gas cylinder, when in use, which may damage the safety device or interfere with the quick closing of the valve.

Fuel gas shall not be used from cylinders through torches or other devices which are equipped with shutoff valves without reducing the pressure through a suitable regulator attached to the cylinder valve or manifold.

Before a regulator is removed from a cylinder valve, the cylinder valve shall always be closed and the gas released from the regulator.

If, when the valve on a fuel gas cylinder is opened, there is found to be a leak around the valve stem, the valve shall be closed and the gland nut tightened. If this action does not stop the leak, the use of the cylinder shall be discontinued, and it shall be properly tagged and removed from the work area. In the event that fuel gas should leak from the cylinder valve, rather than from the valve stem, and the gas cannot be shut off, the cylinder shall be properly tagged and removed from the work area. If a regulator attached to a cylinder valve will effectively stop a leak through the valve seat, the cylinder need not be removed from the work area.

If a leak should develop at a fuse plug or other safety device, the cylinder shall be removed from the work area.

1926.350(j)

For additional details not covered in this subpart, applicable technical portions of American National Standards Institute, Z49.1-1967, *Safety in Welding and Cutting*, shall apply.

ANSI Standard Z49.1-1967, Fire Watch Duties:

Fire watchers shall be trained in the use of fire extinguishing equipment. They shall be familiar with facilities for sounding an alarm in the event of a fire. They shall watch for fires in all exposed areas, try to extinguish them only when obviously within the capacity of the equipment available, and otherwise sound the alarm. A fire watch shall be maintained for at least a half hour after completion of welding or cutting operations to detect and extinguish possible smoldering fires.

Arc Welding and Cutting

1926.351(d) (1) through (5)

Employers shall instruct employees in the safe means of arc welding and cutting as follows:

When electrode holders are to be left unattended, the electrodes shall be removed and the holders shall be so placed or protected that they cannot make electrical contact with employees or conducting objects.

Hot electrode holders shall not be dipped in water; to do so may expose the arc welder or cutter to electric shock.

When the arc welder or cutter has occasion to leave his work or to stop work for any appreciable length of time, or when the arc welding or cutting machine is to be moved, the power supply switch to the equipment shall be opened.

Any faulty or defective equipment shall be reported to the supervisor.

Other requirements, as outlined in Article 630, *National Electrical Code*, NFPA 70-1971; ANSI C1-1971 (Rev. of 1968), *Electric Welders*, shall be used when applicable.

Fire Prevention

1926.352(e)

When the welding, cutting, or heating operation is such that normal fire prevention precautions are not sufficient, additional personnel shall be assigned to guard against fire while the actual welding, cutting, or heating operation is being performed, and for a sufficient period of time after completion of the work to ensure that no possibility of fire exists. Such personnel shall be instructed as to the specific anticipated fire hazards and how the firefighting equipment provided is to be used.

Welding, Cutting, and Heating in Way of Preservative Coatings

1926.354(a)

Before welding, cutting, or heating is commenced on any surface covered by a preservative coating whose flammability is not known, a test shall be made by a competent person to determine its flammability. Preservative coatings shall be considered to be highly flammable when scrapings burn with extreme rapidity.

Ground-Fault Protection

1926.404(b) (iii) (B)

The employer shall designate one or more competent persons [as defined in § 1926.32(f)] to implement the program.

Scaffolding

1926.451(a) (3)

No scaffold shall be erected, moved, dismantled, or altered except under the supervision of competent persons.

1926.451(b) (16)

All wood pole scaffolds 60 feet or less in height shall be constructed and erected in accordance with Tables L-4 to 10. If they are over 60 feet in height, they shall be designed by a qualified engineer competent in this field, and it shall be constructed and erected in accordance with such design.

1926.451(c) (4) and (5)

Tube and coupler scaffolds shall be limited in heights and working levels to those permitted in Tables L- 10, 11, and 12. Drawings and specifications of all tube and coupler scaffolds above the limitations in Tables L- 10, 11, and 12 shall be designed by a qualified engineer competent in this field.

All tube and coupler scaffolds shall be constructed and erected to support four times the maximum intended loads, as set forth in Tables L- 10, 11, and 12, or as set forth in the specifications by a licensed professional engineer competent in this field.

1926.451(d) (9)

Drawings and specifications for all frame scaffolds over 125 feet in height above the base plates shall be designed by a registered professional engineer.

1926.451(g) (3)

Unless outrigger scaffolds are designed by a registered professional engineer competent in this field, they shall be constructed and erected in accordance with Table L-13. Outrigger scaffolds, designed by a registered professional engineer, shall be constructed and erected in accordance with such design.

1926.451(h) (6) and (14)

Where the overhang exceeds six feet six inches, outrigger beams shall be composed of stronger beams or multiple beams and be installed under the supervision of a competent person.

Each scaffold shall be installed or relocated under the supervision of a competent person.

1926.451(k) (10)

Single-point adjustable suspension scaffolds

For additional details not covered in this paragraph, applicable technical portions of American National Standards Institute, A120.1-1970, *Power-Operated Devices for Exterior Building Maintenance Powered Platforms*, shall be used.

ANSI Standard A120.1-1970

"Qualified Operators: Powered platform shall be operated only by qualified persons who have been instructed in the operation and in the inspection, with respect to safe operating condition of the particular powered platform to be operated."

Guarding of Low-Pitched Roof Perimeters During the Performance of Built-Up Roofing Work

1926.500(g)(6)(i) and (ii)(a) through (f) and (iii)

The employer shall provide a training program for all employees engaged in built-up roofing work so that they are able to recognize and deal with the hazards of falling associated with working near a roof perimeter. The employees shall also be trained in the safety procedures to be followed in order to prevent such falls.

The employer shall ensure that employees engaged in built-up roofing work have been trained and instructed in the following areas:

The nature of fall hazards in the work area near a roof edge

The function, use, and operation of the MSS system, warning line, and safety monitoring systems to be used

The correct procedures for erecting, maintaining, and disassembling the systems to be used

The role of each employee in the safety monitoring system when this system is used

The limitations on the use of mechanical equipment

The correct procedures for the handling and storage of equipment and materials

Training shall be provided for each newly hired employee, and for all other employees as necessary, to ensure that employees maintain proficiency in the areas listed in paragraph (g)(6)(ii) of this section.

Fall Protection

1926.503(a) (1) and (2) (i) through (vii)

The employer shall provide a training program for each employee who might be exposed to fall hazards. The program shall enable each employee to recognize the hazards of falling and shall train each employee in the procedures to be followed in order to minimize these hazards.

The employer shall ensure that each employee has been trained, as necessary, by a competent person qualified in the following areas:

The nature of fall hazards in the work area

The correct procedures for erecting, maintaining, disassembling, and inspecting the fall protection systems to be used

The use and operation of guardrail systems, personal fall arrest systems, safety net systems, warning line systems, safety monitoring systems, controlled access zones, and other protection to be used

The role of each employee in the safety monitoring system when this system is used

The limitations on the use of mechanical equipment during the performance of roofing work on low-sloped roofs

The correct procedures for the handling and storage of equipment and materials and the erection of overhead protection

The standards contained in this subpart

Cranes and Derricks

1926.550(a) (1), (5), and (6)

The employer shall comply with the manufacturer's specifications and limitations applicable to the operation of any and all cranes and derricks. Where manufacturer's specifications are not available, the limitations assigned to the equipment shall be based on the determinations of a qualified engineer competent in this field and such determinations will be appropriately documented and recorded. Attachments used with cranes shall not exceed the capacity, rating, or scope recommended by the manufacturer.

The employer shall designate a competent person who shall inspect all machinery and equipment prior to each use, and during use, to make sure it is in safe operating condition. Any deficiencies shall be repaired, or defective parts replaced, before continued use.

A thorough, annual inspection of the hoisting machinery shall be made by a competent person, or by a government or private agency recognized by the U.S. Department of Labor. The employer shall maintain a record of the dates and results of inspections for each hoisting machine and piece of equipment.

1926.550(g) (4) (i) (A)

Personnel platforms (i) *Design criteria*—The personnel platform and suspension system shall be designed by a qualified engineer or a qualified person competent in structural design.

1926.550(g) (5) (iv)

A visual inspection of the crane or derrick, rigging, personnel platform, and the crane or derrick base support or ground shall be conducted by a competent person immediately after the trial lift to determine whether the testing has exposed any defect or produced any adverse effect upon any component or structure.

Material Hoists, Personnel Hoists, and Elevators

1926.552(a) (1)

The employer shall comply with the manufacturer's specifications and limitations applicable to the operation of all hoists and elevators. Where manufacturer's specifications are not available, the limitations assigned to the equipment shall be based on the determinations of a professional engineer competent in the field.

1926.552(b) (7)

All material hoist towers shall be designed by a licensed professional engineer.

1926.552(c) (15) and (17) (i)

Personnel hoists—Following assembly and erection of hoists, and before being put in service, an inspection and test of all functions and safety devices shall be made under the supervision of a competent person. A similar inspection and test is required following major alteration of an existing installation. All hoists shall be inspected and tested at not more than three-month intervals. Records shall be maintained and kept on file for the duration of the job.

Personnel hoists used in bridge tower construction shall be approved by a registered professional engineer and erected under the supervision of a qualified engineer competent in this field.

Material Handling Equipment

1926.602(c) (1)(vi)

Lifting and hauling equipment (other than equipment covered under Subpart N of this part)—All industrial trucks in use shall meet the applicable requirements of design, construction, stability, inspection, testing, maintenance, and operation, as defined in American National Standards Institute B56.1-1969, *Safety Standards for Powered Industrial Trucks.*

ANSI Standard B56.1-1969:

> "Operator Training: Only trained and authorized operators shall be permitted to operate a powered industrial truck. Methods shall be devised to train operators in the safe operation of powered industrial trucks. Badges or other visual indication of the operators' authorization should be displayed at all times during work period."

Site Clearing

1926.604(a) (1)

Employees engaged in site clearing shall be protected from hazards of irritant and toxic plants and suitably instructed in the first-aid treatment available.

Excavations—General Protection Requirements (Excavations, Trenching, and Shoring)

1926.651(c) (1) (i)

Access and egress (1) *Structural ramps*—Structural ramps that are used solely by employees as a means of access or egress from excavations shall be designed by a competent person. Structural ramps used for access or egress of equipment shall be designed by a competent person qualified in structural design, and shall be constructed in accordance with the design.

1926.651(h) (2) and (3)

Protection from hazards associated with water accumulation—If water is controlled or prevented from accumulating by the use of water removal equipment, the water removal equipment and operations shall be monitored by a competent person to ensure proper operation.

If excavation work interrupts the natural drainage of surface water (such as streams), diversion ditches, dikes, or other suitable means shall be used to prevent surface water from entering the excavation and to provide adequate drainage of the area adjacent to the excavation. Excavations subject to runoff from heavy rains will require an inspection by a competent person.

1926.651(i) (1)

Stability of adjacent structures—Where the stability of adjoining buildings, walls, or other structures is endangered by excavation operations, support systems such as shoring, bracing, or underpinning shall be provided to ensure the stability of such structures for the protection of employees.

1926.651(i) (2) (iii)

A registered professional engineer has approved the determination that the structure is sufficiently removed from the excavation so as to be unaffected by the excavation activity; or

1926.651(i) (2) (iv)

A registered professional engineer has approved the determination that such excavation work will not pose a hazard to employees.

1926.651(k) (1) and (2)

Inspections—Daily inspections of excavations, the adjacent areas, and protective systems shall be made by a competent person for evidence of a situation that could result in possible cave-ins, indications of failure of protective systems, hazardous atmospheres, or other hazardous conditions. An inspection shall be conducted by the competent person prior to the start of work and as needed throughout the shift. Inspections shall also be made after every rainstorm or other hazard increasing occurrence. These inspections are only required when employee exposure can be reasonably anticipated.

Where the competent person finds evidence of a situation that could result in a possible cave-in, indications of failure of protective systems, hazardous atmospheres, or other hazardous conditions, exposed employees shall be removed from the hazardous area until the necessary precautions have been taken to ensure their safety.

Concrete and Masonry Construction

1926.701(a)

No construction loads shall be placed on a concrete structure or portion of a concrete structure unless the employer determines, based on information received from a person who is qualified in structural design, that the structure or portion of the structure is capable of supporting the loads.

1926.703(b) (8) (i)

The design of the shoring shall be prepared by a qualified designer, and the erected shoring shall be inspected by an engineer qualified in structural design.

Bolting, Riveting, Fitting-Up, and Plumbing-Up

1926.752(d) (4)

Plumbing-up guys shall be removed only under the supervision of a competent person.

Underground Construction

1926.800(d)

Safety instruction—All employees shall be instructed in the recognition and avoidance of hazards associated with underground construction activities including, where appropriate, the following subjects:

> Air monitoring
> Ventilation
> Illumination
> Communications
> Flood control
> Mechanical equipment
> Personal protective equipment
> Explosives
> Fire prevention and protection
> Emergency procedures, including evacuation plans and check-in/check-out systems

1926.800(g) (2)

Emergency provisions; Self-rescuers—The employer shall provide self-rescuers having current approval from the National Institute for Occupational Safety and Health and the Mine Safety and Health Administration to be immediately available to all employees at work stations in underground areas where employees might be trapped by smoke or gas. The selection, issuance, use, and care of respirators shall be in accordance with paragraphs (b) and (c) of 1926.103.

1926.800(g) (5) (iii) through (v)

Emergency provisions—Rescue team members shall be qualified in rescue procedures, the use and limitations of breathing apparatus, and the use of firefighting equipment. Qualifications shall be reviewed not less than annually.

On jobsites where flammable or noxious gases are encountered or anticipated in hazardous quantities, rescue team members shall practice donning and using self-contained breathing apparatus monthly.

The employer shall ensure that rescue teams are familiar with conditions at the jobsite.

1926.800(j) (1) (i) (A)

Air quality and monitoring—The employer shall assign a competent person who shall perform all air monitoring required by this section.

Where this paragraph requires monitoring of airborne contaminants "as often as necessary," the competent person shall make a reasonable determination as to which substances to monitor and how frequently to monitor taking into consideration location of jobsite; geology of the jobsite; presence of air contaminants in nearby jobsites and changes in levels of substances monitored on prior shifts; and work practices and jobsite conditions including use of diesel engines, explosives, fuel gas, volume and flow of ventilation, visible atmospheric conditions, decompression of the atmosphere, welding, cutting, and hot work, and employees' physical reactions to working underground.

1926.800(j)(1)(vi) (A) and (B)

Air quality and monitoring—When the competent person determines, on the basis of air monitoring results or other information, that air contaminants may be present in sufficient quantity to be dangerous to life, the employer shall

Prominently post a notice at all entrances to the underground jobsite to inform all entrants of the hazardous condition

Ensure that the necessary precautions are taken

1926.800(o) (3) (i) (A)

Ground support Underground areas—A competent person shall inspect the roof, face, and walls of the work area at the start of each shift and as often as necessary to determine ground stability.

1926.800(o) (3) (iv) (B)

Ground support Underground areas—A competent person shall determine whether rock bolts meet the necessary torque and shall determine the testing frequency in light of the bolt system, ground conditions, and the distance from vibration sources.

1926.800(t) (3) (xix) and (xx)

Hoisting unique to underground construction—A competent person shall visually check all hoisting machinery, equipment, anchorages, and hoisting rope at the beginning of each shift during hoist use, as necessary.

Each safety device shall be checked by a competent person at least weekly during hoist use to ensure suitable operation and safe condition.

Compressed Air

1926.803(a) (1) and (2)

There shall be present, at all times, at least one competent person designated by and representing the employer who shall be familiar with this subpart in all respects and responsible for full compliance with these and other applicable subparts.

Every employee shall be instructed in the rules and regulations which concern his safety or the safety of others.

1926.803(b)(1) and (10)(xii)

There shall be retained one or more licensed physician familiar with and experienced in the physical requirements and the medical aspects of compressed air work and the treatment of decompression illness. He or she shall be available at all times while work is in progress in order to provide medical supervision of employees employed in compressed air work. He shall himself be physically qualified and be willing to enter a pressurized environment.

The medical lock shall be in constant charge of an attendant under the direct control of the retained physician. The attendant shall be trained in the use of the lock and suitably instructed regarding steps to be taken in the treatment of employee exhibiting symptoms compatible with a diagnosis of decompression illness.

1926.803(e) (1)

Every employee going under air pressure for the first time shall be instructed on how to avoid excessive discomfort.

1926.803(f) (2) and (3)

In the event it is necessary for an employee to be in compressed air more than once in a 24-hour period, the appointed physician shall be responsible for the establishment of methods and procedures of decompression applicable to repetitive exposures.

If decanting is necessary, the appointed physician shall establish procedures before any employee is permitted to be decompressed by decanting methods. The period of time that the employees spend at atmospheric pressure between the decompression following the shift and recompression shall not exceed five minutes.

1926.803(h) (1)

At all times there shall be a thoroughly experienced, competent, and reliable person on duty at the air control valves as a gauge tender who shall regulate the pressure in the working areas. During tunneling operations, one gauge tender may regulate the pressure in not more than two headings: provided that the gauge and controls are all in one location. In caisson work, there shall be a gauge tender for each caisson.

Preparatory Operations

1926.850(a)

Prior to permitting employees to start demolition operations, a competent person shall make an engineering survey of the structure to determine the condition of the framing, floors, and walls, and possibility of unplanned collapse of any portion of the structure. Any adjacent structure where employees may be exposed shall also be similarly checked. The employer shall have in writing evidence that such a survey has been performed.

Chutes

1926.852(c)

A substantial gate shall be installed in each chute at or near the discharge end. A competent employee shall be assigned to control the operation of the gate and the backing and loading of trucks.

Mechanical Demolition

1926.859(g)

During demolition, continuing inspections by a competent person shall be made as the work progresses to detect hazards resulting from weakened or deteriorated floors, or walls, or loosened material. No employee shall be permitted to work where such hazards exist until they are corrected by shoring, bracing, or other effective means.

General Provisions (Blasting and Use of Explosives)

1926.900(a)

The employer shall permit only authorized and qualified persons to handle and use explosives.

1926.900(k) (3) (i)

There shall be a prominent display of adequate signs warning against the use of mobile radio transmitters on all roads within 1,000 feet of blasting operations. Whenever adherence to the 1,000-foot distance would create an operational handicap, a competent person shall be consulted to evaluate the particular situation, and alternative provisions may be made which are adequately designed to prevent any premature firing of electric blasting caps. A description of any such alternatives shall be reduced to writing and shall be certified as meeting the purposes of this subdivision by the competent person consulted. The description shall be maintained at the construction site during the duration of the work and shall be available for inspection by representatives of the Secretary of Labor.

1926.900(q)

All loading and firing shall be directed and supervised by competent persons thoroughly experienced in this field.

Blaster Qualifications

1926.901(c), (d), and (e)

A blaster shall be qualified, by reason of training, knowledge, or experience, in the field of transporting, storing, handling, and use of explosives, and have a working knowledge of state and local laws and regulations which pertain to explosives.

Blasters shall be required to furnish satisfactory evidence of competency in handling explosives and performing in a safe manner the type of blasting that will be required.

The blaster shall be knowledgeable and competent in the use of each type of blasting method used.

Surface Transportation of Explosives

1926.902(b) and (i)

Motor vehicles or conveyances transporting explosives shall only be driven by, and be in the charge of, a licensed driver who is physically fit. He shall be familiar with the local, state, and federal regulations governing the transportation of explosives.

Each vehicle used for transportation of explosives shall be equipped with a fully charged fire extinguisher, in good condition. An Underwriters Laboratory–approved extinguisher of not less than 10-ABC rating will meet the minimum requirement. The driver shall be trained in the use of the extinguisher on his vehicle.

Firing the Blast

1926.909(a)

A code of blasting signals equivalent to Figure 5.1 shall be posted on one or more conspicuous places at the operation, and all employees shall be required to familiarize themselves with the code and conform to it. Danger signs shall be placed at suitable locations.

WARNING SIGNAL—A one-minute series of long blasts five minutes prior to blast signal.

BLAST SIGNAL—A series of short blasts one minute prior to the shot.

ALL CLEAR SIGNAL—A prolonged blast following the inspection of blast area.

Figure 5.1 Code of Blasting Signals

General Requirements (Power Transmission and Distribution)

1926.950(d) (1)(ii)(a) through (c), (vi), and (vii)

When de-energizing lines and equipment operated in excess of 600 volts and the means of disconnecting from electric energy is not visibly open or visibly locked out, the provisions of subdivisions (i) through (vii) of this subparagraph shall be complied with:

Notification and assurance from the designated employee [a qualified person delegated to perform specific duties under the conditions existing] shall be obtained that:

- All switches and disconnectors through which electric energy may be supplied to the particular section of line or equipment to be worked have been de-energized
- All switches and disconnectors are plainly tagged indicating that men are at work
- Where design of such switches and disconnectors permits, they have been rendered inoperable

When more than one independent crew requires the same line or equipment to be de-energized, a prominent tag for each such independent crew shall be placed on the line or equipment by the designated employee in charge.

Upon completion of work on de-energized lines or equipment, each designated employee in charge shall determine that all employees in his crew are clear, that protective grounds installed by his crew have been removed, and he shall report to the designated authority that all tags protecting his crew may be removed.

1926.950(d) (2) (ii)

When a crew working on a line or equipment can clearly see that the means of disconnecting from electric energy are visibly open or visibly locked-out, the provisions of subdivisions (i) and (ii) of this paragraph shall apply:

Upon completion of work on de-energized lines or equipment, each designated employee in charge shall determine that all employees in his crew are clear, that protective grounds installed by his crew have been removed, and he shall report to the designated authority that all tags protecting his crew may be removed.

1926.950(e) (1) (i) and (ii) and (2)

The employer shall provide training or require that his employees are knowledgeable and proficient in procedures involving emergency situations, and first-aid fundamentals including resuscitation.

Overhead Lines

1926.955(b) (3) (i)

A designated employee shall be used in directing mobile equipment adjacent to footing excavations.

1926.955(b) (8) and (d) (1)

A designated employee shall be utilized to determine that required clearance is maintained in moving equipment under or near energized lines.

Stringing adjacent to energized lines—Prior to stringing parallel to an existing energized transmission line, a competent determination shall be made to ascertain whether dangerous induced voltage buildups will occur, particularly during switching and ground fault conditions.

1926.955(e) (1) and (4)

Employees shall be instructed and trained in the live-line bare-hand technique and the safety requirements pertinent thereto before being permitted to use the technique on energized circuits.

All work shall be personally supervised by a person trained and qualified to perform live-line bare-hand work.

Underground Lines

1926.956(b) (1)

While work is being performed in manholes, an employee shall be available in the immediate vicinity to render emergency assistance as may be required. This shall not preclude the employee in the immediate vicinity from occasionally entering a manhole to provide assistance, other than emergency. This requirement does not preclude a qualified employee [a person who by reason of experience or training is familiar with the operation to be performed and the hazards involved] working alone from entering, for brief periods of time, a manhole where energized cables or equipment are in service, for the purpose of inspection, housekeeping, taking readings, or similar work if such work can be performed safely.

Construction in Energized Substations

1926.957(a) (1)

When construction work is performed in an energized substation, authorization shall be obtained from the designated, authorized person [a qualified person delegated to perform specific duties under the conditions existing] before work is started.

1926.957(d) (1)

Work on or adjacent to energized control panels shall be performed by designated employees.

1926.957(e) (1)

Use of vehicles, gin poles, cranes, and other equipment in restricted or hazardous areas shall at all times be controlled by designated employees.

Ladders

1926.1053(b) (15)

Ladders shall be inspected by a competent person for visible defects on a periodic basis and after any occurrence that could affect their safe use.

1926.1060(a) (1) (i) through (v) and (b)

The employer shall provide a training program for each employee using ladders and stairways, as necessary. The program shall enable each employee to recognize hazards related to ladders and stairways and shall train each employee in the procedures to be followed to minimize these hazards.

The employer shall ensure that each employee has been trained by a competent person in the following areas, as applicable:

> The nature of fall hazards in the work area

> The correct procedures for erecting, maintaining, and disassembling the fall protection systems to be used

> The proper construction, use, placement, and care in handling of all stairways and ladders

> The maximum intended load-carrying capacities of ladders used

> The standards contained in this subpart

Retraining shall be provided for each employee as necessary so that the employee maintains the understanding and knowledge acquired through compliance with this section.

Endnotes:

[1] The first section of this chapter (through Construction Training Requirements on page 158) is adapted from: OSHA, *Publication 2254 (Revised 1995),* http://www.osha.gov/Publications/osha2254.pdf

6

Noise Control

INTRODUCTION

Noise, or unwanted sound, is one of the most pervasive occupational health problems. Approximately 30 million people in the U.S. are occupationally exposed to hazardous noise. About ten million people have noise-induced hearing loss, nearly all of which were caused by occupational exposures. Fortunately, the incidence of noise-induced hearing loss can be reduced, or often eliminated, through the successful application of engineering controls and hearing conservation programs.[1]

OSHA is considering rulemaking to revise the construction noise standards to include a hearing conservation component for the construction industry that provides a similar level of protection to that afforded to workers in general industry 29 CFR 1910.95.

The federal government has recognized the hazardous conditions caused by noise on construction projects for many years. OSHA's current noise standard for construction stems from the Occupational Noise Standard originally published in 1969 by the Bureau of Labor Standards under the authority of the Construction Safety Act (40 U.S.C. 333).

OSHA adopted the construction noise standard in 1971 and later recodified it at 29 CFR 1926.52. Another section of the construction standard (29 CFR 1926.101) contains a provision requiring employers to provide hearing protection devices when needed. Both sections 1926.52 and 1926.101 apply to employers engaged in construction and renovation work when high noise levels are present.

NOISE EXPOSURE IN CONSTRUCTION

Many construction jobs, such as concrete work, site excavation, highway construction, and carpentry, involve high levels of noise. Major noise sources include heavy equipment, such as loaders, dozers, and cranes, as well as tools like jackhammers and chipping guns. Excessive noise at construction sites not only causes hearing loss, but also can create a safety hazard by masking the sounds of oncoming vehicles. Hearing loss and the use of hearing protectors by those with pre-existing hearing loss may further interfere with the workers' ability to hear and perceive the sounds of danger. Although these difficulties occur in many occupational settings, they are a particular problem in construction, where a variety of moving vehicles, back-up alarms, and other signals and activities may occur simultaneously.[2]

Construction workers are among the most affected by industrial deafness. The 1998 International Labour Office encyclopedia lists the construction industry as the fourth noisiest industry sector. The types of workers at risk include

- Users of impact equipment and tools (e.g., piling hammers, concrete breakers, manual hammers)
- Users of explosives (e.g., blasting, cartridge tools)
- Users of pneumatically powered equipment
- Operators of plant powered by internal combustion engines
- Bystanders in the vicinity of the plant
- Operators and bystanders in enclosed spaces where there are noisy activities or a concentration of plant
- Service and equipment maintenance personnel

Occupational Noise Exposure Standards in Construction

OSHA standard 1926.52 requires protection against the effects of noise exposure when eight-hour time-weighted average sound levels exceed a permissible exposure limit (PEL) of 90 decibels (dBA) measured on the A scale of a sound level meter set at slow response. The exposure level is raised 5 dB for every halving of exposure duration as shown in Table 6.1 (Table D-1 of the standard) below.

Table 6.1 Permissible Noise Exposures

Table D-1 -- Permissible Noise Exposures	
Duration per day, hours	Sound level DBA slow response
8	90
6	92
4	95
3	97
2	100
1.5	102
1	105
.5	110
.25 or less	115

OSHA Standard 29 CFR 1926.52 states that when employees are subjected to noise doses exceeding those shown in Table D-1, feasible administrative or engineering controls must be used to lower employee noise exposure. If such controls fail to reduce sound to the levels shown in the table, personal protective equipment must be provided and used to reduce noise exposure to within those levels.

OSHA defines continuous noise as noise levels that occur at intervals of one second or less and requires that a "continuing, effective hearing conservation program" be administered whenever levels exceed those shown in Table D-1. It gives instruction on how to calculate an employee's noise exposure when the employee is exposed to two or more periods of noise at different levels, and states that exposure to impulsive or impact noise should not exceed a peak sound pressure level of 140 dB.

The requirements of 29 CFR 1926.101 include

- Hearing protection devices shall be provided and used wherever it is not feasible to reduce the noise exposure (level times duration) to within the Permissible Exposure Limit (PEL) specified in Table D-2 (see above)
- Hearing protection devices inserted in the ear shall be fitted by competent persons
- Plain cotton is not an acceptable protective device

Construction Noise

By conducting sound surveys, you can identify the construction trades and activities most likely to expose workers to hazardous noise.

Most trades are typically exposed to noise levels greater than 85 dB (A) averaged over an eight-hour shift.

A SAMPLE HEARING CONSERVATION PROGRAM

The objective of the *(Company Name)* Hearing Conservation Program is to minimize occupational hearing loss by providing hearing protection, training, and annual hearing tests to all persons working in areas or with equipment that have noise levels equal to or exceeding an eight-hour time-weighted average (TWA) sound limit of 85 dBA (decibels measured on the A scale of a sound level meter). A copy of this program will be maintained by all affected departments. A copy of OSHA's Hearing Conservation Standard, 29 CFR 1926.52, can be obtained from ***Responsible Person***. A copy of the standard will also be posted in areas with affected employees.

(Enter Name Here) will be responsible for overall direction of the Safety Program. We have adopted this program in order to implement 29 CFR 1926.52, the OSHA standard regulating Occupational Noise exposure.

Employers Covered

This program covers employers those who have information about employees who are exposed to excess noise. Those who do not have such information do not have to measure the noise levels or do anything else.

In the case of employers who determine that they do not have any such information, it will be difficult for OSHA to second-guess them or to prove noncompliance because, without full-shift precise measurements, no one can establish what the eight-hour average noise levels were on any particular day.

The fact that an OSHA inspector measures excess noise on June 1, for example, does not mean that the noise was excessive on the preceding month, the previous July 1, or on any other day. Noise exposures vary from day to day because of production requirements, down time, varying employee work schedules and movement, and many other variables. Even if an OSHA inspection shows excess exposure levels, that is not necessarily determinative either because past experience has shown that OSHA's noise measurements are sometimes inaccurate.

If it is your determination that you do not have the requisite information, and at some time in the future an OSHA inspector asks you for noise measurement records or information, you can tell him that no measurements have been made because there is no employee exposure above the average limits at your plant (or in the particular areas thereof in question).

That could be a correct response at the time even if the OSHA inspector then takes measurements that show the contrary. If the inspector's measurements are correct, the moment when you receive them would become the first information you received of excess noise exposure (assuming that you never previously had any reason to believe otherwise).

Monitoring

Monitoring for noise exposure levels will be conducted by ***Responsible Person***. It is the responsibility of the individual departments to notify ***Responsible Person*** when there is a possible need for monitoring. Monitoring will be performed with the use of sound level meters and personal dosimeters at the discretion of ***Responsible Person***.

Monitoring will also be conducted whenever there is a change in equipment, process or controls that affect the noise levels. This includes the addition or removal of machinery, alteration in building structure, or substitution of new equipment in place of that previously used. The responsible supervisor must inform ***Responsible Person*** when these types of changes are instituted.

If an employer does have information indicating any exposure, the first obligation is to measure the noise to determine whether it equals or exceeds the 85-decibel (dB) level. If you have already done this once, it need not ever be repeated. This noise level trigger is an average for a full eight-hour day. Because the outcome of this measurement determines whether the other requirements of the standard apply, it is important that this measurement be done properly. It should not be left to chance. When, where, and how it is done should be carefully determined.

When Should Monitoring Be Done?

Choose a typical day to take your measurements. If, for example, the principal noise sources have only been operated 50 percent of the time recently, make your measurements on a day when they will be in operation for 50 percent of an eight-hour work shift. If the employees are out of the vicinity of the noise for some part of their workday, noise during that period of time should be excluded in figuring the eight-hour exposure. That time should be counted at zero noise exposure in figuring the average for that eight-hour shift. Be sure to conduct measurements for the full eight-hour day, even though the principal noise sources are not in operation during this full period.

Where Should Monitoring Be Done?

Noise measurements should be made for those areas for which you have the information described above. If that information relates to only one or two places or to only particular jobs within your facility, those areas are all you need to measure

How Should Monitoring Be Done?

You can use a sound level meter, a dosimeter, or a tape recorder (the latter two require an additional instrument to obtain a proper read out). You can place it anywhere in the area to be measured that you deem to be appropriate, or you can attach a dosimeter to an employee who works in such an area. Each employer is authorized to determine what is appropriate. The results obtained by the measuring device will, in many cases, differ depending upon where it is placed on the employee or in the area to be measured. If you decide to place it in a particular spot and leave it there for eight hours, you may want to try several different places before you decide which one you will choose for the full eight-hour measurement. Bear in mind, when making this choice, that your purpose is not to measure the noise made by machines but to determine employee noise exposure.

Who Should Conduct the Monitoring?

You can do the measurements yourself or you can have it done by an outside consultant from either the state government or a private firm. You will need to obtain the proper measuring instrument, either dosimeter, sound level meter, or tape recorder. It will have to be calibrated prior to and subsequent to the measurement.

The integrity of a noise measurement can easily be distorted by horseplay or deliberate employee action; you should be careful to guard against this.

Once a noise measurement is made, make a record of the result and keep it for two years. It does not have to be in any particular form or detail. You can simply write out on a sheet of paper the date on which the noise measurements were made and the result and file it away.

Employee Involvement in Noise Measurements

Those employees who are affected by excess noise are entitled to observe the noise measurements being made and those who are overexposed must be told of the results. It is up to the employer to determine how this is to be done. Some employers, concerned about potential distortion of the results of the measurements caused by employee conduct, handle this notification requirement on a low-key basis. If they use a dosimeter attached to an employee, they will tell him (but no one else) at the moment it is attached that it is for the purpose of obtaining the OSHA-required noise measurement.

If they use an instrument placed in a single location, they will post a notice with it stating that this instrument is being used to measure the noise in this area. No verbal statement or advance notification need to be made. No matter what method is used, however, steps should be taken to ensure that there is no tampering with the instrument and that no abnormal noise is picked up during its eight-hour measuring period. When the results are obtained, there is no particular time frame for notifying the employees who are exposed to excess noise levels and no specified method for so doing. You can tell those employees verbally, post a notice in the area, or put a note in their pay envelope. It is your choice.

Noise Measuring

Once noise levels have been measured as described above, your obligation to measure has been fulfilled for all time unless you later have information that the measured employee exposures may have increased or that there are additional work areas where excess noise exists.

If your results show that no employee in the work area where measurements were taken is exposed to noise at or above an 85-decibel eight-hour time-weighted average, you have no further obligations. If your results do not show employee exposure at or above those levels, you must observe the requirements for those employees and no others.

Hearing Protectors

The standard does not alter the long-existing requirements that hearing protectors shall be provided and used to reduce sound levels to within those specified in Table G-16 (a 90 dBA time-weighted average for an eight-hour day), but it does add a number of additional conditions that trigger a hearing protector requirement.

1. The employer shall make hearing protectors available to all employees exposed to an eight-hour time-weighted average of 85 decibels.
2. The employer shall ensure that hearing protectors are worn by employees so exposed if they have not yet had the baseline audiogram.
3. The employer shall ensure that hearing protectors are worn by employees if any annual audiogram shows a potential hearing loss.
4. The employees are obligated to wear hearing protectors because of the long-existing 90 dBA rule.

Provide Options

Give employees the choice of selecting their hearing protectors from a variety of suitable ones. This doesn't mean that you must let them choose from a catalog listing all such devices on the market. It is sufficient if you choose two or three types and let them pick one of those.

Training

Affected employees will be required to attend training concerning the proper usage and wearing of hearing protection. The training will be conducted by ***Responsible Person***, or a designated representative, within a month of hire and annually thereafter.

Training shall consist of the following components:

- How noise affects hearing and hearing loss
- Review of the OSHA Hearing Protection Standard
- Explanation of audiometric testing
- Rules and procedures
- Locations within company property where hearing protection is required
- How to use and care for hearing protectors

Training records will be maintained by ***Responsible Person***.You should provide training in the hearing protector's use and care. This does not mean a formal classroom session. The type of pamphlet supplied by many distributors of hearing protectors will suffice. Most hearing protectors come in individual packages with use and care instructions printed on them, employers may avail themselves of prepared materials such as pamphlets. This may be more effective and less expensive than individually prepared programs.

Fitting

You should ensure initial fitting and correct use of the hearing protectors. For most employees, this can be accomplished by supplying ear muffs or ear plugs of the self-fitting or malleable type and, if such an employee is later observed to be using them incorrectly, instructing him in proper use at that time. If you should furnish a different kind of hearing protectors, the correct fitting of which is not obvious, then you must show the employee how to wear them.

Individual attention should be provided for those employees who have ear allergies or infections. Perhaps the simplest way to handle this is to provide each employee with an explanatory pamphlet or to include in all distributions of hearing protectors a directive to each employee who receives them that he must notify his supervisor if he does not know how to fit or use the protectors or if they cause him any discomfort. Some hearing protectors come in packages with the advice printed on the package.

Attenuation

You should make sure that the hearing protectors reduce the noise being heard by the employee to permissible levels. For most employees, the eight-hour, 90 dBA level is the permissible level. Reduction to the 85 dBA level is only for those employees who fall within the first two categories.

Meeting this requirement is very simple because all hearing protectors are accompanied by their attenuation rating (known as the NRR—Noise Reduction Rating). If it is not printed right on the package, contact your supplier for the appropriate NRR number. Simply deduct this number from the noise decibel level you measured for the area where the employee works. If this calculation shows a result less than 90 (or 85, where appropriate), this requirement is satisfied.

In the event that this calculation results in a number higher than permitted, get hearing protectors with an NRR high enough to produce the required attenuation.

Training

In addition to the hearing protector use and care instruction discussed above, you should provide training to those employees who are exposed to the 85 dB (or higher) average noise levels. It must cover the effects of noise; the purpose, advantages, disadvantages and attenuation of various types of hearing protectors; and the purpose and procedures of audiometric testing. You can decide how you want to do this. Perhaps the simplest way is to obtain or make a printed pamphlet containing this information and distribute it to each such employee.

Potential Hearing Loss Cases

When any annual audiometric test shows that the STS criterion has been met or exceeded (and that is confirmed by the re-test, if you decide to exercise this option), you must provide that employee with written notification within 21 days of a determination. If a physician determines that the STS are neither work-related nor aggravated by occupational noise exposure, then the further obligations to this employee are at an end when you supply this notification.

In the absence of a medical opinion you should do the following:

1. If the employee is not already wearing hearing protectors, make sure that matters discussed are observed even if that employee does not work in an excess noise level area.
2. If the employee is already wearing hearing protectors, he or she must be refitted and retrained in their use. If necessary, he or she must be provided with hearing protectors that provide greater attenuation.
3. In certain cases, the employee must be referred to a clinical audiological evaluation or an otological examination, whichever is appropriate. This simply means that you should advise the employee that the hearing tests indicate that he or she should have the test in order to preserve his or her hearing.

You are not required to pay for the test

- Where additional testing is necessary
- Where the employer suspects that a medical pathology of the ear is caused or aggravated by the wearing of hearing protectors

The employee must be informed of the need for an ontological examination if a medical pathology of the ear that is unrelated to the use of hearing protectors is suspected. This simply means that the employee should be told this (either orally or in writing). There is no obligation to determine what the employee does with this information.

The standard does not require that cases of hearing loss be recorded on the OSHA 300 Form (record of job injuries and illnesses). OSHA's explanation of the standard states that such an entry may be imposed by the existing general regulations governing records of work injuries and illnesses.

OSHA has no jurisdiction to impose any requirement for non-work-related matters; hearing loss cases should not be entered on the OSHA 300 unless it is clear that the loss was the result of employment. Be very careful in this regard because recording a hearing loss entry on the OSHA 300 is an admission that the loss was caused by the employee's job. That could very well be used by the employee to support future compensation claims.

There will be cases where one annual audiogram shows an STS that does not show upon a later one. That is not an unusual situation because industrial audiometric testing is far less exact than clinical testing and because different technicians administer the tests in different years, often with different equipment.

The standard recognizes these variables by making it clear that you may discontinue the follow-up obligations discussed above for an employee (previously classified as having a potential hearing loss) when a subsequent audiogram fails to show any STS. When that happens, the standard requires that the employee be so advised and permits you to discontinue the required use of hearing protectors so long as the emplyee works in an area where noise levels are within the permissible 90 dBA limits.

Employee Information

You must make available to employees and unions who request them a copy of the Standard and any explanatory pamphlets given to you by OSHA; post a copy of the Standard in the workplace, and provide the OSHA pamphlets upon request along with all materials relating to your hearing conservation training and education programs. If you post a copy of the Standard in the workplace, that should satisfy the make-available provision.

All requirements can be satisfied by keeping a copy of all OSHA standards at some place on the premises and attaching a statement like this to the OSHA poster that is required to be displayed:

> Copies of all OSHA standards that are required to be posted are available on the premises for you to examine. Contact your immediate supervisor and tell him or her which standard you want to see.

That is not an accurate statement of the legal requirement. Nevertheless, hundreds, if not thousands, of employers have been cited for failure to post copies of OSHA standards. You can avoid citations of that kind as well as unnecessary bulletin board clutter by following this suggestion.

Records

There are only two hearing-conservation record-keeping requirements:

1. Keep a record of your noise measurements (in any form you wish) for two years
2. Keep the employee audiometric test results for as long as the person tested remains your employee

The Audiometric Test record must contain the following:

The employee's name and job classification; the date the test was taken; the name of the technician who administered the test; and calibration data on the test equipment, and the noise level in the testing booth. This is a rather easy requirement to follow but it is often overlooked.

The same provision also states that your most recent noise measurement (monitoring) of the area where the tested employee works must be kept. It can be listed right on the employee's audiometric test result or it can be kept separately.

It goes without saying that if you are not required to make noise measurements or perform audiometric testing, the Standard imposes no record-keeping requirement on you. There is no requirement in the Standard to record hearing-loss cases on the OSHA 300 Form.

If you conduct audiometric testing, you should obtain and keep a record of each tested employee's non-work activities that can contribute to his hearing acuity, such as recreational hunting, firing weapons, listening to rock music, or similar activities. When each test is given, the employee should be given a form to fill out.

Endnotes:

[1] OSHA, http://www.osha.gov/SLTC/constructionnoise/

[2] *Federal Register*, Vol.67, No.150, 5 August 2002.

Any employee found to have a noise exposure on the 80 to 130dB scale of greater than or equal to (50 percent with 2dBA error factor) will be included in a Hearing Conservation Program that meets the requirements established by our company.

Employee's Name: _____

Date Sampled: _____

Hearing Conservation Program Review

After completion of this review, or if you have any questions about compliance with any part of the "Hearing Conservation Program" requirements, consult with the Industrial Hygienist or Health Specialist.

Noise Exposure Assessment

YES / NO Has the employee's noise exposure been evaluated? If the operator has included the miner in a hearing conservation program without assessing exposure to the "Action Level," a citation may not be warranted.

YES / NO Has the mine operator informed the miner in writing within the last twelve months of exposure determination?

Hearing Protectors

YES / NO Has the mine operator provided the miner with a selection of hearing protection at no cost?

Audiometric Testing

YES / NO Has the employer offered to the employee an audiometric test at no cost?

YES / NO Has audiometric testing been conducted every 12 months?

YES / NO Is an audiometric test record maintained for the employee that documents l) name and job classification, 2) copy of all audiograms, 3) evidence that the audiogram is scientifically valid, 4) any exposure determination, and 5) results of follow-up exams?

YES / NO Has the employee been provided with a written record of the results of the audiogram?

YES / NO Has a reportable hearing loss been incurred by the employee (25dBA reduction) and been filed?

Date of Baseline Audiometric Test: _____

Date of Last Audiometric Test: _____

Training

YES / NO Has the employee received training for 1) effects of noise on hearing; 2) use, care, fitting of, advantages, disadvantages, and types of hearing protection devices; 3) requirements of noise controls; and 4) purpose and value of audiometric testing?

YES / NO Has training been provided within 12 months of last HCP training?

YES / NO Has the employee been certified of the date and type of training?

Date of Training: _____

Figure 6.1 Hearing Conservation Program CheckList for a Mine Operation

HEARING CONSERVATION TRAINING LOG

Training Date:_____

Topic:_____

Training Conducted by:_____

Employee Name (printed)	Employee Signature	Job Title

Figure 6.2 Hearing Conservation Training Log

Record of Hearing Protection Needs

Company Name **Personnel in Hearing Conservation Program** *Date*				
Hearing protection is required for and has been issued to the following personnel:				
Employee Name	**Department**	**Job Description/ Equipment Being Used**	**Type of Hearing Protection Issued**	**Date Issued**

Figure 6.3 Record of Hearing Protection Needs

7

Signs, Signals, and Barricades

Safety Standards for Signs, Signals, and Barricades[1]

INTRODUCTION

The Occupational Safety and Health Administration (OSHA) has revised the construction industry safety standards to require that traffic control signs, signals, barricades or devices protecting workers conform to either the 1988 Edition of the Federal Highway Administration (FHWA) Manual on Uniform Traffic Control Devices (MUTCD) or the Millennium Edition of the FHWA MUTCD (Millennium Edition). This final rule became effective December 11, 2002.

This final rule addresses the types of signs, signals, and barricades that must be used to protect construction employees from traffic hazards. The vast majority of road construction in the United States is funded through federal transportation grants. As a condition to receiving federal funding, the U.S. Department of Transportation's (DOT's) Federal Highway Administration requires compliance with MUTCD.

OSHA's statutory mandate is to protect the health and safety of employees. OSHA also requires employers that are within the scope of its authority to comply with the MUTCD. However, OSHA's current standard incorporates the 1971 version of the MUTCD, which FHWA has since updated. The purpose of this final rule is to update OSHA's standard.

BACKGROUND

Currently, under 29 CFR part 1926 subpart G—Signs, Signals, and Barricades, OSHA requires that employers comply with the 1971 MUTCD. Specifically, employers must ensure that the following conform to the 1971 MUTCD: traffic control signs or devices used to protect construction workers (29 CFR 1926.200(g)(2)); signal directions by flagmen (29 CFR 1926.201); and barricades for the protection of workers (29 CFR 1926.202).

In contrast, a DOT rule, 23 CFR 655.601 through 655.603, requires that such traffic control signs or devices conform to a more recent version of the MUTCD. DOT regulations provide that the MUTCD is the national standard for all traffic control devices on streets, highways, and bicycle trails. DOT's rule requires that traffic control devices on roads for which federal funds were involved be in substantial conformance with the MUTCD.

OSHA was asked to update subpart G because of the outdated OSHA requirements in addition to the DOT rule. OSHA pointed out that the Millennium Edition reflected updated standards and technical advances based on 22 years of experience in work zone traffic control design and implementation, as well as human behavior research and experience.

The National Committee on Uniform Traffic Control Devices (NCUTCD)—which consists of various national associations and organizations interested in highway construction or highway safety including the American Road and Transportation Builders Association, the Association of American Railroads, the American Automobile Association, the National Association of Governor's Highway Safety Representatives, and the National Safety Council—unanimously resolved in January 1999 to request that OSHA adopt Revision 3 in place of the 1971 MUTCD. In May 2000, OSHA's Advisory Committee on Construction Occupational Safety and Health (ACCSH) also expressed support for adopting a more recent edition of the MUTCD as the OSHA standard for the construction industry.

OSHA reviewed the differences between the 1971 version, Revision 3, and the Millennium Edition and concluded that compliance with the more recently published manuals would provide all the safety benefits (and more) of the 1971 version. The differences between OSHA's regulations that reference the 1971 MUTCD and DOT's modern regulations create potential industry confusion and inefficiency, without advancing worker safety. Accordingly, in an interpretation letter dated 16 June 1999 to Cummins Construction Company, Inc., OSHA stated that it would accept compliance with Revision 3 in lieu of compliance with the 1971 MUTCD referenced in section 1926.200(g) through its *de minimis* policy.

The numerous and various changes to the 1971 MUTCD reflected in Revision 3 and the Millennium Edition stem from over 20 additional years of experience in temporary traffic control zone design, technological changes, and contemporary human behavior research and experience. Revision 3 and the Millennium Edition provide highway work zone planners more comprehensive guidance and greater flexibility in establishing effective temporary traffic control plans based on type of highway, traffic conditions, duration of project, physical constraints, and the nature of the construction activity. Revision 3 and the Millennium Edition, accordingly, better reflect current practices and techniques to best ensure highway construction worker safety and health.

OSHA is amending the safety and health regulations for construction to adopt and incorporate Revision 3 (and the option to comply with the Millennium Edition), instead of the 1971 MUTCD, and to make certain editorial changes. The amendment deletes the references in 29 CFR 1926.200(g)(2) and 1926.202 to the 1971 MUTCD and inserts references to Revision 3 (and the option to comply with the Millennium Edition). The amendment clarifies and abbreviates 29 CFR 1926.201(a) by simply adopting the requirements of Revision 3 (and the option to comply with the Millennium Edition) with regard to the use of flaggers. The amendment also makes certain editorial corrections, replacing the term workers for the term workmen and the term flaggers for the term flagmen in 29 CFR 1926.200(g)(2) and 1926.201(a).

Updating OSHA's rule eliminates the technical anomaly of having to meet both OSHA's outdated requirement to comply with the 1971 version and DOT's more modern requirements.

Instead, OSHA's final rule requires compliance with Revision 3 (or, at the option of the employer, the Millennium edition). In addition to harmonizing OSHA's requirements with those of DOT, the final rule's additional safety measures (described below) will be enforceable as OSHA requirements. With the current emphasis on rebuilding the nation's highways and improving safety in work zone areas, OSHA's update is particularly appropriate.

Temporary Traffic Control Elements

The 1971 MUTCD does not discuss traffic control plans (TCPs), which are used by industry to describe traffic controls that are to be implemented in moving vehicle and pedestrian traffic through a temporary traffic control zone. Revision 3 emphasizes the importance of TCPs in facilitating safe and efficient traffic flow. Revision 3 recognizes that different TCPs are suitable for different projects and does not detail specific requirements. The Millennium Edition offers expanded guidance and options for TCPs, but it adds no requirements. In both Revision 3 and the Millennium Edition, a TCP is recommended but not required. Revision 3 and the Millennium Edition also discuss the "temporary traffic control zone," comprised of several areas known as the "advance warning area," "transition area," "activity area," and "termination area." In addition, Revision 3 and the Millennium Edition explain the need for differing traffic control measures in each control zone area.

The 1971 MUTCD only briefly describes "tapers" and provides a formula for calculating the appropriate taper length. However, Revision 3 defines and discusses five specific types of tapers used to move traffic in or out of the normal path of travel. It illustrates each of them, and sets out specific formulae for calculating their appropriate length. In all three editions, information relating to tapers is limited to guidance and contains no mandatory requirements.

All versions of the MUTCD require the coordination of traffic movement, when traffic from both directions must share a single lane. Revision 3 and the Millennium Edition describe five means of "alternate one-way traffic control," adding the "Stop or Yield Control Method" to the methods described in the 1971 MUTCD. The "Stop or Yield Control Method" is appropriate for a low-volume two-lane road where one side is closed and the other side must serve both directions. It calls for a stop or yield sign to be installed on the side that is closed. The approach to the side that is not closed must be visible to the driver who must yield or stop.

Pedestrian and Worker Safety

The key elements of traffic control management that should be considered in any procedure for ensuring worker safety are training, worker clothing, barriers, speed reduction, use of police, lighting, special devices, public information, and road closure. Revision 3 recommends that these traffic control techniques be applied by qualified persons exercising good engineering judgment. The Millennium Edition makes this recommendation a requirement. The Millennium Edition also requires advance notification of sidewalk closures.

Hand Signaling or Flagger Control

A flagger should wear an orange, yellow, or "strong yellow green" (called "yellow-green" in Millennium Edition) vest, shirt, or jacket, instead of an "orange vest and/or an orange cap," as directed in the 1971 ANSI standard. For nighttime work, Revision 3 requires that the outer garment be retro-reflective orange, yellow, white, silver, strong yellow-green, or a fluorescent version of one of these colors. This clothing must be designed to clearly identify the wearer as a person, and the clothing must be visible through the full range of body motions. For nighttime work, the Millennium Edition requires that the colors noted above be retro-reflective but does not mandate that the clothing be visible through the full range of body motions. Both Revision 3 and the Millennium Edition allow the employer more flexibility in selecting colors.

Under the 1971 ANSI standard, the flagger was required to be visible to approaching traffic at a distance that would allow a motorist to respond appropriately. Revision 3 and the Millennium Edition contain more specific requirements. Under both versions, flaggers must be visible at a minimum distance of 1,000 feet. In addition, Revision 3 and the Millennium Edition list training in "safe traffic control practices" as a minimum flagger qualification.

Revision 3 and the Millennium Edition depart significantly from the 1971 ANSI standard by requiring that "Stop/Slow" paddles, not flags, be the primary hand-signaling device. The paddles must have an octagonal shape on a rigid handle, and be at least 18 inches wide with letters at least six inches high. The 1971 ANSI standard recommended a 24-inch width. Revision 3 and the Millennium Edition require that paddles be retro-reflective when used at night.

Flags would still be allowed in emergency situations or in low-speed and/or low-volume locations. Revision 3 and the Millennium Edition differ in that Revision 3's recommendations for flag and paddle signaling practice are requirements in the Millennium Edition. In addition, the Millennium Edition applies several new requirements when flagging is used. The flagger's free arm must be held with the palm of the hand above shoulder level toward approaching traffic and the flagger must motion with the free hand for road users to proceed. These requirements were guidance in Revision 3, and options in the 1971 ANSI standard.

Devices

There are numerous differences in the design and use of various traffic control devices such as signs, signals, cones, barricades, and markings used in temporary traffic control zones. Several signs or devices are described in Revision 3 and the Millennium Edition that are not mentioned in Part VI of the 1971 ANSI standard.

Also, Revision 3 and the Millennium Edition offer expanded options for the color of temporary traffic control signs. Signs that, under the 1971 ANSI standard, were required to have orange backgrounds may now have fluorescent red-orange or fluorescent yellow-orange backgrounds.

The 1971 ANSI standard required that signs in rural areas be posted at least five feet above the pavement; signs in urban areas were required to be at least seven feet above the pavement. Revision 3 eliminated the distinction between urban and rural areas, and downgraded the requirement to a recommendation. It recommended that signs in all areas have a minimum height of seven feet. In the Millennium Edition, the FHWA returned to the 1971 ANSI requirements.

The Millennium Edition also introduced the requirement that signs and sign supports be crash-worthy.

The Millennium Edition introduced and clarified mandatory requirements for the design of the following signs: Weight Limit, Detour, Road (Street) Closed, One Lane Road, Lane(s) Closed, Shoulder Work, Utility Work, Blasting Area, Shoulder Drop-Off, Road Work Next [XX] km (Miles), and Portable Changeable Message.

The dimensions, color, or use of certain channelizing devices have also changed. "Channelizing devices" include cones, tubular markers, vertical panels, drums, barricades, temporary raised islands, and barriers. The 1971 ANSI standard required that traffic cones and tubular markers be at least 18 inches in height and that the cones be predominantly orange. Revision 3 raised the minimum height for traffic cones and tubular markers to 28 inches "when they are used on freeways and other high-speed highways, on all highways during nighttime, or whenever more conspicuous guidance is needed" (6F-5b(1), 5c(1)). Revision 3 also expanded the color options for cones to include fluorescent red-orange and fluorescent yellow-orange. The Millennium Edition maintained these requirements.

Revision 3 and the Millennium Edition require that vertical panels be eight to 12 inches wide, rather than the six to eight inches required by the 1971 ANSI standard. Under Revision 3 and the Millennium Edition, drums must be made of lightweight, flexible, and deformable materials; at least 36 inches in height; and at least 18 inches in width. Steel drums may not be used. The Millennium Edition adds the requirement that each drum have a minimum of two orange and two white stripes with the top stripe being orange. Revision 3 and the Millennium Edition require that delineators only be used in combination with other devices; be white or yellow, depending on which side of the road they are on; and be mounted approximately four feet above the near roadway edge.

The 1971 ANSI standard required warning lights to be mounted at least 36 inches high. Revision 3 and the Millennium Edition reduced the minimum height to 30 inches and introduced new requirements for warning lights. Type A low-intensity flashing warning lights and Type C steady-burn warning lights must be maintained to allow a nighttime visibility of 3,000 feet. Type B high intensity flashing warning lights must be visible on a sunny day from a distance of 1,000 feet.

Revision 3 and the Millennium Edition contain an additional requirement not found in the 1971 ANSI standard that requires employers to remove channelizing devices that are damaged and have lost a significant amount of their retro-reflectivity and effectiveness. Revision 3 and the Millennium Edition also specifically prohibit placing ballast on the tops of drums or using heavy objects such as rocks or chunks of concrete as barricade ballast.

Revision 3 and the Millennium Edition address in greater detail the appearance and use of pavement markings and devices used to delineate vehicle and pedestrian paths. They require that after completion of the project, pavement markings be properly obliterated to ensure complete removal and a minimum of pavement scars. Whereas Revision 3 requires that all temporary broken-line pavement markings be at least four feet long, the Millennium Edition sets the minimum at two feet.

Temporary Traffic Control Zone Activities

This section, not found in the 1971 ANSI standard, provides information on selecting the appropriate applications and modifications for a temporary traffic control zone. The selection depends on three primary factors: work duration, work location, and highway type. Section 6G in both Revision 3 and the Millennium Edition emphasizes that the specific typical applications described do not include a layout for every conceivable work situation and that typical applications should, when necessary, be tailored to the conditions of a particular temporary traffic control zone.

Among the specific new requirements in Revision 3 and the Millennium Edition are the following: retro-reflective and/or illuminated devices in long-term (more than three days) stationary temporary traffic control zones; warning devices on (or accompanying) mobile operations that move at speeds greater than 20 mph; warning signs in advance of certain closed paved shoulders; a transition area containing a merging taper in advance of a lane closure on a multi-lane road; temporary traffic control devices accompanying traffic barriers that are placed immediately adjacent to the traveled roadway; and temporary traffic barriers or channelizing devices separating opposing traffic on a two-way roadway that is normally divided.

The Millennium Edition includes several additional requirements in Section 6G. It requires the use of retro-reflective and/or illuminated devices in intermediate-term stationary temporary traffic control zones. A zone is considered intermediate-term if it is occupying a location more than one daylight period up to three days or if there is nighttime work in the zone lasting more than one hour. The Millennium Edition also requires a transition area containing a merging taper when one lane is closed on a multi-lane road. When only the left lane on undivided roads is closed, the merging taper must use channelizing devices and the temporary traffic barrier must be placed beyond the transition area channelizing devices along the centerline and the adjacent lane. In addition, when a directional roadway is closed, inapplicable "WRONG WAY" signs and markings and other existing traffic control devices at intersections within the temporary two-lane two-way operations section must be covered, removed, or obliterated.

RELATIONSHIP TO EXISTING DOT REGULATIONS

Through this rule, OSHA requires that traffic control signs, signals, barricades, or devices conform to Revision 3 or Part VI of the Millennium Edition, instead of the ANSI MUTCD. The ANSI MUTCD was issued in 1971. In 1988 the FHWA substantially revised and reissued the MUTCD. Since that time, FHWA has published several updates, including a 1993 revision to Part VI—Revision 3. In December 2000, FHWA published a Millennium Edition of the MUTCD that changed the format and revised several requirements. Employers that receive federal highway funds are currently required to comply with Revision 3 and have up until January 2003 to bring their programs into compliance with the Millennium Edition.

This is a significant regulatory action and has been reviewed by the Office of Management and Budget under Executive Order 12866. OSHA has determined that this action is not an economically significant regulatory action within the meaning of Executive Order 12866. Revision 3 of the MUTCD adds to the ANSI requirements some new alternative traffic control devices and expanded provisions and guidance materials, including new typical application diagrams that incorporate technology advances in traffic control device application. Part VI of the Millennium Edition includes some alternative traffic control devices and only a very limited number of new or changed requirements. However, the activities required by compliance with either Revision 3 or the Millennium Edition would not be new or a departure from current practices for the vast majority of work sites. All of these requirements are now or have been part of DOT regulations that cover work-related activities on many public roadways.

According to DOT regulations, the MUTCD is the national standard for streets, highways, and bicycle trails. While OSHA's *de minimus* policy is applied to situations in which there is failure to comply with the 1971 ANSI MUTCD, when there is compliance with Revision 3, this action will reduce any confusion created by the current requirement for employers to comply both with the 1971 ANSI MUTCD and DOT's MUTCD.

PERCENTAGE OF ROADS COVERED UNDER OSHA'S STANDARD VERSUS THE DOT STANDARD

The majority of U.S. roads are currently covered by DOT regulations and their related state MUTCDs. DOT regulations cover all federal-aid highways, which carry the majority of traffic. Moreover, many states extend MUTCD coverage to non-federal-aid and private roads. Thus, the requirements imposed by this OSHA final rule will be new only for the small percentage of the work that is not directly regulated by DOT or state transportation agencies.

FEDERAL-AID HIGHWAYS

Employers must comply with Revision 3 for all construction work respecting federal-aid highways. Although federal-aid highways constitute a minority of all public highways as measured by length, these highways carry the great majority of traffic. According to OSHA's analysis, 84 percent of vehicle-miles are driven on federal-aid highways. Though not a perfect measure, vehicular use corresponds more directly than length of road to the need for construction, repair, and other work activities addressed by the MUTCD. This suggests that most of these activities occur with respect to federal-aid highways. Conforming to the standards of the MUTCD during these work activities is a clear requirement of receiving federal highway funds and is therefore regulated by DOT.

State, Local, County, and Municipal Roads Not Receiving Federal Aid

The available data suggest that work respecting most non-federal-aid roads are required to comply with the MUTCD. Many states choose to regulate public roadways that are not federal-aid highways and thereby extend the coverage of the MUTCD. For example, OSHA reviewed the practices of nine states (Alabama, Arkansas, Colorado, Connecticut, Delaware, Kentucky, Michigan, North Carolina, and Texas), which include 23 percent of all U.S. public roads. In conducting this review, OSHA found that eight of the states require MUTCD standards on all state roads, while the ninth state requires MUTCD standards on state roads if the state contracts the work to be done. Five of these states also require that MUTCD standards be met on all county and municipal roads. For the sample of nine states, individual state coverage of public roads by state MUTCDs ranges from 12 percent to 100 percent. OSHA found that, on average, MUTCD coverage of all public roads in these nine states is 84 percent. OSHA computed the average across the nine states by weighting by total highway miles.

Private Roads

OSHA also examined MUTCD coverage of private roads. Although data on the extent of private roads is very limited, the best available information indicates that about 20 percent of the total mileage is accounted for by private roads. Some of these private roads are covered by state MUTCD standards. Of the nine states examined by OSHA, one state included private roads under the MUTCD standards if the state enforced traffic laws on these roads (e.g., roads in gated communities). Another state extended MUTCD standards to private roads if the state was involved in road design or approval. A third state deferred coverage to municipal ordinances, which may require meeting MUTCD standards on private roads. Thus, although it is clear that some local governments extend coverage to private roads, no data are available to specify with precision the extent to which this is the case.

INCENTIVES TO COMPLY WITH THE MUTCD

The estimates of the percentage of roads and highways covered by the MUTCD presented above are conservative. States, localities, and their contractors have additional incentives to comply with the MUTCD when it is not required. OSHA policy reinforces these incentives because OSHA does not enforce compliance with the ANSI MUTCD when there is compliance with Revision 3.

States must have highway safety programs that are approved by the Secretary of Transportation. The Secretary is directed to promulgate guidelines for establishing these programs. Those guidelines state that programs "should" conform with the MUTCD. DOT does not have the authority to require compliance with the MUTCD on roads that do not receive federal aid but recommends it. In light of this, and the statement that the MUTCD is "the national standard for all traffic control devices," the MUTCD has become the standard of care for litigation purposes. Thus, when a state or local government engages in a road construction project, it will likely seek to meet a reasonable standard of care (i.e., compliance with a recent edition of the MUTCD). If it does not, it could face substantial liability if the construction on its roads is a contributing factor in an accident. While compliance with the MUTCD does not insulate a state or locality from liability, it significantly reduces its exposure.

Moreover, many of the contractors who conduct work on covered roads are likely to conduct work on non-covered roads as well. In the interest of efficiency, these contractors are likely to consistently apply the current version of the MUTCD to all work, rather than switch back to the ANSI version for a small percentage of their overall business.

Finally, as is discussed below, signs and devices meeting 1993 specifications are often less expensive than signs meeting 1971 ANSI specifications. This has provided contractors involved in road construction and repair operations with a natural incentive to replace old and worn signs with signs meeting the more up-to-date standard.

Section 18 of the Occupational Safety and Health (OSH) Act (29 U.S.C. 651 et seq.) expresses Congress's intent to preempt state laws where OSHA has promulgated occupational safety and health standards. Under the OSH Act, a state can avoid preemption on issues covered by federal standards only if it submits and obtains federal approval of a plan for the development of such standards and their enforcement (29 U.S.C. 667). Occupational safety and health standards developed by such plan states must, among other things, be at least as effective in providing safe and healthful employment and places of employment as the federal standards. Subject to these requirements, State Plan states are free to develop and enforce their own requirements for road construction safety.

Although Congress has expressed a clear intent for OSHA standards to preempt state job safety and health rules in areas involving the safety and health of road construction workers, this final rule has only a minimum impact on the states. DOT requires compliance with the MUTCD for "application on any highway project in which federal highway funds participate and on projects in federally administered areas where a federal department or agency controls the highway or supervises the traffic operations" (23 CFR 655.603(a)). For this work, which represents the majority of road construction work in every state, all states must require compliance with the current edition of the MUTCD or another manual that substantially conforms to the current edition. States have been required to enforce Revision 3 or their own substantially conforming manual since 1994. DOT regulations allow states until January 2003 to adopt the Millennium Edition or another manual that substantially conforms to the Millennium Edition. (See 23 CFR 655.603(b).)

States must also have highway safety programs that are approved by the Secretary of Transportation, even for roads that do not receive federal aid. The Secretary is directed to promulgate guidelines for establishing these programs. Those guidelines and programs should conform with the current edition of the MUTCD. Most states require compliance with the latest edition of the MUTCD even on roads that receive no federal funding.

The requirements described in this document are new requirements only for the very small percentage of employers that are not already covered by the DOT regulations or corresponding state requirements. Therefore, the required state plan adoption of the provisions of Revision 3 or the Millennium Edition or an equivalent standard will also effectively impose a new regulation only on that small percentage of employers. OSHA concludes that this action does not have a significant impact on the states.

STATE PLAN STANDARDS

The 26 states or territories with OSHA-approved occupational safety and health plans must adopt an equivalent amendment or one that is at least as protective for employees within six months of the publication date of this final standard. These states are Alaska, Arizona, California, Connecticut (for state and local government employees only), Hawaii, Indiana, Iowa, Kentucky, Maryland, Michigan, Minnesota, Nevada, New Mexico, New Jersey (for state and local government employees only), New York (for state and local government employees only), North Carolina, Oregon, Puerto Rico, South Carolina, Tennessee, Utah, Vermont, Virginia, Virgin Islands, Washington, and Wyoming.

The rule, which applies to employers involved in road construction and repair operations, requires compliance with either the Millennium Edition or Revision 3 of the Federal Highway Administration's Manual on Uniform Traffic Control Devices (MUTCD).

Among the specific changes the revised standard would require retro-reflective and illuminated devices at intermediate and long-term stationary temporary traffic control zones; warning devices for mobile operations at speeds above 20 mph; advance warning signs for certain closed paved shoulders; a transition area containing a merging taper when one lane is closed on a multi-lane road; temporary traffic control devices with traffic barriers that are immediately adjacent to an open lane; and temporary traffic barriers separating opposing traffic on a two-way roadway.

ACCIDENT PREVENTION SIGNS AND TAGS—§1926.200

General

Signs and symbols required by this Subpart G shall be visible at all times when work is being performed and shall be removed or covered promptly when the hazards no longer exist.

Danger Signs

Danger signs shall be used only where an immediate hazard exists.

Danger signs shall have red as the predominating color for the upper panel, black outline on the borders, and a white lower panel for additional sign wording (see Figure 7.1).

Figure 7.1 Danger Sign

Caution Signs

Caution signs shall be used only to warn against potential hazards or to caution against unsafe practices.

Caution signs shall have yellow as the predominating color, black upper panel and borders, yellow lettering of "caution" on the black panel, and the lower yellow panel for additional sign wording. Black lettering shall be used for additional wording.

Standard color of the background shall be yellow; the panel, black with yellow letters. Any letters used against the yellow background shall be black. The colors shall be those of opaque glossy samples as specified in Table 1 of American National Standard ANSI Z53.1-1967 (see Figure 7.2).

Figure 7.2 Caution Sign

Exit Signs

Exit signs, when required, shall be lettered in legible red letters, not less than six inches high, on a white field, and the principal stroke of the letters shall be at least three-fourths inch in width (see Figure 7.3).

Figure 7.3 Exit Sign

Safety Instruction Signs

Safety instruction signs, when used, shall be white with green upper panel with white letters to convey the principal message. Any additional wording on the sign shall be black letters on the white background (see Figure 7.4).

Figure 7.4 Safety Instruction Sign

Directional Signs

Directional signs, other than automotive traffic signs specified in the paragraph below, shall be white with a black panel and a white directional symbol. Any additional wording on the sign shall be black letters on the white background.

Traffic Signs

Construction areas shall be posted with legible traffic signs at points of hazard.

All traffic control signs or devices used for protection of construction workers shall conform to American National Standards Institute ANSI D6.1-1971, *Manual on Uniform Traffic Control Devices for Streets and Highways*.

Construction areas shall be posted with legible traffic signs at points of hazard.

Figure 7.5 Traffic Sign

Accident Prevention Tags

Accident prevention tags shall be used as a temporary means of warning employees of an existing hazard, such as defective tools, equipment, etc. They shall not be used in place of, or as a substitute for, accident prevention signs.

Specifications for accident prevention tags similar to those shown below shall apply.

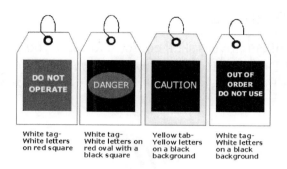

Figure 7.6 Accident Prevention Tags

SIGNALING — § 1926.201

Flagmen

When operations are such that signs, signals, and barricades do not provide the necessary protection on or adjacent to a highway or street, flagmen or other appropriate traffic controls shall be provided.

Signaling directions by flagmen shall conform to the American National Standards Institute ANSI D6.1-1971, *Manual on Uniform Traffic Control Devices for Streets and Highways.*

Hand signaling by flagmen shall be by use of red flags at least 18 inches square or sign paddles and, in periods of darkness, red lights.

Figure 7.7 Hand signaling

Flagmen shall be provided with and shall wear a red or orange warning garment while flagging. Warning garments worn at night shall be of reflective material.

Crane and Hoist Signals

Regulations for crane and hoist signaling will be found in applicable American National Standards Institute Standards.

BARRICADES — §1926.202

Barricades for protection of employees shall conform to the portions of the American National Standards Institute ANSI D6.1-1971, *Manual on Uniform Traffic Control Devices for Streets and Highways*, relating to barricades.

Figure 7.8 Barricade

Endnotes

[1] Excerpts from *Federal Register*, Vol. 67, No. 177, 12 September 2002.

8

Ergonomics

Preventing Musculoskeletal Injuries at Your Worksite

INTRODUCTION

Ergonomics is a science new to most of us, but it actually dates back to 1949. It was used by the automotive industry in the early 1970s to design car interiors that were more functional and easier to use. Also referred to as human engineering, ergonomics is concerned with helping people interact more comfortably and efficiently with their environment, whether at home, work or on the playing field.

Constructing a new building is, by its very nature, a problem in ergonomics. Installing floors and ceilings requires work at floor and ceiling height, which by definition is ergonomically hazardous since ceilings have to be above shoulder level and floors below knee height. Building materials are necessarily heavy and present manual materials-handling problems. Today, implementing ergonomics in the construction industry is recognized as one of the best ways to minimize on-the-job stress and strain and to prevent Musculoskeletal Disorders (MSDs).

In the workplace, ergonomics helps adapt the job to fit the person, rather than force the person to fit the job. Adapting the job to fit the worker can help reduce stress and eliminate many potential injuries and disorders associated with the overuse of muscles, bad posture, and repetitive motion. The objective of ergonomics is to adapt the job and workplace to the worker by designing tasks, work stations, controls, displays, safety devices, tools, lighting, and equipment to fit the worker's physical capabilities and limitations.

The extensive problems related to workplace ergonomic hazards were brought to OSHA's attention in the meatpacking industry, which has a cumulative trauma rate at least 75 times higher than that of the construction industry. Even though there is no specific safety or health regulation governing ergonomic hazards in the workplace, OSHA will continue to use its authority to cite companies under the General Duty Clause.

Section 5(a)(1), more commonly known as the General Duty Clause, specifically states: "Each employer shall furnish to each of his employees employment which is free from recognized hazards that are causing or are likely to cause death or serious physical harm to his employees." The intent of the general duty clause was simply to fill in gaps that may exist in the OSHA standards; it has been held from the very beginning that an employer cannot be held in violation of Section 5(a)(1) when OSHA had adopted an occupational safety and health standard that

211

addresses the subject matter of an alleged violation. For example, a general duty clause violation cannot be based upon an employer's failure to provide his employees with a safe means of access to a scaffold because there is an OSHA standard, 1926.451(a)(13), which provides that. OSHA inspectors will cite a company whenever they feel obligated to issue a citation but cannot identify an OSHA standard that regulates the condition or practice that they feel should be penalized.

About half of OSHA's ergonomic inspections are triggered by employee complaints of ergonomic problems in the workplace. Other ergonomic inspections occur after OSHA has reviewed an employer's OSHA 300 Log for injuries and illnesses during a normal inspection and has identified a general pattern of muscle injuries in the workplace. If it is determined that the company has a high incidence of ergonomic related injuries, another OSHA inspection focusing specifically on ergonomic hazards will be made. OSHA has also sent new guidelines to its inspectors to cite employers for failing to record work-related cumulative trauma disorders on the OSHA 300 Log where the injury is evidenced by at least one "physical finding" or a subjective symptom in combination with medical treatment or lost work days.

OSHA ACTIVITY RELATING TO ERGONOMICS[1]

During the last days of the Clinton Administration, an ergonomic regulation was passed. It would have required many employers to implement an Ergonomics program. But when George W. Bush became president, Congress passed and President George W. Bush signed Senate Joint Resolution 6, which rescinded the original ergonomics rule and, under the Congressional Review Act, prohibits the agency from issuing a rule that is substantially the same.

There are a number of reasons why guidelines are preferable to a rule. OSHA must follow certain criteria in establishing a rule—any rule. In terms of ergonomics, there are factors that make establishing a rule very difficult:

- There are a variety of different hazards and combinations of different hazards to be addressed.
- Exposure to the hazards is not readily measured in some cases.
- The exposure-response relationship is not well understood.
- Cost and feasibility of abatement measures may be uncertain and may be very high in some cases.
- It is very difficult, except in the most general terms, to prescribe remedies for abating such hazards in a single rule.

These considerations make it very difficult to develop simple criteria for compliance that can apply to a broad range of industries.

On the other hand, industry and task specific guidelines can be developed more quickly and are more flexible, and can provide specific and helpful guidance for abatement to assist employees and employers in minimizing injuries. Guidelines are the most effective method available for reducing injuries.

HOW OSHA GUIDELINES CAN REDUCE INJURIES AND ILLNESSES RELATED TO MSDS[2]

Injuries and illnesses related to MSDs have consistently declined over the last ten years, even though there has not been a standard addressing them. Guidelines, such as OSHA's Meatpacking Guidelines, and voluntary industry efforts have been successful in reducing the injury and illness rates for these disorders. For example, on a national basis, rates for carpal tunnel injuries with days away from work have gone down by 39 percent from 1992 to 1999. For the same time period, rates for strains and sprains with days away from work have also gone down by 39 percent, and rates for back injuries with days away from work have gone down by 45 percent. In the meatpacking industry, with industry-specific guidelines and focused OSHA enforcement, rates of carpal tunnel injuries with days away from work have gone down 47 percent from 1992 to 1999.

Over the same time period, rates of strains and sprains with days away from work have gone down by 61 percent, and rates for back injuries with days away from work have gone down by 64 percent. OSHA expects that industry-or-task-specific guidelines will further reduce injuries and illnesses as they are completed and implemented. OSHA's VPP (Voluntary Protection Program) participants, who have implemented safety and health programs, have injury and illness rates 53 percent below the average for their respective SIC codes.

Problem Recognition[3]

What is an "ergonomic injury"?

Input from the recent ergonomics forums demonstrated to OSHA that there are a wide variety of opinions on how the Agency should define an ergonomic injury and that the definition adopted by OSHA depends on the context. Ergonomic injuries are often described by the term "musculoskeletal disorders" or "MSDs." This is the term of art in scientific literature that refers collectively to a group of injuries and illnesses that affect the musculoskeletal system; there is no single diagnosis for MSDs. As OSHA develops guidance material for specific industries, the agency may narrow the definition as appropriate to address the specific workplace hazards covered. OSHA will work closely with stakeholders to develop definitions for MSDs as part of its overall effort to develop guidance materials.

Are all MSDs work-related?

No. MSDs can and do develop outside the workplace.

How do you determine whether MSDs are work-related?

The determination of whether any particular MSD is work-related may require the use of different approaches tailored to specific workplace conditions and exposures. Broadly speaking, establishing the work-relatedness of a specific case may include:

- Taking a careful history of the patient and the illness
- Conducting a thorough medical examination
- Characterizing factors on and off the job that may have caused or contributed to the MSD

How do I look for conditions that may contribute to musculoskeletal disorders?

Both work-related and non-work related conditions can, either individually or by interacting with each other, give rise to musculoskeletal disorders (MSDs). There are several approaches that may be used to determine whether conditions in the workplace might be contributing to employees developing MSDs, including reviewing and analyzing injury and illness records and analyzing job and work tasks. These approaches can be used individually or in combination.

REVIEW AND ANALYSIS OF INJURY AND ILLNESS RECORDS[4]

Step 1: Review Injury and Illness Records

An easy and straightforward approach to assessing potential ergonomic-related problems is to conduct a review of various injury and illness records. The records can be different for every company but may include

- OSHA 200 forms for years prior to 2002
- OSHA 300 Log forms for years after 2002
- Workers' Compensation claims
- Group health insurance records
- First aid logs
- Absentee and turnover records
- Records of employee complaints or grievances

Step 2: Identify Potential MSD Cases

Using all these information sources, try to identify entries which may indicate the presence of Musculoskeletal Disorders (MSDs). Musculoskeletal disorders (MSDs) is a term of art in scientific literature that refers collectively to a group of injuries and illnesses that affect the musculoskeletal system. While there is no single diagnosis for MSDs, the conditions identified below are commonly included in the range of conditions that potentially constitute MSDs:

- Tendinitis
- Tenosynovitis
- Epicondylitis
- Carpal tunnel syndrome
- Bursitis
- deQuervain's disease
- Ganglion cyst
- Thoracic outlet syndrome
- Less precise entries such as sprains, strains, tears, or even just pain

Step 3: Categorize

Once all of the records have been examined, categorize MSD cases by job, department, division, work task, shift, and so on, to determine whether patterns or trends exist. For example, look for instances where

- one unit (e.g., job, department, or division) has a higher number of MSDs than other units in the company
- one unit has more severe MSDs than other units in the company
- one unit's rate of MSDs has increased each year for several years
- one unit's rate of MSDs is higher than other companies doing the same job
- MSDs increase during a particular shift or time of the year
- MSDs increase when producing a particular product or when performing a particular work task

ANALYSIS OF JOBS OR WORK TASKS[5]

Not all MSDs are related to work activities. Other factors such as personal characteristics and societal factors have also been associated with ergonomic related injuries and illnesses. When analyzing jobs or work tasks that may be associated with MSDs, conditions to consider may include, but are not limited to

Awkward Postures—which might include prolonged work with hands above the head or with the elbows above the shoulders; prolonged work with the neck bent; squatting, kneeling, or lifting; handling objects with back bent or twisted; repeated or sustained bending or twisting of wrists, knees, hips or shoulders; forceful and repeated gripping or pinching

Forceful Lifting, Pushing, or Pulling—which might include handling heavy objects; moving bulky or slippery objects; assuming awkward postures while moving objects.

Prolonged Repetitive Motion—which might include keying; using tools or knives; packaging, handling, or manipulating objects.

Contact Stress—which might include repeated contact with hard or sharp objects, like desk or table edges.

Vibration—which might include overuse of power hand tools.

Defining the Problem

In OSHA's view, ergonomics means more than learning. It means redesigning workstations and job requirements in order to make them less stressful and less prone to result in injury or illness to those who perform the jobs.

The ergonomic injuries and illnesses that concern OSHA are. In his address to the meatpackers, OSHA's former deputy assistant Secretary of Labor Alan C. McMillan stated that MSDs include things like carpal tunnel syndrome and tendinitis. And that it is not confined to one or two industries but that it "exists in every type of workplace." That is true, of course. But it also exists outside of the workplace. OSHA is aware of that, but it believes that MSDs are also produced by the manner in which certain jobs are performed and it thinks there are figures to prove it. Some reliable surveys, for example, report that MSDs are the basis for fully one-half of all worker compensation claims. The most recent injury/illness report of the Bureau of Labor Statistics shows that almost half the reported occupational illnesses were CTD cases.

Correcting the Problem

In any event, MSDs have caught OSHA's attention, and OSHA's resulting enforcement activities, which have resulted in citations for violation of the General Duty Clause (Section 5(a)(1)) coupled with multimillion dollar fines, have caught the attention of business and industry. OSHA is pleased with the result of its initial efforts and is determined to keep pushing ahead. Those General Duty Clause citations against some of the giants of American industry have produced settlement agreements under which firms like Chrysler, IBP, Sara Lee, and others have instituted comprehensive ergonomics programs. OSHA is convinced that those programs will be beneficial to employees by reducing the MSD incidence rate.

Disorders and Injuries Associated with MSDs

Technological advances in recent years have greatly enhanced products and services, but at the same time, they have extracted a physical price from workers in this society. New technology has played a major role in the rapid evolution of musculoskeletal disorders (MSDs). MSDs are recognized by OSHA as the major occupational health hazard in the workplace and account for the largest share of occupational illnesses known as "repeated trauma" disorders, according to the Bureau of Labor Statistics.[6]

MSDs are disorders of the musculoskeletal and nervous systems which may be caused or aggravated by repetitive motions, forceful exertions, vibration, mechanical compression (hard and sharp edges), sustained or awkward postures, or by exposure to noise over extended periods of time. MSDs can affect nearly all tissues, the nerves, tendons, tendon sheaths, and muscles, with the upper extremities being the most frequently affected. These painful and sometimes crippling injuries develop gradually over periods of weeks, months, and years and result from repeated actions such as twisting and bending of the hands, arms, and wrists. A common risk factor among these disorders is the use of force combined with repetitive motion over time.

The anatomical areas of most concern, the upper extremities and the lower back, have subcategories such as upper extremity MSDs, carpal tunnel syndrome, tendinitis, low-back syndrome, lumbago, disc degeneration, and herniated nucleus pulposus (HNP). The main underlying factor for this kind of injury is accumulated trauma caused by repetitive events that leads to inflammatory responses, restrictive movements, and temporary or permanent tendon, ligament, muscle, or nerve injury. Studies conducted by NIOSH reveal that musculoskeletal injury rates (number of injuries per worker-hours on job) and severity rates (number of hours lost because of injury per worker-hours on job) increase significantly when heavy or bulky objects are lifted, objects are lifted from the floor, or objects are frequently lifted.

The most common MSDs in the workplace are tendon disorders such as tendinitis, tenosynovitis, De Quervain's disease, trigger finger, Raynaud's syndrome, and carpal tunnel syndrome. Tendon disorders often occur at or near the joints where the tendons rub against ligaments and bones. The most frequently noted symptoms are a dull aching sensation over the tendon, discomfort with specific movements, and tenderness to the touch. Recovery is usually slow and the condition may easily become chronic if the cause is not eliminated. Some of the most common disorders associated with MSDs are as follows:

- **Tendinitis**—This is a form of tendon inflammation that occurs when a muscle or tendon is repeatedly tensed from overuse or unaccustomed usage of the wrist and shoulder. With further exertion, some of the fibers that make up the tendon can actually fray or tear apart. The tendon becomes thickened, bumpy, and irregular in certain areas of the body (e.g., the shoulder), and the injured area may calcify. Without rest and sufficient time for the tissues to heal, the tendon may be permanently weakened. Tendinitis is common among power press operators, welders, and painters.

- **Tenosynovitis**—This is an inflammation or injury to the synovial sheath surrounding the tendon. These sheaths secrete synovial fluid which acts as a lubricant to reduce friction during movement. Repetitive motion using the hands and wrists may provoke an excessive secretion of synovial fluid, with the sheath becoming swollen and painful. Repetitions exceeding 1,500 to 2,000 per hour are known to produce symptoms associated with tendon sheath irritation in the hands. Tenosynovitis affects workers in jobs such as core packing, poultry processing, and meatpacking.

- **De Quervain's Disease**—With this disorder, the tendon sheath of the thumb is inflamed. De Quervain's Disease is attributed to excessive friction between two thumb tendons and their common sheath. Twisting and forceful gripping motions with the hands, similar to a clothes-wringing movement, can place sufficient stress on the tendons to cause De Quervain's disease. Tasks involving these kinds of motions are frequently performed by butchers, housekeepers, packers, seamstresses, and cutters.

- **Trigger Finger**—This is another tendon disorder and is attributed to the creation of a groove in the flexing tendon of the finger. If the tendon becomes locked in the sheath, attempts to move that finger will cause snapping and jerking movements. The palm side of the finger is the usual site for trigger finger. This disorder is often associated with using tools that have handles with hard or sharp edges. Carpenters are at risk of developing trigger finger.

- **Raynaud's Syndrome**—This disorder, also known as white finger, occurs when the blood vessels of the hand are damaged as a result of repeated exposure to vibration for long periods of time. The skin and muscles are unable to get the necessary oxygen from the blood and eventually die. Common symptoms include intermittent numbness and tingling in the fingers; skin that turns pale, ashen, and cold; and eventual loss of sensation and control in the fingers and hands. This condition is also intensified when the hands are exposed to extremely cold temperatures. Raynaud's syndrome is associated with the use of vibrating tools over time, such as pneumatic hammers, electric chain saws, and gasoline powered tools. After long-term exposure, perhaps ten or fifteen years working six to seven hours a day with vibrating tools, the blood vessels in the fingers may become permanently damaged. There is no medical remedy for white finger. If the fingers are fairly healthy, the condition may improve if exposure to vibration stops or is reduced. Construction workers working with power drills are at risk of developing this Raynaud's Syndrome.

- **Carpal Tunnel Syndrome**—When early Stone Age Man began chip-chip-chipping away while using ill-designed tools and poor job techniques to perform a repetitive task, he may have become the first victim of Carpal Tunnel Syndrome (CTS). Eons later, Carpal Tunnel Syndrome is still with us, and may be increasing for those same three basic reasons: 1) ill-designed tools, 2) poor job techniques, and 3) repetitive work processes involving the hands. An October 1988 article in *Men's Health* magazine provided a description of the problems associated with carpal tunnel syndrome:

  > CTS is a disease of the times, a clash between anatomy and labor-saving technologies that require less sweat but more repetitive hand work. It hits carpenters, musicians, butchers and meatpackers, auto workers, gardeners, construction workers, supermarket checkers, assembly-line workers, writers using computer keyboards—anyone who uses his hands repetitively with the wrists bent.

Today, more than half of all U.S. workers are susceptible to developing CTS. Anyone whose job demands a lot of repetitive wrist, hand, and arm motion—which need not always be forceful or strenuous—might be a potential victim of CTS. Throughout time, CTSs ramifications have extended into such varied trades and crafts as stone cutting, weaving, garment-cutting, hand-sewing, and electronics assembly work. Indeed, the condition can affect virtually anyone involved with repetitive motions of the wrist area, whether one is performing a simple, direct, hands-on work process such as sorting and flipping mail or whether one is holding a tool (customarily a pair of pliers, cutters, or screwdriver, but for some workers the tool may be a control button, a pencil, a telephone, or an assembly part).

A person may first notice CTS as a tingling, numb, pins-and-needles sensation, usually occurring some hours after work or during sleep. While this may happen to everyone occasionally, it persists in the individual with CTS. The affected person may awaken regularly and have to shake his or her hand to help regain normal feeling. Then, as the syndrome progresses, this sensation becomes more frequent, eventually occurring in the daytime, too.

Many people who have CTS attribute the numbness and tingling to arthritis or aging. Others resign themselves to the fact that when they go to work, their hands will hurt. But, while numbness, tingling, and loss of function in the hands can be caused by arthritis, diabetes, or other ailments, researchers are finding that more frequently a person's work or hobby involves repetitive activities which eventually cause CTS. For example, a hand surgeon at St. Luke's Medical Center in Chicago, recently reported that about 30 percent of his patients' cumulative trauma syndrome injuries are caused by work, while another 30 percent are caused by natural factors, such as arthritis. The cause of the remaining 40 percent is unknown. Many of his patients work in construction and undertake many repetitive motions each day. Those having surgery for carpal tunnel syndrome usually are out of work from eight to ten weeks.

The major problem with CTS is the fact that employees are often unaware of the causes of CTS and what to do about them. They may not associate their pain with their work because symptoms may only occur during evening or off-duty hours. When workers finally seek medical help, they may be given the wrong diagnosis and find the road to recovery takes more time and money than had been anticipated.

The medical community has taken a variety of positions in the dialogue over whether such injuries are related to the workplace. These positions, as with most medical determinations, have evolved along with findings about MSDs. Some of the early medical reviews were slow to recognize that problems such as cumulative trauma disorders and carpal tunnel syndrome even existed. Some experts in the field made an initial judgment that there were no such injuries and that wear and tear was due to the normal aging process. But this did not explain disorders in younger workers, particularly if the inability to use their extremities prevented them from working.

Other practitioners, while recognizing the existence of CTS, have been slow to accept that there are industrial problems. They have expressed concern about CTS, and MSDs in general, only as some form of worker mass hysteria, griping, malingering, or fraud. Some have undertaken studies to show that such injuries are not primarily due to workplace activities but represent a combination of work and home activities, with no way to determine how much work activity is responsible for the injury. Yet, while some credence must be given to these experts, it may turn out that many of them lack expertise regarding the workplace and the ergonomic problems that led the worker to be injured.

While disagreement exists within the medical community, most physicians and medical researchers agree that cumulative trauma disorders are caused by repeated motion of the same body parts over a long period of time. Cumulative trauma can be triggered by using vibrating tools such as chippers and grinders which can lead to back problems, tendinitis, bursitis, and many other medical problems. MSDs can be extremely painful, and they often cripple victims for life, leaving them unable to perform simple tasks.

One reason disagreements exist in the medical field concerning MSDs is that ergonomics is a new science and no long-term studies of the problem have been conducted as have been for other illnesses such as heart disease. Tracking MSDs in this manner is difficult because few workplaces have stable populations that can be studied over time. Another reason why data on the prevalence of MSDs in the workforce is insufficient is because of problems in sorting out occupational and non-occupational contributing factors. Thus, millions of American workers are at risk of injuries classified as cumulative trauma disorders, but the medical community is having difficulty diagnosing or even defining these conditions.

We have adopted this Ergonomics Program in order to assist in preventing ergonomic disorders that can produce physical pain and disabilities in our employees through sound safety and health programs. It is to be adapted to the needs and resources of each workstation, each job and each employee as necessary. The Program will be administered by a team approach under the overall coordination of_____.

This is *(Company Name)* Ergonomics Written Program. It meets all OSHA requirements and applies to all our work operations.

(Enter Name Here) will be responsible for overall direction of the Safety Program.

INTRODUCTION

The purpose of this program is to inform our employees that *(Company Name)* is committed to preventing and reducing musculoskeletal disorders (MSDs). Our Ergonomics Program is the most effective way to reduce risk, decrease exposure, and protect our workers against MSDs. This program applies to all work operations. However, it does not address injuries caused by slips, trips, falls, vehicle accidents, or similar accidents.

We have adopted this Ergonomics Program in order to assist in preventing ergonomic disorders that can produce physical pain and disabilities in our employees through sound safety and health programs. It is to be adapted to the needs and resources of each workstation, each job, and each employee as necessary. The program will be administered by a team approach under the overall coordination of *(Enter Name Here)*.

The Ergonomics Program consists of six main elements. They are the following:

1. Management Commitment and Employee Involvement
2. Job Hazard Analysis
3. Hazard Prevention and Control
4. MSD Management
5. Training and Education
6. Program Evaluation

Management Commitment and Employee Involvement

The first step in implementing our Ergonomics Program will involve establishing a management commitment. Commitment and involvement are complementary and essential elements of a sound Ergonomics Program. Commitment by management provides the organizational resources and motivating force necessary to deal effectively with ergonomic hazards.

Employee involvement and feedback through clearly established procedures are likewise essential, both to identify existing and potential hazards and to develop and implement an effective way to abate such hazards.

Figure 8.1 Sample Ergonomics Program

The company recognizes that the implementation of an effective Ergonomics Program requires a commitment to provide the visible involvement of top management so that all employees, from management to line workers, fully understand that management has a serious commitment to the program. Our program is a team approach, with top management as the team leader.

Management's involvement is demonstrated through our personal concern for employee safety and health and by the priority we place on preventing MSDs. Our policy places the safety and health of our employees on the same level of importance as production. It integrates production processes and safety and health protection to ensure that this protection is part of the daily production activity.

Our program assigns and communicates the responsibility for the various aspects of the Ergonomics Program so that all managers, supervisors, and employees involved know what is expected of them. We have provided adequate authority and resources to all responsible parties so that assigned responsibilities can be met. Each manager, supervisor, and employee responsible for the Ergonomics Program in the workplace is accountable for carrying out those responsibilities.

Our written program is communicated to all personnel because it encompasses the total workplace, regardless of the number of workers employed or the number of work shifts. It establishes clear objectives to meet those goals. We will do whatever is necessary to ensure that they are communicated to and understood by all company employees and supervisors.

Our program encourages employee involvement in the following:

1. An employee complaint and suggestion procedure that allows workers to bring their concerns to management and provide feedback without fear of reprisal.
2. Any employee who believes that he has sustained or is suffering from any work-related MSD must immediately notify his supervisor.
3. Any employee who believes that his work method or pace, his tools or equipment, his workstation or anything else connected with his job is producing or could produce an MSD must immediately report that to his supervisor.
4. Any employee who believes that there is a way to change the way his work is done that will result in less strain or pain and reduce the potential for an MSD should report those suggestions to his supervisor.
5. Each supervisor shall promptly transmit each such reports to the appropriate company person and shall be responsible for seeing to it that the employee making any such report is informed of the action that has been taken in response to the report.
6. No supervisor or member of management shall discharge or in any manner discriminate against any employee because he has made any report or supplied any information relating to ergonomics, MSDs, or this program.
7. We encourage the prompt and accurate reporting of signs and symptoms of MSDs by employees so that they can be evaluated and, if warranted, treated.

Figure 8.1 Sample Ergonomics Program, continued

221

8. We will select ergonomic teams or monitors with the required skills to identify and analyze jobs for ergonomic stress and to recommend solutions.
9. Procedures and mechanisms have been developed to evaluate the implementation of the Ergonomics Program and to monitor the progress. Management will review the program at least annually to evaluate success in meeting our goals and objectives. Evaluation techniques will include the following:

~ Analyses of trends in injury/illness rates
~ Employee surveys
~ Evaluations of job/workstation changes
~ Review of results of those evaluations
~ Up-to-date records or logs of job improvements tried or implemented

The results of the reviews will be a progress report and program update that will include whatever new or revised goals may arise from the review, identification, and correction of any deficiencies that are disclosed.

Job Hazard Analysis

The second step in implementing our Ergonomics Program involves conducting a job hazard analysis. If an MSD incident has occurred, _____ *(Enter Name Here)* will be responsible for conducting all job hazard analyses.

If an employee's job does not meet the Action Trigger, our company does not need to take further action for that MSD incident.

If it is determined that there is an MSD hazard in the job, the job will be termed a "problem job." However, if it is determined that MSD hazards pose a risk only to the employee who reported the MSD, our company may limit our job controls, training, and evaluation to that individual employee's job by the Quick Fix Option.

Quick Fix Option

A Quick Fix Option is a way to quickly fix a job when an MSD incident has occurred in that job. The benefit to using the Quick Fix Option is that the hazards are controlled quickly and more informally; employees in that job are safer.

We will use the Quick Fix Option for a job if either of the following occurs:

• Our employees have experienced no more than one MSD incident in that job
• There have been no more than two MSD incidents in our company

Our company will observe the following Quick Fix Options:

1. The company makes MSD management available.
2. Our company talks with employees in the job and their representatives about the tasks they perform that may relate to the MSD incident, ob-

Figure 8.1 Sample Ergonomics Program, continued

serves the employees performing the job to identify which risk factors are likely to have caused the MSD incident, and asks employees performing the job and their representatives to recommend measures to reduce exposure to the MSD hazards identified.

3. Our company implements quick fix controls. For each problem job, we will use feasible engineering, administrative controls, or any combination of these to reduce MSD hazards in the job.
4. Our company will train employees in the use of the selected quick fix controls.
5. Our company will check all problem jobs within 30 days after controls are implemented. If our company determines that MSD hazards have been reduced, no further action except to maintain controls for that job will be required.

Hazard Prevention and Control

The third step in implementing our Ergonomics Program will involve controlling and reducing hazards. Job hazard controls are engineering and administrative controls used to reduce or eliminate MSD hazards. While engineering controls, where feasible, are the preferred methods, administrative controls also may be important in addressing MSD hazards. Personal protective equipment (PPE) may also be used to supplement engineering and administrative controls, but may only be used alone where other controls are not feasible. Where PPE is used our company provides it at no cost to employees.

There are three types of controls our company will consider using for controlling ergonomic hazards. They are:

- Engineering Controls
- Administrative Controls
- Personal Protective Equipment (PPE)

Engineering Controls

The preferred approach to prevent and control MSDs is to design the job to include the workstation layout, selection and use of tools, and work methods to take account of the capabilities and limitations of the workforce. Examples of Engineering Controls are as follows:

- Changing the way products can be transported. For example, using mechanical assist devices to relieve heavy load lifting and using handles in packages requiring manual handling.
- Changing the process or product to reduce worker exposures to risk factors. Examples could include maintaining the fit of plastic molds to reduce the need for manual removal of flashing, or using easy-connect electrical terminals to reduce manual forces.
- Modifying containers such as height-adjustable material bins.
- Changing workstation layout, which might include using height-adjustable workbenches or locating tools and materials within short reaching distances.

Figure 8.1 Sample Ergonomics Program, continued

- Changing the way parts, tools, and materials are to be manipulated. Examples include using fixtures (clamps, vise-grips, etc.) to hold work pieces to relieve the need for awkward hand and arm positions to reduce weight and allow easier access.
- Changing tool designs. For example, pistol handle grips for knives reduce wrist bending postures required by straight-handle knives, and squeeze-grip-actuated screwdrivers replace finger-trigger-actuated screwdrivers.

Administrative Controls

Administrative controls include changes in job rules and procedures such as scheduling more rest breaks, rotating workers through jobs that are physically tiring, and training workers to recognize ergonomic risk factors and to learn techniques for reducing the stress and strain while performing their work tasks.

Although engineering controls are preferred, administrative controls can be helpful as temporary measures until engineering controls can be implemented or when engineering controls are not technically feasible. Since administrative controls do not eliminate hazards, management must ensure that the practices and policies are followed.

Examples of Administrative Controls include the following:

- Providing protective equipment to employees, such as vibration-absorbing gloves or rubber mats
- Rotating workers from one job to another to provide time for relief from certain risk factors
- Matching workers to jobs based on job requirements and worker capabilities
- Providing training programs in safe work methods
- Enforcing safety practices
- Changing pay practices to encourage appropriate rest breaks
- Scheduling exercise/stretch breaks
- Arranging work to interrupt static postures
- Training in the recognition of risk factors for musculoskeletal disorders

Personal Protective Equipment

One of the most controversial questions in the prevention of MSDs is whether the use of personal equipment worn or used by the employee (such as wrist supports, back belts, or vibration attenuation gloves) is effective. In the field of occupational safety and health, PPE generally provides a barrier between the worker and the hazard source. Respirators, earplugs, safety goggles, chemical aprons, safety shoes, and hard hats are all examples of PPE. Whether braces, wrist splints, back belts, and similar devices can be considered as personal protection against ergonomic hazards remains open to question. Although these devices may, in some situations, reduce the duration, frequency, or intensity of exposure, evidence of their effectiveness in injury reduction is inconclusive. In some instances they may de-

Figure 8.1 Sample Ergonomics Program, continued

crease one exposure but increase another because the worker has to "fight" the device to perform his or her work.

Implementing Hazard Controls

Testing and evaluation verify that the proposed solution actually works and identifies any additional enhancements or modifications that may be needed. Employees who perform the job can provide valuable input into the testing and evaluation process. Worker acceptance of the changes put into place is important to the success of the intervention.

After the initial testing period the proposed solution may need to be modified. If so, further testing should be conducted to ensure that the correct changes have been made followed by full-scale implementation. Designating the personnel responsible, creating a timetable, and considering the logistics necessary for implementation are elements of the planning needed to ensure the timely implementation of the controls.

A good idea in general is that ergonomic control efforts start small, targeting those problem conditions which are clearly identified through safety and health data and job analyses information. Control actions should be directed to those conditions that appear easy to fix. Early successes can build the confidence and experience needed in later attempts to resolve more complex problems.

Our company will implement job hazard controls in the following ways:

- Make sure MSD hazards are controlled.
- Make sure hazards are reduced in accordance with or to levels below those in the hazard identification tool.
- Make sure MSD hazards are reduced to the extent feasible. Then at least every three years, our company will assess the job to determine whether additional feasible controls would control or reduce MSD hazards.

Evaluating Hazard Controls

A follow up evaluation is necessary to ensure that the controls reduced or eliminated the ergonomic risk factors and that new risk factors were not introduced. This follow up evaluation should use the same risk factor checklist or other method of job analysis that first documented the presence of ergonomic risk factors. If the hazards are not substantially reduced or eliminated, the problem-solving process is not finished.

The follow up may also include a symptom survey that can be completed in conjunction with the risk-factor checklist or other job analysis method. The results of the follow up symptom survey can then be compared with the results of the initial symptom survey (if one was performed) to determine the effectiveness of the implemented solutions in reducing symptoms.

Because some changes in work methods and the use of different muscle groups may actually make employees feel sore or tired for a few days, follow up should occur no sooner than a couple of weeks after implementation.

Figure 8.1 Sample Ergonomics Program, continued

MSD Management

The fourth step in implementing our Ergonomics Program involves developing a program to manage MSD incidents when they occur.

A MSD Management Program should spell out how your company provides for prompt and appropriate management when an employee has experienced an MSD incident. MSD management will include access to a healthcare professional. MSD management is important largely because it helps ensure that employees promptly report MSDs as well as signs and symptoms of MSDs.

Our company will ensure that all employees receive timely attention for the reported MSD. Whenever it is determined that an employee has suffered an MSD incident, our company will provide the employee with prompt and effective MSD management.

MSD management will include the following:

- Access to a Healthcare Professional (HCP)
- Evaluation and follow up of the MSD incident.

Whenever an employee consults an HCP for MSD management, we will provide the HCP with a description of the employee's job and information about the physical work activities and risk factors in the job.

The HCP will provide the following:

- The HCP's assessment of the employee's medical condition as related to the physical work activities, risk factors, and MSD hazards in the employee's job
- A statement that the HCP has informed the employee of the results of the evaluation, the process to be followed to effect recovery, and any medical conditions associated with exposure to physical work activities, risk factors, and MSD hazards in the employees job
- A statement that the HCP has informed the employee about work-related or other activities that could impede recovery from the injury

Evaluations

Our company has the option to request the assistance of a healthcare professional to evaluate our employees, determine the employees' functional capabilities, and prepare opinions regarding the employee jobs and job tasks. With specific knowledge of the physical demands involved in various jobs and the physical capabilities or limitations of employees, the healthcare professional can match the employee's capabilities with the appropriate jobs. Being familiar with employee jobs not only assists the healthcare professional in making informed case management decisions but also assists with the identification of ergonomic hazards and alternative job tasks.

Figure 8.1 Sample Ergonomics Program, continued

If we decide to use a healthcare professional, he or she will become familiar with the jobs and job tasks by a periodic plant walk through. Once familiar with plant operations and job tasks, the healthcare professional will periodically revisit the facility to remain knowledgeable about changing working conditions. Other approaches that may help the healthcare professional to become familiar with jobs and job tasks include reviewing job analysis reports, detailed job descriptions, job safety analyses, and photographs or videotapes that are accompanied by written descriptions of the jobs.

Early Reporting

Our employees will report as early as possible any signs or symptoms of musculoskeletal disorders. Early reporting allows corrective measures to be implemented before the effects of a job problem worsen. Individual worker complaints that certain jobs cause undue physical fatigue, stress, or discomfort may be signs of ergonomic problems. Following up on these reports—particularly reports of MSDs—is essential. Such reports indicate a need to evaluate the jobs to identify any ergonomic risk factors that may contribute to the cause of the symptoms or disorders.

Employees reporting any signs or symptoms of potential MSDs will have the opportunity for prompt evaluation by a healthcare provider. The earlier symptoms are identified and treatment is initiated, the less likely a more serious disorder will develop. Our company will not discourage employees from reporting and signs or symptoms. Employees need not fear any discipline or discrimination on the basis of such reporting.

Training and Education

The fifth step in implementing our Ergonomics Program involves developing a training program. Our company will train each employee who works at a job with exposure to specific risk factors and each employee in a job where a work-related musculoskeletal disorder has been recorded.

Training and education for employees potentially exposed to ergonomic hazards is a critical component of our ergonomics-training program. Training allows managers, supervisors, and employees to understand ergonomic and other hazards associated with a job or production process and their prevention, control, and medical consequences.

The following individuals will be included in our training and education program:

- All affected employees
- Engineers and maintenance personnel
- Supervisors
- Managers
- Healthcare professionals

Figure 8.1 Sample Ergonomics Program, continued

227

The program will be presented in a language and at a level of understanding that is appropriate for the individuals being trained. The company training program includes an opportunity for employees to ask questions and receive answers. This allows employees to fully understand the material presented to them. It will provide an overview of the potential risk of illnesses and injuries, their causes, symptoms, and the means of prevention and treatment.

The program will also include a means for adequately evaluating its effectiveness, such as employee interviews, testing, and observing work practices, in order to determine if those who received the training understand the material and the work practices to be followed.

General training will consist of the following:

- How to recognize risk factors associated with work-related musculoskeletal disorders and ways to reduce them
- The signs and symptoms of work-related musculoskeletal disorders, the importance of early reporting, and medical management procedures
- Reporting procedures and the person to whom the employee is to report work-related musculoskeletal disorders
- The process our company is taking to address and control workplace risk factors, each employee's role in the process, and how to participate in the process
- Opportunity to practice and demonstrate proper use of implemented control measures and safe work methods which apply to the job

Job-specific training. New employees and reassigned workers will receive an initial orientation and hands-on training when first placed in a full-production job. Training lines may be used for this purpose. Each new hire will be taught the proper use of and procedures for all tools and equipment that they will use.

The initial training program will include the following:

- Care, use and handling techniques for tools and equipment
- Use of special tools and devices associated with individual workstations, as necessary
- Use of appropriate guards and safety equipment, including personal protective equipment
- Use of proper lifting techniques and devices
- On-the-job training that will emphasize employee development and use of safe and efficient techniques

Training for supervisors. Supervisors are responsible for ensuring that employees follow safe work practices. They will receive appropriate instruction and training to enable them to do that. Supervisors must have training comparable to that of the employees and any additional training as may be necessary to enable them to recognize the signs and symptoms of MSDs, to recognize hazardous work practices, to correct the practices, and to reinforce our ergonomic program, especially through the ergonomic training of employees as may be needed.

Figure 8.1 Sample Ergonomics Program, continued

Training for managers. Managers must be aware of their safety and health responsibilities. We will, therefore, make sure that they have sufficient instruction and training pertaining to ergonomic issues at each workstation and in the production process as a whole, so that they can effectively carry out their responsibilities.

PROGRAM EVALUATION

The final step in implementing our Ergonomics Program involves performing a program evaluation. If any problems arise in our Ergonomics Program, although we may not be able to eliminate all problems, we will try to reduce exposures to and eliminate as many problems as possible to improve employee protection and encourage safe practices. *(Enter Name Here)* will thoroughly evaluate and, as necessary, promptly take action to correct any compliance deficiencies in our program so we can eliminate problems effectively. *(Enter Name Here)* will also evaluate the Ergonomics Program annually.

Our Ergonomics Program will have a system for long-term implementation, feedback, and review to assess the progress and success of the program. Components of the ergonomics plan will include accountability for implementation, ongoing assessment of effectiveness, and identification of new problems.

The following elements will be part of our Ergonomics Program:

- Employee input procedures
- Annual review and update of objectives and actions
- Corrective actions and dates of implementation
- Workplace monitored for ongoing ergonomic hazards
- Injury claim trends monitored
- OSHA 300 reviews monitored
- Employee suggestion mechanisms in place
- Ergonomics plan annually reviewed

Figure 8.1 Sample Ergonomics Program, continued

CONSTRUCTION ERGONOMICS CHECKLIST

To be filled out and updated jointly by contractors and union reps every 2 weeks or as a site changes. This document is intended to help develop an "eye" for ergonomic problems and prevent injuries.

Date:_____ Site: _____ General contractor:_____

Union rep:_____ Subcontractor:_____ Signature:_____

Materials Handling

What heavy materials or equipment are being handled on site—drywall, rebar, concrete forms, anything over 20 pounds?	
Do any workers have to lift more than 50 pounds at one time without help?	Yes_____ No_____
Do workers have to lift more than 20 pounds often? **If yes**, how can this be changed?	Yes_____ No_____
Are there handles to help carry materials?	Yes_____ No_____
If yes, are the handles easy to use and comfortable?	Yes_____ No_____
Are workers told to get someone's help to lift heavy materials?	Yes_____ No_____
Are there carts, dollies, or other aids readily available for moving materials?	Yes_____ No_____
If yes, are the carts being used?	Yes_____ No_____
If no, why not?	
If no, is the site clear enough to permit the use of carts?	Yes_____ No_____
Are materials delivered as close as possible to where they will be used?	Yes_____ No_____
If no, how can this be changed?	
On what jobs do workers have to lift overhead?	
How can this lifting be avoided?	
Are materials stored at floor or ground level?	Yes_____ No_____
If yes, do workers have to bend down to lift materials?	Yes_____ No_____
Can the materials be stored at waist height?	Yes_____ No_____
On which tasks do workers have to stretch to pick up or lift materials?	
Can the materials be kept closer?	Yes_____ No_____

Figure 8.2 Construction Ergonomics Checklist

Tools

Are tools sharp and in good condition?	Yes_____ No_____

Which tools are very heavy or not well balanced?

Which tools vibrate too much?	
Which tools must be used while in a difficult position?	

Which tools have poor handle design?
<ul><li>Grips too big or too small?</li><li>Handles that are too short and dig into hands?</li><li>Handles with ridges that dig into hands?</li><li>Slippery handles?</li></ul>

Which tools require bending of wrists to use?

Do gloves ever make it hard to grip tools?	Yes_____ No_____
Are there other tools with a better design?	Yes_____ No_____

If yes, what are they?

Repetitive Work

Which tasks or jobs use the same motion dozens of times an hour for more than one hour per day?

What are the motions?	
Can the number of repetitions be reduced by job rotation or rest breaks?	Yes_____ No_____

Figure 8.2 Construction Ergonomics Checklist, continued

Awkward Postures

Which tasks or jobs involve work above the shoulder more than one hour per day?	
Can scaffolds, platforms, or other equipment cut down on the need to work overhead?	Yes_____ No_____
Which tasks or jobs involve work at floor level or on knees for more than one hour a day?	
Are knee pads or cushions available and are they used?	Yes_____ No_____
Can equipment be used to reduce kneeling?	Yes_____ No_____
Which jobs require workers to stay in one position for a long time?	
Can rotation or rest breaks be used to reduce time in awkward postures?	Yes_____ No_____
Which jobs require a lot of twisting or turning?	
Which jobs require a lot of bending?	
How can the need to twist or bend be reduced?	

Standing

What jobs require workers to stand all day, especially on concrete floors?	
Can anti-fatigue matting be used?	Yes_____ No_____
Is it possible to use adjustable stools to allow workers to rest periodically?	Yes_____ No_____

Figure 8.2 Construction Ergonomics Checklist, continued

Surfaces for Walking and Working

Are working and walking surfaces clean and dry?	Yes_____ No_____
Are the surfaces unobstructed?	Yes_____ No_____
Are the surfaces even?	Yes_____ No_____

Seating

What jobs require sitting all day?	
Are the seats well-designed, easy to adjust and comfortable?	Yes_____ No_____
In heavy equipment, do workers have to lean forward to see/do their work?	Yes_____ No_____
Does the seating in any heavy equipment vibrate a lot?	Yes_____ No_____

Weather

Do workers have enough protection from heat, cold, rain, wind, and sun?	Yes_____ No_____

Lighting

Are work areas well lit to prevent tripping and falling?	Yes_____ No_____
Is there enough light to do the work?	Yes_____ No_____
Are there areas where glare is a problem?	Yes_____ No_____

Production Pressures

Do any workers work piece rate?	Yes_____ No_____
Have supervisors or workers been under production pressures that could lead to shortcuts and injuries?	Yes_____ No_____

How could this problem be reduced?
        More rest breaks?
        More safety meetings?
        A special safety rep on site?
        Other	

Figure 8.2 Construction Ergonomics Checklist, continued

Training

What training have workers had on ergonomics and preventing musculoskeletal disorders?	
What training have supervisors had on ergonomics and preventing musculoskeletal disorders?	

Musculoskeletal Symptoms

Do workers feel free to report symptoms?	Yes_____ No_____
Have any workers been reporting muscle pain, tingling, numbness, loss of strength, or loss of joint movement?	Yes_____ No_____
If yes, where? • Back • Neck • Shoulder • Arm • Wrist • Knee	
Which trades have the most problems?	
And what may be the main cause(s)? • Repetitive motion • Awkward postures • Fixed postures • Heavy lifting • Not enough rest breaks • Other	
Do workers often appear exhausted at the end of the day?	Yes_____ No_____

Figure 8.2 Construction Ergonomics Checklist, continued

Solutions

What jobs on site are the most hazardous for musculoskeletal injuries?

Most hazardous jobs for musculoskeletal injuries

1.
2.
3.
4.
5.

What has been done to get worker ideas to help reduce musculoskeletal injuries on the job?

What can be done working together to reduce these injuries?

What can be done to reduce the hazards or make the jobs easier?

Proposed solutions

Most Effective	Easiest to Implement	Least Expensive
1. 2. 3. 4. 5.		
Least Effective	**Hardest to Implement**	**Most Expensive**

Figure 8.2 Construction Ergonomics Checklist, continued

Endnotes:

[1] http://www.osha.gov/SLTC/ergonomics/faqs.html

[2] *Ibid.*

[3] *Ibid.*

[4] http://www.osha.gov/SLTC/ergonomics/review_of_records.html

[5] http://www.osha.gov/SLTC/ergonomics/job_analysis.html

[6] See Occupational Injuries and Illnesses in the United States, U.S. Department of Labor, BLS (1995).

9

Personal Protective Equipment

Protecting Your Workers from Head to Toe

INTRODUCTION

The purpose of a company's Personal Protection Equipment (PPE) Program is to protect employees from workplace hazards through use of personal protection equipment.

In general, the safety of workers depends upon a thorough knowledge of their operations and the hazards posed. A written personal protection program is designed with these objectives:

- To provide a reference document for any employee with questions concerning the proper application of PPE and how our company is complying with the relevant OSHA regulation
- To provide managers and employees with clear guidance on their responsibilities in the overall PPE Program

Personal protective equipment includes all clothing and other work accessories designed to create a barrier against workplace hazards. The basic element of any program for personal protective equipment should be an in-depth evaluation of the equipment needed to protect against the hazards of the workplace. Management dedicated to the safety and health of employees should use that evaluation to set a standard operating procedure for personnel, then train employees on the protective limitations of personal protective equipment and on its proper use and maintenance.

Using personal protective equipment requires hazard awareness and training on the part of the user. Employees must be aware that the equipment does not eliminate the hazard. If the equipment fails, exposure will occur. To reduce the possibility of failure, equipment must be properly fitted and maintained in a clean and serviceable condition.

Selection of the proper personal protective equipment for a job is important. Employers and employees must understand the equipment's purpose and its limitations. The equipment must not be altered or removed even though an employee may find it uncomfortable. (Sometimes equipment may be uncomfortable simply because it does not fit properly.)

A personal protection program covers may types of equipment most commonly used for protection of the head, including eyes and face; arms; hands; and feet. The use of equipment to protect against life-threatening hazards also is discussed. Information on respiratory protective equipment may be found in 29 CFR 1926.103. The standard should be consulted for more information on specialized equipment.

GENERAL REQUIREMENTS

Protective equipment, including PPE for eyes, face, head, and extremities; protective clothing; respiratory devices; and protective shields and barriers must be provided, used, and maintained in a sanitary and reliable condition. PPE must be provided whenever it is necessary by reason of hazards of processes or environment, chemical hazards, radiological hazards, or mechanical irritants (i.e., flying chips or sparks, abrasive moving parts) encountered in a manner capable of causing injury or impairment in the function of any part of the body through absorption, inhalation, or physical contact.

In order to assess the need for personal protective equipment in your company, conduct a walk-through survey of the areas in our workplace that may need PPE. Consideration will be given to the following hazard categories:

- Impact
- Penetration
- Compression (roll-over)
- Chemical
- Heat
- Harmful dust
- Light (optical) radiation

During the walk-through survey, the safety manager should observe:

- Sources of motion, i.e., machinery or processes where any movement of tools, machine elements, or particles could exist; or movement of personnel that could result in collision with stationary objects
- Sources of high temperatures that could result in burns, eye injury, or ignition of protective equipment, etc.
- Types of chemical exposures
- Sources of harmful dust
- Sources of light radiation, i.e., welding, brazing, cutting, furnaces, heat-treating, high intensity lights, etc.
- Sources of falling objects or potential for dropping objects
- Sources of sharp objects which might pierce the feet or cut the hands
- Sources of rolling or pinching objects which could crush the feet
- Layout of workplace and location of co-workers
- Any electrical hazards.

In addition, injury/accident data should be reviewed to help identify problem areas.

Following the walk-through survey, decide if PPE is required for certain areas of work. The objective is to prepare for an analysis of the hazards in the environment to enable proper selection of protective equipment.

Having conducted a thorough survey of the workplace, an estimate of the potential for injuries should be made and a determination made as to the type and level of risk and any potential injuries from each of the hazards found in the area.

All PPE must be of safe design and construction for the work that is to be performed.

All equipment that is damaged or defective will be taken out of service and thrown away.

Hazard Assessment

The employer must assess the workplace to determine if hazards are present or are likely to be present that will necessitate the use of PPE. If hazards are present or likely to be present

- Select proper PPE and require the use of PPE
- Communicate the selection decision to each affected employee
- Select PPE that properly fits each affected employee

A written certification of the workplace hazard assessment must be made. It is not necessary to prepare and retain a formal written hazard assessment, only a certification of the assessment. It must include

- The workplace evaluated
- The person certifying that the evaluation has been performed
- The date(s) of the hazard assessment

Equipment Selection

In order for employers to make correct selections, they must survey for hazards, organize, and analyze the data and then make the proper selection based on the types of hazards. It will be the safety manager's responsibility to use common sense and fundamental techniques to accomplish these tasks. This process is somewhat subjective due to the variety of situations where PPE may be required.

The following list is a guide to possible personal protective equipment requirements:

1.	Wire brush wheels	Aprons, gloves, face shields, and goggles
2.	Grinding stones	Face shield and safety glasses
3.	Metal working machines	Impact goggles, safety glasses with shields
4.	Compressed air	Impact goggles, safety glasses with side shields
5.	Woodworking machines	Safety goggles, face shields if dusts and no flying objects; safety glasses with side shield, abdominal guard or anti-kickback apron
6.	Handling wood, metal,	Kevlar (or other comparable material) leather, etc. gloves, or hand pads

7.	Landscaping tools	Eye protection with side shields
8.	Maintenance	Breast pockets sewn closed or removed or with flaps, tool belt with tools on side, gloves, harness and lifeline, goggles
9.	Material handling	Gloves, hardhats, eye protection
10.	Cold weather	Hardhat liners, Gloves
11.	Close quarters work	Hard hats, Bump caps
12.	Sparks, hot metals	Flame resistant caps, aprons, hoods, Nomex (or other comparable material) Canvas Spats
13.	Hair protection	Cool, lightweight cap with long visor (hair under cap), hairnets
14.	Acids, Alkalis	Acid suits, hoods (large quantities), face shield and splash hazard-proof goggles (small quantities), gloves, aprons.
15.	Limited direct splash from alkalis, acids	Face shield, chemical goggle
16.	Lifting 15 lbs. solid objects once a day one foot or more; rolling rolls of paper, steel, hogsheads	Toe protection
17.	Chain saws	Chaps, eye protection, hearing protection

Training

The employer must train each employee who is required to use PPE. Each employee must know at least the following:

- When PPE is necessary
- What PPE is necessary
- How to properly wear and adjust PPE
- The limitations of PPE
- The proper care, maintenance, useful life, and disposal of the PPE

Retraining is required when the employer has reason to believe that any employee who has been previously trained does not have the understanding or skill to use PPE properly, such as when

- Changes in the workplace render previous training obsolete
- Changes in the types of PPE to be used render previous training obsolete
- The employee has not retained the understanding or skill to use PPE properly

The employer must verify in writing through a certification record that employee received and understood the training. The record must contain the following:

- Name of each employee trained
- Date(s) of training
- Subject of the certification

EYE AND FACE PROTECTION

Employers must ensure that the proper eye and/or face protection is used when the employee is exposed to hazards from flying particles, molten metal, liquid chemicals, acid or caustic liquids, chemical gases or vapors, or potentially injurious light radiation. Care should be taken to recognize the possibility of multiple and simultaneous exposure to a variety of hazards. Adequate protection against the highest level of each of the hazards should be taken.

Each employee will wear eye protection that provides side protection when there is a hazard from flying objects. Detachable (clip-on or slide-on) side shields are acceptable.

Employees who must wear prescription lenses while engaged in activities requiring use of protection must be provided with eye protection which has the prescription incorporated into it or protection that can be worn effectively over the prescription lenses. Wearers of contact lenses must wear appropriate eye and face protection devices in hazardous environments. Dusty or chemical environments may represent an additional hazard to contact lens wearers.

Each eye and face PPE must be marked to identify the manufacturer.

For protection against potentially injurious light radiation, employees must use equipment with filter lenses with the appropriate shade number for the work being done. Tinted and shaded lenses are not filter lenses unless marked or identified as such.

Some occupations (not a complete list) for which eye protection should be routinely considered are carpenters, electricians, machinists, mechanics and repairers, millwrights, plumbers and pipe fitters, sheet metal workers and tinsmiths, assemblers, sanders, grinding machine operators, lathe and milling machine operators, sawyers, welders, laborers, chemical process operators and handlers, and timber cutting and logging workers.

Each eye, face, or face-and-eye protector is designed for a particular hazard. In selecting the protector, consideration should be given to the kind and degree of hazard, and the protector should be selected on that basis. Where a choice of protectors is given and the degree of protection required is not an important issue, worker comfort may be a deciding factor.

Persons using corrective spectacles and those who are required by OSHA to wear eye protection must wear face shields, goggles, or spectacles of one of the following types:

- Spectacles with protective lenses providing optical correction
- Goggles worn over corrective spectacles without disturbing the adjustment of the spectacles
- Goggles that incorporate corrective lenses mounted behind the protective lenses

When limitations or precautions are indicated by the manufacturer, they should be transmitted to the user and strictly observed. Over the years, many types and styles of eye and face-and-eye protective equipment have been developed to meet the demands for protection against a variety of hazards.

Fitting of goggles and safety spectacles should be done by someone skilled in the procedure. Prescription safety spectacles should be fitted only by qualified optical personnel.

Inspection and Maintenance for Eye Protection

It is essential that the lenses of eye protectors be kept clean. Continuous vision through dirty lenses can cause eye strain-often an excuse for not wearing the eye protectors. Daily inspection and cleaning of the eye protector with soap and hot water, or with a cleaning solution and tissue, is recommended. DAILY...

Pitted lenses, like dirty lenses, can be a source of reduced vision. They should be replaced. Deep scratches or excessively pitted lenses are apt to break more readily.

Goggles should be kept in a case when not in use. Spectacles, in particular, should be given the same care as one's own glasses, since the frame, nose pads, and temples can be damaged by rough usage.

Personal protective equipment that has been previously used should be disinfected before being issued to another employee.

When each employee is assigned protective equipment for extended periods, it is recommended that such equipment be cleaned and disinfected regularly.

Several methods for disinfecting eye-protective equipment are acceptable. The most effective method is to disassemble the goggles or spectacles and thoroughly clean all parts with soap and warm water. Carefully rinse all traces of soap, and replace defective parts with new ones. Swab thoroughly and immerse all parts in a solution of germicidal deodorant fungicide for ten minutes. Remove parts from solution and suspend in a clean place for air drying at room temperature or with heated air. Do not rinse after removing parts from the solution because this will remove the germicidal residue, which retains its effectiveness after drying.

The dry parts or items should be placed in a clean, dust-proof container, such as a box, bag, or plastic envelope, to protect them until reissue.

HEAD PROTECTION

Each affected employee must wear a protective helmet (hard hat) when working in areas where there is a potential injury to the head from falling objects.

Where falling object hazards are present, helmets must be worn. Some examples include working below other workers who are using tools and materials which could fall; working around or under conveyor belts which are carrying parts or materials; working below machinery or processes which might cause material or objects to fall; and working on exposed energized conductors. Some examples of occupations for which head protection should be routinely considered are carpenters, electricians, linemen, mechanics and repairers, plumbers and pipe fitters, assemblers, packers, wrappers, sawyers, welders, laborers, freight handlers, timber cutters and loggers, stock handlers, and warehouse laborers.

Head protection is required where there is a risk of injury from moving, falling, or flying objects, or for work near high-voltage equipment.

Hard hats are designed to protect from impact and penetration caused by objects hitting workers' heads, and from limited electrical shock or burns. The shell of the hat is designed to absorb some of the impact. The suspension, which consists of the headband and strapping, is even more critical for absorbing impact. It must be adjusted to fit the wearer and to keep the shell a minimum distance of one and one-fourth inches above the wearer's head.

Hard hats are tested to withstand the impact of an eight-pound weight dropped five feet-that's about the same as a two-pound hammer dropped 20 feet and landing on your head. They also must meet other requirements including weight, flammability, and electrical insulation.

Materials used in helmets should be water-resistant and slow burning. Each helmet consists essentially of a shell and suspension. Ventilation is provided by a space between the headband and the shell. Each helmet should be accompanied by instructions explaining the proper method of adjusting and replacing the suspension and headband.

The wearer should be able to identify the type of helmet by looking inside the shell for the manufacturer, ANSI designation, and class. For example

Manufacturer's Name

ANSI Z89.1-1969 (or later year)

Class A

All head protection is designed to provide protection from impact and penetration hazards caused by falling objects. Head protection is also available which provides protection from electric shock and burn. When selecting head protection, knowledge of potential electrical hazards is important.

Each type and class of head protectors is intended to provide protection against specific hazardous conditions. An understanding of these conditions will help in selecting the right hat for the particular situation. Protective hats are made in the following types and classes:

- Type 1-helmets with full brim, not less than one and one-fourth inches wide
- Type 2-brimless helmets with a peak extending forward from the crown

Be sure to issue and wear the right hard hat for the job. Hard hats come in three classes:

- Class A hard hats or helmets, in addition to impact and penetration resistance, are made from insulating material to protect from falling objects and electric shock by voltages of up to 2,200 volts. Class A hats are used in general service work and for limited voltage protection. They are used in mining, construction, shipbuilding, tunneling, lumbering, and manufacturing.

- Class B hard hats or helmets, in addition to impact and penetration resistances, are made from insulating material to protect from falling objects and electric shock by voltages up to 20,000 volts. Class B hats are used in utility service work and provide protection against high voltage. They are used extensively by electrical workers.

- Class C hard hats or helmets provide impact and penetration resistance. They are designed to protect workers from falling objects but are not designed for use around live electrical wires or where corrosive substances are present. The safety hat or cap in Class C is designed specifically for lightweight comfort and impact protection. This class is usually manufactured from aluminum and offers no dielectric protection. Class C helmets are used in certain construction and manufacturing occupations, oil fields, refineries, and chemical plants where there is no danger from electrical hazards or corrosion. They also are used on occasions where there is a possibility of bumping the head against a fixed object.

Headbands are adjustable in one-eighth size increments. When the headband is adjusted to the right size, it provides sufficient clearance between the shell and the headband. The removable or replaceable sweatband should cover at least the forehead portion of the headband. The shell should be of one-piece seamless construction and designed to resist the impact of a blow from falling material. The internal cradle of the headband and sweatband forms the suspension. Any part that comes into contact with the wearer's head must not be irritating to normal skin.

Inspection and Maintenance for Head Protection

Manufacturers should be consulted with regard to paint or cleaning materials for their helmets because some paints and thinners may damage the shell and reduce protection by physically weakening it or negating electrical resistance.

A common method of cleaning shells is dipping them in hot water (approximately 140 0F) containing a good detergent for at least a minute. Shells should then be scrubbed and rinsed in a clear hot water. After rinsing, the shell should be carefully inspected for any signs of damage.

All components, shells, suspensions, headbands, sweatbands, and any accessories should be visually inspected daily for signs of dents, cracks, penetration, or any other damage that might reduce the degree of safety originally provided.

Users are cautioned that if unusual conditions occur (such as higher or lower extreme temperatures than described in the standards), or if there are signs of abuse or mutilation of the helmet or any component, the margin of safety may be reduced. If damage is suspected, helmets should be replaced or representative samples tested in accordance with procedures contained in ANSI Z89.1-1986. This booklet references national consensus standards, for example, ANSI standards, that were adopted into OSHA regulations. Employers are encouraged to use up-to-date national consensus standards that provide employee protection equal to or greater than that provided by OSHA standards.

Helmets should not be stored or carried on the rear window shelf of an automobile because sunlight and extreme heat may adversely affect the degree of protection.

FOOT PROTECTION

Safety shoes or boots will provide employees both impact and compression protection. Where necessary, safety shoes can be obtained which provide puncture protection.

Safety footwear is required for employees who regularly handle solid objects weighing 15 pounds or more which can fall on their toes.

For protection of feet and legs from falling or rolling objects, sharp objects, molten metal, hot surfaces, and wet slippery surfaces, workers should use appropriate footguards, safety shoes, or boots and leggings.

Aluminum alloy, fiberglass, or galvanized steel footguards can be worn over usual work shoes, although they may present the possibility of catching on something and causing workers to trip. Heat-resistant soled shoes protect against hot surfaces like those found in the roofing, paving, and hot metal industries.

Safety shoes should be sturdy and have an impact-resistant toe. In some shoes, metal insoles protect against puncture wounds. Additional protection, such as metatarsal guards, may be found in some types of footwear. Safety shoes come in a variety of styles and materials, such as leather and rubber boots and oxfords.

Employees who work around exposed electrical wires or connections need to wear metal-free non-conductive shoes or boots. Rubber or synthetic footwear is recommended when working around chemicals. Avoid wearing leather shoes or boots when working because chemicals can eat through the leather right to your foot.

Safety shoes or boots with impact protection will be required for carrying or handling materials, such as packages, objects, parts or heavy tools, which could be dropped and for other activities in which objects might fall onto the feet.

Safety shoes or boots with compression protection will be required for work activities involving skid trucks (manual material handling carts) around bulk rolls (such as paper rolls) and around heavy pipes, all of which could potentially roll over an employee's feet.

Safety shoes or boots with puncture protection will be required where sharp objects such as nails, wire, tacks, screws, large staples, scrap metal, etc., could be stepped on by employees, causing a foot injury.

Some occupations (not a complete list) for which foot protection should be routinely considered are shipping and receiving clerks, stock clerks, carpenters, electricians, machinists, mechanics and repairers, plumbers and pipe fitters, structural metal workers, assemblers, drywall installers, packers, wrappers, craters, punch and stamping press operators, sawyers, welders, laborers, freight handlers, gardeners and grounds-keepers, timber cutting and logging workers, stock handlers and warehouse laborers.

HAND PROTECTION

Hand protection is required for employees who are exposed to hazards such as those from cuts, abrasions, burns, and skin contact with chemicals that are capable of causing local or systemic effects following dermal exposure. OSHA is unaware of any gloves that provide protection against all potential hand hazards, and commonly available glove materials provide only limited protection against many chemicals. Therefore, it is important to select the most appropriate glove for a particular application and to determine how long it can be worn and whether it can be reused.

It is also important to know the performance characteristics of gloves relative to the specific hazard anticipated; e.g., chemical hazards, cut hazards, flame hazards. These performance characteristics should be assessed by using standard test procedures. Before purchasing gloves, the gloves must meet the appropriate test standard(s) for the hazard(s) anticipated.

Employers need to determine what hand protection their employees need. The work activities of the employees should be studied to determine the degree of dexterity required; the duration, frequency, and degree of exposure to hazards, and the physical stresses that will be applied.

★ Fingers, hands, and arms are injured more often than any other parts of the body. Be especially careful to protect them by wearing the proper hand protection.

Gloves are the most common protectors for the hands. When working with chemicals, gloves should be taped at the top, or folded with a cuff to keep liquids from running inside your glove or on your arm.

Vinyl, rubber, or neoprene gloves are sufficient when working with most chemicals. However, if you work with petroleum-based products, a synthetic glove will be needed.

Leather or cotton knitted gloves are appropriate for handling most abrasive materials. Gloves reinforced with metal staples offer greater protection from sharp objects.

It is dangerous to wear gloves while working on moving machinery. Moving parts can easily pull your glove, hand, and arm into the machine.

Do not wear metal-reinforced gloves when working with electrical equipment.

As long as the performance characteristics are acceptable, in certain circumstances it may be more cost effective to regularly change cheaper gloves than to reuse more expensive types.

With respect to selection of gloves for protection against chemical hazards, the toxic properties of the chemical(s) must be determined.

Generally, any "chemical resistant" glove can be used for dry powders.

For mixtures and formulated products (unless specific test data are available), protective gloves should be selected on the basis of the chemical component with the shortest breakthrough time because it is possible for solvents to carry active ingredients through polymeric materials.

Employees must be able to remove the gloves in such a manner as to prevent skin contamination.

RESPIRATORY PROTECTION   *USE APPROPRIATE MASK WHEN NEEDED.*

Feasible engineering controls are the primary measures used to control employee exposure to harmful dust, fog, fumes, mists, gasses, smokes, sprays, or vapors. Such engineering controls include, but are not limited to, enclosures and confinement, general and local ventilation, and substitution of less toxic materials.

When effective engineering controls are not feasible, or while they are being instituted, appropriate respirators are used as specified by the following requirements. Applicable and suitable respirators are provided when necessary to protect employee health.

A respiratory protection program has been established and is properly maintained to protect employees from atmospheric contamination and hazards. Key elements of the program include

- A written standard operating procedure governing the selection and use of respirators.
- Selection of respirators based on hazardous exposure.
- Instruction and training of users concerning proper respirator use and their limitations.
- Regular cleaning and disinfection of respirators, and thorough cleaning and disinfection before use by another employee.
- Respirators stored in a convenient, clean, and sanitary location.
- Routine inspection of respirators during cleaning, and replacement of worn or deteriorated parts. Respirators for emergency use such as self-contained breathing apparatus, are thoroughly inspected at least monthly and after each use. Records are maintained of these inspections.
- Work areas are routinely surveyed to review work area conditions and degree of employee exposure or stress.
- Regular inspections and evaluations are conducted to determine continued program effectiveness. A formal annual evaluation is conducted and a written report prepared.
- A determination must be made and recorded that employees are physically able to wear respiratory protection and are able to perform the work and use the equipment prior to assigning them to wear respirators.
- The respirator user's medical status is reviewed at least annually.
- Only approved respirators are worn which provide adequate respiratory protection against the particular hazard. Recognized authorization for respirator approval includes ANSI and NIOSH.

Procedures have been written covering the safe use of respirators in dangerous atmospheres that might be encountered in normal operations or in emergencies. These procedures are located in the work areas where respirators are used, and employees have been informed of the procedures and the available respirators.

At least one additional person is required to be present in areas where the wearer, with respirator failure, could be overcome by a toxic or oxygen-deficient atmosphere. Communications, including visual, voice, or signal line, are maintained between the respirator user and the attendant. Plans are provided such that one individual will be unaffected by any likely incident and will have the proper rescue equipment necessary to assist the others in an emergency.

When self-contained breathing apparatus or hose masks with blowers are used in atmospheres immediately dangerous to life and health (IDLH), an attendant is required outside the work area with suitable rescue equipment.

Persons using air line respirators in IDLH atmospheres are equipped with a safety harness and safety lines for lifting or removing them from the hazardous atmosphere or other equivalent provisions for rescue used. The attendant(s) and/or standby person shall have suitable self-contained breathing apparatus and be stationed at the nearest fresh air base for emergency rescue. All confined space entry and rescue must comply with OSHA standard 1910.146.

Respiratory Protection Inspections

Frequent random respiratory protection inspections are conducted by the respiratory protection program coordinator to ensure that respirators are properly selected, used, cleaned, and maintained.

All respirators are routinely inspected before and after use by the user to ensure they meet their original effectiveness. Any defects or possible defects detected are reported to supervision so the necessary evaluations and maintenance can be performed prior to reuse.

Respirators not routinely used but kept ready for emergency use are inspected after each use and at least monthly to ensure they are in satisfactory working condition. A record is maintained of these inspections showing the date of the inspection and findings. Self-contained breathing apparatus are inspected monthly to ensure

- The breathing air cylinder is fully charged according to the manufacturer's instructions
- The regulator and warning devices function properly
- Connections are tight
- Face piece, headband, valve, connecting tubes, and canister are in good condition
- Rubber or elastomer parts are pliable, and not deteriorated, and are kept pliable by massaging to prevent a set during storage

Education and Training

Supervisors and employees are properly instructed by competent persons in the selection, use, and maintenance of respirators. During the training program, respirator users are provided an opportunity to handle the respirator, have it fitted properly, test its face piece-to-face seal, wear it in normal air for a long familiarity period, and wear it in a test atmosphere.

Fitting and Maintenance of Respirators

Every respirator wearer receives fitting instructions including demonstration and practice in how the respirator should be worn, how to adjust it, and how to determine if it fits properly.

Respirators must not be worn when conditions prevent a good face seal including growth of a beard, sideburns, a skull cap projecting under the face piece, temple pieces on glasses, or absence of dentures.

Worker diligence in observing respirator fit factors is evaluated by periodic checks. Also, the respirator wearer has been instructed to check the respirator face piece fit each time the respirator is put on as prescribed by the manufacturer instructions.

A respirator maintenance and care program is provided which covers the type of operations, working conditions, and hazards involved. The program includes

- Inspection for defects (including leak checks)
- Cleaning and disinfecting
- Repair
- Storage

Routinely used respirators are collected, cleaned, and disinfected as frequently as necessary to ensure proper wearer protection. Emergency use respirators are cleaned and disinfected after each use.

Respirator replacement and repairs are performed with parts designed for the respirator only by authorized and experienced persons approved by the respiratory protection program coordinator and per the manufacturer's recommendations. Reducing or admission valves or regulators are returned to the manufacturer or to a trained technician for adjustment or repair. Trained technicians must be authorized by the respiratory protection program coordinator to perform repairs.

Respirators are stored to protect against dust, sunlight, heat, extreme cold, excessive moisture, or damaging chemicals. Routinely used respirators may be placed in plastic bags.

Emergency respirators placed at stations and in work areas for quick accessibility are stored in special compartments built for that purpose. These compartments are clearly marked.

Storage of respirators in lockers or tool boxes are prohibited unless they are in carrying cases or cartons.

LIMITATIONS OF PPE

You should know the limitations of your PPE. It won't protect workers from everything. Employees must follow all the safety rules in your workplace and know how personal protective equipment fits into your company's safety program.

Find out the limitations of your equipment. For example, gloves may protect from the chemicals you work with, but may dissolve if they come into contact with chemicals used in the shop next door.

It is the employer's responsibility to teach workers what personal protective equipment is needed. However, it is the employee's responsibility to wear it.

PPE must be used correctly to protect your employees. Learn how to use PPE, but most importantly, use it.

Personal protective equipment can be effective only if the equipment is selected based on its intended use, employees are trained in its use, and the equipment is properly tested, maintained, and worn.

PERSONAL PROTECTIVE EQUIPMENT SAFETY CHECKLIST

This personal protective equipment (PPE) safety checklist will help employees and supervisors follow safety practices. This list is not meant to be comprehensive or form part of any official self-assessment practice. Relevant references are noted after each question.

Responsibilities	Yes	No
Is proper PPE provided, used, and maintained in a sanitary and reliable condition?	_____	_____
Is PPE used to protect personnel against hazards after engineering and administrative controls have been applied?	_____	_____
Do supervisors inform each employee of the hazards to which they might be exposed, the PPE required, and how to use and maintain PPE?	_____	_____
Do supervisors ensure that PPE is readily available, clearly identified, used for the purpose intended, fits correctly, and is properly worn?	_____	_____
Does the employee PPE not interfere with the task?	_____	_____
Do employees ensure that the PPE is cleaned, repaired, stored, and disposed of as applicable?	_____	_____
Do personnel who wear contact lenses notify their supervisors of this fact, and do supervisors know which employees wear contact lenses?	_____	_____

Figure 9.1 PPE Checklist

Head Protection	Yes	No
Do employees wear safety hats when required?	_____	_____
Are safety hats with any defects or damage immediately removed from service?	_____	_____
Are personnel who work around chains, belts, rotating devices, suction devices, and blowers required to cover long hair?	_____	_____

Eye and Face Protection	Yes	No
Is protective eye and face equipment required and provided where there is a reasonable probability of injury that can be prevented by such equipment?	_____	_____
Does the facility provide proper eye and face protection for the work to be performed?	_____	_____
If needed, do employees use eye and face protection?	_____	_____
Does eye protection meet the following minimum requirements:		
Provide adequate protection against the particular hazards for which they are designed?	_____	_____
Fit snugly and not unduly interfere with the movements of the wearer?	_____	_____
Look durable, clean, and in good repair?	_____	_____

Are employees provided with goggles or spectacles of one of the following methods when their vision requires corrective lenses or spectacles?	Yes	No
Spectacles whose protective lenses provide optical correction?	_____	_____
Safety goggles that can be worn over corrective spectacles without disturbing the adjustment of the spectacles?	_____	_____
Safety goggles that incorporate corrective lenses mounted behind the protective lens?	_____	_____
Is eye protection distinctly marked to identify the manufacturer?	_____	_____
Are face shields used as primary eye and face protection in areas where splashing or dust, rather than impact resistance, is the problem?	_____	_____

Figure 9.1 PPE Checklist, continued

Is it prohibited for personnel to wear contact
lenses when the work environment exposes the
individual to toxic or chemical fumes and vapors,
splash hazards, intense heat, molten metal, or highly
particulate atmospheres?   _____ _____

Is it prohibited for employees who work on or near
energized electrical circuits or in flammable/
explosive atmospheres to wear conductive frame eye/
face protection?   _____ _____

Hand Protection

Is proper hand protection required for employees
whose work may involve hand injury or impairment?   _____ _____

Are sleeves worn outside glove gauntlets when
caustic substances are being poured?   _____ _____

Foot and Leg Protection   Yes   No

Is protective foot wear provided and worn when
needed?   _____ _____

Are requirements for protective footwear determined
by supervision with assistance from safety?   _____ _____

Are conductive shoes provided to protect employees
against the buildup of static electricity?   _____ _____

Are electrical-hazard safety shoes provided for
employees exposed to electrical hazards?   _____ _____

Are spark-resistant shoes provided in working
areas for explosive storage and petroleum tank cleaning?   _____ _____

Light Reflective Products   Yes   No

Are traffic directors and others exposed to
vehicular traffic equipped with reflective vests or
light-reflective clothing?   _____ _____

Safety Belts, Lifelines, and Lanyards

Are lifelines, safety belts, and lanyards used only
for their intended purpose?   _____ _____

Are lifelines secured above the point of operation
to an anchorage or structural member capable of
supporting a minimum dead weight of 5,400 pounds?   _____ _____

Figure 9.1 PPE Checklist, continued

Are lifelines that are subjected to cutting or abrasion a minimum of 7/8-inch (2.2-centimeter) wire core manila rope? _____ _____

Are safety belt lanyards a minimum of 0.5-inch nylon, with a maximum length to provide for a fall of no greater than six feet (1.8 meters)? _____ _____

Does the cushion part of the body belt meet the following conditions?

                                                                            Yes   No

Contain no exposed rivets on the inside _____ _____

Be at least three inches (7.6 centimeters) in width _____ _____

Be at least 5/32 inch (0.39 centimeter) thick, if made of leather _____ _____

Figure 9.1 PPE Checklist, continued

10

Fall Protection

Preventing Falls at Your Worksite

INTRODUCTION

Fall hazards create the greatest exposure to injury on construction sites. These hazards include falls from elevation, falls on the same level, and being struck by falling objects. The most common hazard, falls from elevation, generally includes falls from ladders, from elevated workplaces, through floor openings, and from leading edges. Falls on the same level are usually caused by slipping and tripping hazards. Accidents involving falling objects are caused by objects that are improperly stored, disposed of, or mishandled at elevation.

If any of your employees are exposed to a fall of six feet or more, you are affected by OSHA's Fall Protection Standard (29 CFR 1926.500) that went into effect on February 6, 1995.

A fall hazard is created when a construction activity exposes an employee to an unprotected fall which may result in injury or death.

Fall Hazard Locations

To prevent fall hazards, it is necessary to recognize where falls are most likely to occur. Construction activities in the following locations require increased fall hazard awareness:

- Ladders
- Scaffolds
- Work platforms
- Form work
- Bridges
- Structural steel
- Openings
- Reinforcing steel
- Stay-in-place decking
- Cluttered/congested areas
- Excavations

Falls from any of these locations may occur while accessing, climbing, traveling, or working at elevation. Once the fall hazard locations have been recognized, the next step in the fall prevention process involves identifying the type of fall hazards that may occur at these locations.

Identifying Fall Hazards

Effectively preventing falls requires identifying, evaluating and controlling the fall hazards most likely to cause accident/injury. This effort can only be accomplished by organizing a team of participants (committee) representing each department of the construction company. This committee should be comprised of personnel representing production, engineering, estimating, and safety. These members should participate in a collaborative effort to identify, evaluate and control all fall hazards.

Interviews

The best way to identify fall hazards is to talk to the construction workers. Attempt to identify the construction activities performed at elevation and determine how those activities are carried out. By asking the right questions, one can gain a better understanding of the work and its inherent hazards. The best employees to interview are those who have the most experience working at elevation (carpenters, laborers). Their insights can be invaluable not only in recognizing fall hazards but also in identifying practical means of control. The interview process may be conducted on an individual basis or in groups at safety meetings.

Job Site Inspections

Another means of identifying fall hazards is to conduct a job site inspection of construction activities. Select both workers and their supervisors to identify elevated work locations, construction activities performed at those locations, and what is required to get the job done. Videotaping fall hazards is an excellent means of identifying and evaluating work at elevation. Taking still photos is another useful means of evaluating fall hazards. Job site inspections can also be performed by insurance carrier representatives.

FALL PREVENTION

Fall prevention is any means used to reasonably prevent employee exposure to fall hazards, either by eliminating work at elevation or using aerial lifts, scaffolds, work platforms with guardrails, or similar protection.

The key to fall prevention is to develop a plan, prior to construction, to eliminate exposure to fall hazards. Fall prevention planning involves an in-depth analysis of each construction activity that exposes employees to potential fall hazards. Based on this analysis, the contractor must select the most appropriate method of protection (elimination, prevention, or control). The primary objective during the planning phase of any construction activity should be disposing of the need for fall arrest equipment in favor of fall prevention measures that eliminate fall hazards.

Fall Prevention Measures

To be effective, fall prevention measures must be initiated in the planning and scheduling phase. The most desirable measures are those which provide an alternative approach to the work by implementing engineering controls and eliminating any fall hazard potential.

When elimination is not possible, prevention measures become necessary. Selecting the appropriate fall prevention measure(s) can only be determined after assessing the location of work and the type of construction activity performed.

Preventing falls involves providing proper access to the work location, protecting unguarded openings and leading edges, practicing good housekeeping, and engineering out hazardous exposures. The following is a list of fall prevention measures that should be implemented to avoid fall hazards:

Engineering Controls

Implement engineering controls into the design process by interacting with designers, fabricators, and material suppliers to build safety measures into the structure, material, or equipment utilized during the construction process.

Maximize the pre-assembly of structural components on the ground. Use mechanical pin extractors to disconnect rigging from the ground. Design holes and/or attachments for stanchions, lifelines, and retractable devices on structural components to permit assembly on the ground and provide protection at elevation.

Determining Alternative Approaches to the Work

Question whether the construction activity can be performed alternatively without employee exposure to a fall hazard.

Use radios to signal cranes so that employees can remain on the ground rather than signaling unprotected from atop walls, forms, and rebar.

Providing Proper Access

Providing proper access includes correct installation and proper use of ladders, scaffolds, stair towers, and stairways. Attach fall arrest systems (lifelines) to bridge steel and form work before erection to accommodate safe access.

Providing Guardrail Protection

Designate work locations requiring guardrail protection. Include

- Elevated work platforms (such as on form work)
- Scaffolds
- Openings/holes in bridge decks, floors, or other unprotected surfaces
- Unprotected sides of ramps/stairways/platforms

Additional fall prevention measures include using elevating equipment, performing work on the ground, practicing good housekeeping, protecting openings/holes, isolating areas below elevated work, and discussing fall prevention measures with employees. All of these measures are intended to prevent the fall before its occurrence.

FALL PROTECTION

Fall protection involves using personal fall arrest equipment to prevent the completion of a fall and to reduce the possibility of resulting injuries.

Fall protection measures are taken when employee exposure to a fall hazard cannot be eliminated by using fall prevention measures. For example, climbing forms or rebar requires complete and continuous protection. Sometimes the only way to provide this protection includes using a self-retracting device anchored overhead and attached to the employee. This fall arrest system protects an employee while climbing by preventing a complete fall.

Fall Arrest Systems

A fall arrest system includes the proper anchorage, body support (harness/belt), and connecting means (lanyards/lifelines) interconnected and rigged to arrest a free fall. The primary function of a fall arrest system is to minimize the consequences of a fall rather than prevent its occurrence. The use of fall arrest equipment should be recognized as a means of minimizing injuries sustained from a fall. It does not prevent the fall.

A fall arrest system could be as basic as a safety belt connected to a six-foot lanyard attached to a safe anchorage point. A more intricate system may include a vertical lifeline attached above the employee to an independent anchorage point and a rope grab connected to the vertical lifeline with a lanyard. The lanyard is then attached to a harness, making a complete fall arrest system. This system is typically used on a two-point suspension scaffold operation such as for finishing concrete walls.

A variety of fall arrest equipment is available to establish an effective fall arrest system. This equipment includes the use of full body harnesses, safety belts, lanyards, lifelines, retractable devices, rope grabs, safety nets, and associated hardware. The various components connected together make up a fall arrest system. Compatibility of these components is the key to their effectiveness.

Fall arrest systems are useful but are the least desirable fall protection method and should only be used as a last resort.

Selecting Proper Equipment

Establishing an effective fall arrest system begins with selecting the proper personal fall arrest equipment to protect the employee in the event of a fall.

The proper equipment should be selected based on

- The task being performed
- Requirements for worker mobility
- The number of employees requiring protection
- The distance of potential fall
- The presence of concrete, dirt, moisture, solvents, and other environmental factors

The proper selection and purchase of safety equipment does not constitute a fall protection program. This is only one step. A complete, effective program requires that 1) appropriate anchorage points are established; 2) proper inspection and maintenance procedures are enforced; and 3) workers are trained and supervised in the proper use and application of all equipment.

COMPONENTS OF A FALL ARREST SYSTEM

The purpose of a fall arrest system is two-fold: 1) to stop the fall, and 2) to distribute the impact energy experienced during fall arrest. Properly installed and used, fall arrest systems can prevent or minimize possible injury. The three basic components of a fall arrest system include

- Anchorage—Component and Structure
- Body Support—Full Body Harness, Safety Belt
- Connecting—Lanyards, Lifelines, Devices

A fall arrest system requires identifying the proper anchorage, wearing the appropriate body support, and using a connecting means capable of attaching the worker to the anchorage. This system is designed to control a free fall and minimize injury. A critical factor in the design of any fall arrest system is ensuring that all components (anchorage, body support, and correcting means) are compatible.

Anchorage

An anchorage is a secure point of attachment for lanyards, lifelines, or deceleration devices capable of withstanding the anticipated forces applied during a fall. Anchorage planning is the key to designing fall arrest systems.

The anchorage point should be positioned on an independent structure and used for securing a lifeline or lanyard. An anchorage point should be located above the worker to avoid unnecessary swing in the event of a fall. The anchorage point should be capable of supporting 5,400 pounds minimum strength for fall protection systems allowing free falls up to six feet. Alternatively, retracting lifelines permitting free falls of two feet or less require anchorage points capable of supporting only 3,000 pounds.

Anchorage points must be engineered by a qualified person. This individual must be capable of determining the required strength, location, and design of the selected anchorage to meet the requirements of the construction activity. Each anchorage point must be carefully planned into the job to provide continuous and complete protection during the work task.

Selecting anchorage points requires evaluating the following characteristics:

- Strength. The strength of an anchorage point is its most important characteristic because failure of any anchorage is likely to result in an unprotected fall. The required strength for a fall arrest system ultimately depends on the potential forces applied and the integrity of the anchorage component selected.
- Independence. Anchorage points for fall arrest systems should be independent of the working platform and its anchorage.
- Height. The primary consideration in determining anchorage point height is to minimize free fall to the shortest distance possible. The shorter the free fall, the less impact force experienced during fall arrest.
- Clearance. The total fall distance must be determined to ensure the height and location of the anchorage is sufficient to prevent collision injury with the ground or other objects.
- Identification. Anchorage points must be identified by a qualified person. Employees should be educated about what is (and what is not) considered acceptable anchorage. When practical, anchorages should be labeled by painting the approved locations so workers know exactly where to secure for proper anchorage.

Anchoring Devices/Points	Non-Anchorages
~ Structural Members	~ Railings
~ Anchors/Fasteners	~ Guardrails
~ Eyebolts	~ Rungs
~ Imbeds	~ Pipe Vents
~ Turnbuckles	~ C-Clamps
~ Shackles	~ Bolt Holes
~ Slings	
~ Retractables	
~ Cross Arm Straps	

Body Support

Body support is the means of fall protection worn by an employee to minimize the consequences of a fall. Body support can take the form of either a body (safety) belt or a full body harness. Body supports are equipped with D-rings for the attachment of a connecting means, such as a lanyard or retractable device. The location of these attachment points is designed to meet the anticipated use of the body support. Specifically, full body harnesses are used to provide fall arrest while body belts provide work positioning and fall restraint protection.

Proper selection of a body support is determined by its anticipated use in conjunction with the construction activity or a fall arrest system. The two types of body supports, the full body harness and the body belt, have different uses and associated hazards.

The full body harness is preferable over the body belt because of its ability to distribute the forces over the body as opposed to the vulnerable mid-section, thereby minimizing injury. Additionally, the harness has a sliding D-ring which helps avoid excessive whipping of the neck during fall arrest. Another advantage of harness use involves permitting prolonged suspension in an upright position. This makes rescue easier and allows an employee to be suspended for prolonged periods of time (more than 30 minutes) without breathing restrictions or extreme discomfort. These advantages make the full body harness the most desirable body support.

Connecting Means

Connecting means secure an employee to an anchorage and include the use of lanyards, lifelines, and devices such as retractables and rope grabs. Selecting the appropriate connecting means requires matching the capability of the fall arrest equipment with the requirements of the work task. For example, where an employee is erecting form work, the work task requires vertical and horizontal mobility while climbing the forms. The appropriate connecting means in this instance might include a retractable device which moves along a horizontal lifeline. However, if the employee is finishing a concrete wall from a two-point suspension scaffold, then a rope grab attached to an independent lifeline would provide the necessary connecting means.

The key to selecting connecting means is determining the workers' mobility during the performance of the work task and providing the most practical means of protection. Whether to use a lanyard, lifeline, or device is dictated by the nature of the work. This choice is critical to controlling a fall hazard.

Most fall arrest systems utilize one or more connecting means, including lifelines (horizontal and vertical), lanyards, retractable devices, rope grabs, hardware (snap hooks, d-rings, or shackles).

Lifelines. Lifelines may be either vertical or horizontal lines.

A vertical lifeline extends from a fixed overhead anchorage point to the ground and is attached to a lanyard, usually by a rope grab. The lifeline should be long enough to reach the ground and prevent the rope grab from running off of it. Vertical lifelines are commonly used on suspended scaffolds and/or where the construction activity requires only vertical mobility. On suspension scaffolds, using rope grabs, shock absorbing lanyards, and full body harnesses is advisable.

Horizontal lifelines serve as anchoring lines that are rigged between fixed anchor points located at the same elevation. The purpose of a horizontal lifeline is to minimize potential swing falls by providing a constant overhead support. Horizontal lifelines serve as anchoring lines that are rigged between fixed positions and move horizontally. The lifeline serves as an attachment point for lanyards or retractable devices. Horizontal lifelines are commonly using during bridge steel erection and other construction activities where horizontal mobility is required. These lifelines should be positioned above the waist and designed by a qualified person knowledge-able in fall arrest loads, anchorage needs, and other erection requirements.

Inspection of vertical and horizontal lifelines for chafing, cuts, abrasions, and other defects is critical for longevity and operating efficiency. Synthetic lifelines of 5/8-inch nylon, polyester, or equivalent capability are preferable. Wire rope lifelines should be a minimum one-half inch extra improved plow-steel with a minimum breaking strength of 5,400 pounds.

Some common hazards associated with lifeline use include:

- Improper installation of the lifeline
- Insufficient anchorage strength
- Overloading the lifeline with employees
- Improper positioning of the lifeline
- Lack of protection over sharp edges
- Knotting of lines at anchorage points (which can reduce lifeline capacity in rope by 50 percent)

Lanyards. A lanyard connects a harness to an anchorage point or to a device such as a retractable. A lanyard is a short, flexible rope or strap webbing. OSHA requires at least one-half-inch diameter synthetic rope to be used as a lanyard.

A lanyard is designed to permit limited freedom of movement during work (two to six feet). Minimum attachment height should be at or above the level of the D-ring to ensure the free fall distance will be less than six feet. The worker should be exposed to as little slack as possible to limit free fall distances.

Shock absorbing lanyards can be used to reduce the impact resulting from a fall and should be incorporated into the fall arrest system, especially when safety belts are used.

Lanyards are commonly used during work from aerial lifts, two-point suspension scaffolds, and other fall hazard locations where movement is limited. Lanyards are sometimes used to restrain a worker from approaching an unprotected opening or a leading edge.

Some common hazards associated with lanyard use include:

- Disregard for manufacturer's instructions
- Incorrect length
- Chokering the lanyard back onto itself
- Tying knots to shorten the lanyard
- Looping the lanyard over sharp edges
- Weakening due to exposure to concrete, solvents, acids, or welding/cutting operations

Retractable Devices. A retractable device attaches the body support (preferably a harness) to a lifeline or anchorage point. The retractable device functions as a deceleration device and features a locking mechanism designed to instantaneously arrest a free fall. The device contains a self-retracting lifeline (cable, rope, webbing) which extends and automatically retracts as the worker climbs up or down. The retractable device operates on the same principle as an automobile seat belt. The device is activated the moment a fall occurs, limiting the workers' free fall to approximately 18 inches and thereby reducing the forces applied at impact.

Retractable devices are used during climbing operations or in conjunction with horizontal lifeline systems. These devices are useful for climbing protection on form work, reinforcing steel, structural steel, and pile driving leads. Retractables should be installed minimally above shoulder height at the highest point of climb and can be accessed by a tag line. For example, retractable devices can be attached to form work on the ground. Then when the form is raised into position the employee can use a tagline to access the device's cable and hook up at ground level for continuous protection while climbing.

Some hazards associated with the use of retractable devices include:

- Cable kinking and accumulating dirt/concrete
- Failure to properly maintain and service the device according to the manufacturer's recommendations
- Improperly using or installing the device
- Not removing the device from service once a fall has been arrested

Rope Grabs. The rope grab connects the lanyard to the lifeline and allows employees vertical movement. Rope grabs are designed to arrest a fall mechanically, bringing the worker to a complete stop. Rope grabs may be manually operated or travel freely on a lifeline. They should be located at the highest possible elevation on the lifeline to minimize the fall distance. A shock absorbing lanyard (three feet maximum) can be used with the rope grab to increase lateral movement and minimize the arresting forces on the employee and fall arrest system.

Rope grabs are typically used on two-point suspension scaffold operations such as finishing concrete walls and during pile driving operations which require the use of bosun-chairs.

ROPE SHOULD CONNECT TO GROUND.

Some common hazards associated with rope grab use include:

- Incompatible lifelines, snap hook attachments, or lanyards
- Worn locking arms or other mechanical parts
- Using the rope grab upside down

Hardware. Hardware consists of snap hooks, D-rings, carabiners, shackles, and other rigging components used to connect the various components of a fall arrest system.

Snap hooks are a self-closing connecting device with a gatekeeper latch that remains closed until manually released. They are part of a lanyard or device which are commonly attached to a D-ring. Double-locking snap hooks should be used to eliminate the risk of "roll-out." Roll-out occurs when the snap hook disengages at the D-ring by a twisting force.

D-rings are attachment points on the harnesses or the safety belt for a lanyard or device.

A carabiner is an oblong snap hook which is commonly secured to an anchorage to support a device such as a retractable. Shackles and similar rigging hardware are also commonly used as connectors.

All hardware must be compatible and properly used and maintained to ensure maximum efficiency as a fall protection device.

DEVELOPING A FALL PROTECTION PROGRAM

Developing a fall prevention program is initiated by establishing a committee comprised of managers, field supervision, labor, engineering personnel, and a safety representative. This committee should work in a collaborative effort to develop the essential elements of a program and become familiar with the methods of protection and available technology for controlling fall hazards.

Once these objectives have been achieved, a fall hazard awareness program should be initiated by providing education and training.

The actual development of a fall prevention program involves the following six steps:

> Step 1: Write Policy/Define Scope. An effective fall protection program begins with a policy statement that expresses the intent of the company regarding all elevated work.

Policy

Employment at (Company Name) job sites shall be free from all recognized fall hazards.

This statement should be signed by the president of the company.

Scope

All employees exposed to a potential free fall greater than _____ feet are required to be protected. This _____-foot requirement applies to all openings, leading edges, work platforms, and surfaces under construction. Emphasis is also placed on providing ground-level fall protection and reducing the potential of falling objects.

The key to an effective fall prevention work plan is establishing a minimum height for protection, determining the appropriate method of protection, and selecting the proper fall arrest equipment.

> Step 2: Identify Fall Hazards. Identification of job site fall hazards begins by analyzing construction operations requiring elevated work (e.g., building concrete walls, placing structural steel). The next step involves defining each elevated work task required to perform that construction operation. The final step involves identifying the specific fall hazards which can occur during each work task as well as during access and egress at the work location.

The key to identifying fall hazards is breaking down the construction operation into the required work tasks and listing the associated fall hazards that occur during each task.

It is also important to identify the hazards associated with access/egress at work locations and the required work task mobility.

The primary objective is to minimize potential fall hazards by controlling the most frequent work task with the greatest probability for injury.

Step 3: Determine Appropriate Methods of Protection. The three primary methods of protecting workers from a fall include

- Eliminating the fall hazard
- Preventing employee exposure
- Controlling the fall

The most critical and perhaps difficult decision in designing a fall arrest system is selecting the appropriate method of protection. The selection process for controlling fall hazards must be carried out with the guidance of qualified personnel who are experienced in a wide range of methods and equipment. Preplanning is the key to selecting and implementing the most appropriate method of protection. The fundamental principle underlying the selection process is that it is more reliable to depend on engineering controls, which provide automatic protection, than to depend on human behavior to control hazards.

Elimination. The elimination of fall hazards is the first and primary means of preventing falls from elevation. This method of protection requires careful assessment of the work to be done and how to accomplish that work safely. By evaluating the location of work and the type of construction activity, one can determine the inherent fall hazards associated with the work and its location. Once the fall hazards have been identified, practical means of eliminating the hazard must be preplanned into the construction process. This may include developing a different way of performing the work while maintaining or even enhancing productivity. The idea is to design safety into the construction process rather than address it as an afterthought.

Examples of eliminating fall hazards include:

- Performing as much work on the ground as possible to eliminate the hazard of climbing/working at elevation
- Assembling guardrail systems and fall arrest systems on form work and/or structural steel at ground level rather than at elevation

Prevention. Preventing employee exposure to fall hazards should be given consideration when a hazard cannot be entirely eliminated. This method also requires assessment of the construction activities and the hazardous work locations. Consideration must be given to planning effective, permanent means of fall prevention and to making changes to the elevated work locations. Taking these measures precludes the need to rely on the employees behavior and the use of fall arrest equipment.

Examples of fall prevention methods include guardrail systems, aerial lifts and stairs, ramps, runways.

Control. Controlling a fall is the least desirable method of protection because it minimizes the consequences of a fall rather than preventing its occurrence. It should only be considered after determining that the fall hazard cannot be eliminated or the possibility of falling prevented. This method of protection requires the use of personal fall protection equipment to prevent a complete fall and to reduce the risk of injury. Preplanning considerations should include the selection of fall arrest equipment and the proper installation and use of the equipment.

> **Step 4: Conduct Education and Training.** Education and training are the key to making employees aware of the need for and use of fall prevention and protection methods. Education provides employees with the knowledge to understand the appropriate methods of protection (elimination/prevention/control) and the capabilities and limitations of fall arrest equipment/systems. Training provides employees with a skill: namely, applying fall protection principles, techniques, and equipment operation.

Adopt an employee training program so that all employees who will be exposed to fall hazards will acquire the understanding, knowledge, and skills necessary for the safe performance of their assigned duties.

The training must be conducted in a manner that will establish employee proficiency in the duties required by the OSHA standard. It will cover the various systems and procedures we have adopted. It also will introduce new or revised procedures, as necessary, in order to accomplish full compliance with the fall protection standard.

Upon its completion, execute a written certificate that the training required by the OSHA Standard has been accomplished. The certificate will contain each employee's name, the signatures or initials of the trainers, and the dates of training. The certificate will be available for inspection by employees and their authorized representatives at the local jobsite office.

Training will be provided to each employee. We will ensure that each employee has been trained by a competent person qualified in the following areas:

- The nature of fall hazards in the work area
- The correct procedures for erecting, maintaining, disassembling, and inspecting the fall protection system to be used
- The use and operation of guardrail systems, personal fall arrest systems, safety net systems, warning line systems, safety monitoring systems, controlled access zones and other protection used
- The limitations on the use of mechanical equipment during the performance of roofing work on low-sloped roofs
- The correct procedures for handling and storage of equipment and materials

Step 5: Perform Inspection and Maintenance. Fall protection systems and equipment must be inspected and maintained regularly. Inspection and maintenance procedures apply to guardrail protection, aerial lifts, covers and barricades for openings, and the use of personal fall arrest equipment and devices including safety nets. Because these systems and equipment are designed to prevent serious or fatal injury, visual inspections before each use and regular inspection and maintenance are vital. This should be performed in accordance with the manufacturers guidelines and written company procedures.

Each worker must be instructed how to use, inspect, and maintain equipment properly. This instruction should address the following measures for using fall arrest equipment and devices:

- Understand how the equipment/device works
- Inspect it frequently
- Use it properly
- Keep it clean
- Store it properly and securely

Step 6: Administer/Audit Program. Administration of the program should be assigned to job site supervisors with designated responsibilities (e.g., who selects/distributes fall arrest equipment, who provides training, who designates anchorage points). To enhance program effectiveness, the job site superintendent should be given the authority to implement the company program, delegate responsibility, and ensure compliance.

Auditing a program involves evaluating the effectiveness of the fall prevention program. It is an opportunity to determine if previously established policy and objectives have been met and to decide what changes (if any) are required. Audits are required to continually improve any program and are based on feedback from supervisors and employees.

This is the company's Fall Protection Plan. It has been adopted to meet the requirements of 29 CFR 1926.500, the OSHA standard that was adopted on Feb. 6, 1995, in order to protect employees in the construction industry from falls whenever workers are exposed to a fall of six feet or more. All of our affected employees have been advised of this written plan. The plan will be available for inspection by our employees and their authorized representatives upon request.

This plan applies to all of our employees who find they cannot use conventional fall protection equipment. The purpose of this plan is to ensure that all our employees are protected against fall hazards while working in leading edge work, precast concrete erection work, or residential construction work. The company is committed to a safe and accident-free jobsite.

The appendix to this program contains definitions for many of the terms defined in the OSHA standard. To avoid any misunderstanding, readers should refer to these definitions.

Policy

A new written Fall Protection Plan must be developed for each site and amended to reflect the type of work being performed.

This fall protection plan is for (name of project)

_____

Location:_____

Date of Plan:_____

Prepared by:_____

Approved by:_____

Plan supervised by:_____

Figure 10.1 Sample Fall Protection Program

Controlling Fall Hazards

The company emphasizes three topics:

- Elimination of fall hazards
- Prevention of falls
- Control of falls

The following will be observed in our fall protection program:

Elimination of Fall Hazards

Elimination of fall hazards is the first and best line of defense against falls from heights. This task requires careful assessment of the workplace and the work itself. The "who, what, when, where, why, how, and how much" of each exposure is considered. Often, pre-consideration of the work and site not only leads to elimination of the hazard but also identifies alternative approaches to the work that can measurably enhance productivity. The idea is to design safety directly into the work process and not simply try to add safety as an afterthought to an inherently unsafe work procedure. Examples include but are not limited to servicing a pile hammer when laid down; backfilling abutments, walls, etc., before employees access structures; using radios for signaling instead of employees hanging over the edge giving signals; and other mechanical devices that can be controlled from the ground.

Prevention of Falls

Preventing falls is the second line of defense when fall hazards cannot be entirely eliminated. This also requires assessment of the workplace and work process. It involves making changes to the workplace so as to preclude the need to rely on the worker's behavior and personal protective equipment to prevent falls. Examples include but are not limited to use of stairs, guardrails, barriers, and travel restriction systems to prevent the worker from direct and unprotected exposure to the fall hazard. These techniques deal with preventing the fall before the onset.

Control of Falls

Controlling falls is the last line of defense. It should be considered only after determining that the fall hazard cannot be eliminated or the possibility of falling prevented. This is the domain of fall protection and calls for equipment such as safety nets or harnesses, lanyards, shock absorbers, fall arresters, lifelines, and anchorage connectors. It deals with reducing the risk of injury in falling after onset of the fall. This fall protection also necessitates workplace and work process assessment and planning in order to select the proper equipment, installation, and proper use of gear.

Figure 10.1 Sample Fall Protection Program, continued

Accident/Incident Record Review

Accident/incident records that give a description of how an accident or incident occurred can be helpful. You may find the work and condition that led to a previous fall is still being performed in the same way at that same location. Or, you may realize that this work is being performed at other locations in the same hazardous way.

Canvas Surveys

Canvas surveys have the advantage of being able to obtain information from a large number of workers relatively easily. Although surveys sometimes give incomplete information, they may also reveal a lack of awareness in the work force which is useful to know in planning future training, instruction, and warning steps.

Interviews

The best way to identify fall hazards is to talk to the workers themselves. Sometimes the worker will not actually recognize the hazard and may not appreciate the risk of injury or the likelihood of the fall taking place, but they will know the work that they do at heights and how they do it. Consequently, we can access their knowledge by asking the right questions.

Fall Hazard Inspection Surveys

Another effective way to identify hazards is to invite experienced workers to assist with a walk-through tour of operations. Workers and their supervisors can point out the various places they have to work and can explain what they do to get the job done.

Evaluation of Fall Hazards

The first consideration in the process of evaluating the list of fall hazards is to analyze and determine the likely consequence of a fall for each hazard. Segregate the hazards most likely to result in death or serious physical harm from those less likely to result in death or serious injury.

The next consideration is the probability or likelihood of an accident occurring. After selecting hazards which will likely result in death or serious physical harm, segregate those most likely to occur from those least likely to occur. Some of the factors affecting probability are

Workers who must traverse or perform their work at the edge or within three feet of the edge of the fall hazard.

Workers traversing or working on ice, snow, oily surfaces, surfaces with trip hazards near the edge, and surfaces not recently inspected for capacity verification.

Figure 10.1 Sample Fall Protection Program, continued

Workers who must push or pull tools or material are more likely to lose their balance and fall, as are workers who cannot maintain three-point contact (two feet and one hand or two hands and one foot).

Dangerous Work

Dangerous work is that done by workers who must maintain good balance while walking I-beams; workers who must jump across floor openings or across edges; workers exposed to high winds; workers in poorly lit areas; and workers who work over water.

Exposure Time

The longer a worker is exposed to a hazard, the greater the likelihood of an accident occurring. Thus, exposure time is a function of the frequency of exposure, the duration of the exposure, and the number of workers exposed.

Feasibility of Effective Controls

Some hazards can be simply and inexpensively eliminated or controlled. Other hazards require relatively large expenditures and greater difficulty from an engineering and design point of view.

In any event, the goal is to provide the greatest amount of protection in the shortest amount of time. We can do this by first isolating the hazards most likely to result in death or serious injury, as well as those easiest to eliminate or prevent falls. However, always keep in mind that hazards of less than six feet can, and have, resulted in death, paralysis, brain damage, etc. Also, hazards which may appear to be less likely to result in accidents can produce more than their share of injury. Unfortunately, there is no simple way to predict with certainty where and how an accident will occur. Therefore, we must establish plans to eliminate, prevent or control to the extent feasible all fall hazards which have been identified.

Figure 10.1 Sample Fall Protection Program, continued

11

Excavation, Trenching, and Shoring[1]

Avoiding Cave-Ins and Other Accidents

Excavating is recognized as one of the most hazardous construction operations. Employees in excavations must be protected from cave-ins, except when the excavation is in stable rock or less than five-feet deep and an examination by a competent person provides no evidence that a cave-in should be expected. Employees in excavations must also be protected from falling rock, soil, or material by use of an adequate system including scaling for loose rock or soil, installation of protective barricades, or other equivalent protection.

Material or equipment which might fall or roll into an excavation must be kept at least two feet from the edge of the excavation, or have retaining devices, or be prevented from falling-in with a combination of precautions.

Daily inspections of excavations, adjacent areas, and protective systems should be made by a competent person and may be necessary after every rainstorm and during periods of freeze/thaw cycles. If there is evidence of possible cave-in, failure of protective systems, hazardous atmospheres, or other hazardous conditions, employees shall be removed from danger until the necessary precautions have been taken. Daily inspections must include a visual and manual test.

EXCAVATION, TRENCHING, AND SHORING PROGRAM

This is *[Company Name]*'s Excavation, Trenching and Shoring Program. _____ is responsible for implementing this program and to ensure the safety of employees who work in or around excavations as part of their job duties. This program is designed to protect employees who work or travel in the vicinity of excavations.

This program establishes guidelines and procedures for use on excavations and compliance with OSHA Excavation Standard at 29 CFR 1926.650. This program applies to all work involving excavations. Work associated with electric power generation, transmission and distribution systems, and fall hazards are covered under the Fall Protection Program.

The program will cover the following:

- Employees who work in or around excavations must be provided training according to their work.

- The excavation or trench must either be sloped or supported as required to comply with OSHA requirements.
- Traffic around the site must be controlled, and barricades, signs, and/or flag persons must be used as needed to control both vehicular and pedestrian traffic.
- Utilities on the site must be protected and suitable precautions taken if any utility will be disturbed by the work.
- Employees must use required personal protective equipment (PPE).
- Each department covered by this program must appoint one or more competent person(s) to ensure compliance with this program.

Excavation Responsibilities

_____ will monitor the overall effectiveness of the program through audits and annual reviews. _____ can also assist with atmospheric testing, other technical assistance, or equipment selection as needed. _____ will also provide or assist with arranging site worker training and competent person training. _____ will conduct an annual audit of the program and will maintain records relating to training and audits.

Each division that conducts work involving excavations must designate a department Competent person and ensure that these persons receive training as such. The department must ensure that a Competent person performs the responsibilities as described in this program.

Competent Person

Each competent person is responsible for ensuring that procedures described in this program are followed including employee training, personal protective equipment usage, site inspections, tests, and recordkeeping.

Each employee has the responsibility to follow established procedures and enter an excavation only after receiving training. Employees must demonstrate a complete understanding of the safe work practices to be followed while working in an excavation.

Contractors performing excavation work on the property must coordinate their work to ensure that related activities such as utility shutdown are addressed.

In order to comply with this program:

1. An initial inspection to evaluate site hazards must be conducted by a competent person.
2. All hazards must be eliminated, controlled, or employees must be provided appropriate personal protective equipment.
3. Protection must be provided against potential cave-ins using either sloping or benching systems or support systems.
4. Inspections must be conducted daily or as conditions occur that may affect or create hazards.

Training

All personnel involved in excavation work must be trained in accordance with the requirements of this program. Training must be provided before the employee is assigned duties.

Retraining will be provided every three years or as necessary to maintain knowledge or skills to safely work within or in the vicinity of excavations.

Personnel who conduct work within or in the vicinity of excavations must receive training prior to beginning work at the site.

The training must include:

- Requirements of the OSHA Excavations standard
- Requirements of the Excavation Safety Program
- Work practices
- Hazards relating to excavation work
- Methods of protection for excavation hazards
- Use of Personal Protective Equipment
- Procedures regarding hazardous atmospheres
- Emergency and non-entry rescue procedures hazards

Underground Installations

The location of sewer, telephone, fuel, electric, and water lines as well as any other underground installations that may be encountered during excavation work must be located and marked prior to opening the excavation. The Competent person must make arrangements as necessary with the appropriate utility agency for the protection, removal, shutdown, or relocation of underground installations.

If it is not possible to establish the exact location of underground installations, the work may proceed with caution, provided detection equipment or other safe and acceptable means (e.g., using hand tools) are used to locate the utility as the excavation is opened and each underground installation is approached.

Excavation work will be conducted in a manner that does not endanger underground installations or employees engaged in the work. Utilities left in place must be protected by barricades, shoring, suspension, or other means as necessary to protect employees.

Access and Egress

Stairs, ladders, or ramps must be provided where employees are required to enter trench excavations four feet or more in depth. The maximum distance of travel in an excavation to a means of egress shall not exceed 25 feet.

Vehicular Traffic

Employees exposed to vehicular traffic must be provided with and wear warning vests or other suitable garments marked with or made of reflectorized or high-visibility material. Warning vests worn by flagmen must be red or orange and be of reflectorized material if worn during night work.

Falling Loads

No employee will be permitted underneath loads handled by lifting or digging equipment. Employees will be required to stand away from any vehicle being loaded or unloaded. Vehicle operators may remain in the cabs of vehicles being loaded or unloaded when the vehicle provides adequate protection for the operator during loading and unloading operations.

Mobile Equipment

When mobile equipment is operated adjacent to the edge of an excavation, a warning system will be used when the operator does not have a clear and direct view of the edge of the excavation. The warning system must consist of barricades, hand or mechanical signals, or stop logs. If possible, the surface grade will slope away from the excavation.

Hazardous Atmospheres

Atmospheric testing must be conducted in excavations over four feet deep where hazardous atmospheres could reasonably be expected to exist (e.g., landfill areas, near hazardous substance storage, gas pipelines).

Adequate precautions will be taken to prevent employee exposure to atmospheres containing less than 19.5 percent oxygen or other hazardous atmospheres. These precautions include providing appropriate respiratory protection or forced ventilation. Forced ventilation or other effective means will be used to prevent exposure to an atmosphere containing a flammable gas in excess of 10 percent of the lower flammable limit. Where needed, respiratory protection will be used in accordance with the Respiratory Protection Program.

Atmospheric monitoring will be performed using a properly calibrated direct reading instrument with audible and visual alarms. Monitoring will be continuous where controls are used to reduce the level of atmospheric contaminants. Monitors will be maintained and calibrated in accordance with manufacturer's specifications.

Water Accumulation

Employees will not work in excavations that contain or are accumulating water unless precautions have been taken to protect employees from hazards posed by water accumulation. The precautions taken could include, for example, special support or shield systems to protect from cave-ins, water removal to control the level of accumulating water, or use of safety harnesses and lifelines.

If water is controlled or prevented from accumulating by the use of water removal equipment, the water removal equipment and operation must be monitored by a person trained in the use of the equipment.

If excavation work interrupts the natural drainage of surface water (such as streams), diversion ditches, dikes, or other suitable means will be used to prevent surface water from entering the excavation. Precautions will also be taken to provide adequate drainage of the area adjacent to the excavation. Excavations subject to runoff from heavy rains must be re-inspected by the Project Manager to determine if additional precautions should be taken.

Support systems (such as shoring, bracing, or underpinning) will be used to ensure the stability of structures and the protection of employees where excavation operations could affect the stability of adjoining buildings, walls, or other structures.

Excavation below the level of the base or footing of any foundation or retaining wall that could be reasonably expected to pose a hazard to employees will not be permitted except when:

- A support system, such as underpinning, is provided to ensure the safety of employees and the stability of the structure; or
- The excavation is in stable rock; or
- A registered professional engineer has approved the determination that the structure is sufficiently removed from the excavation so as to be unaffected by the excavation activity; or
- A registered professional engineer has approved the determination that such excavation work will not pose a hazard to employees.

Sidewalks, pavements, and appurtenant structures will not be undermined unless a support system or other method of protection is provided to protect employees from the possible collapse of such structures.

Where review or approval of a support system by a registered professional engineer is required, the Department will secure this review and approval in writing before the work is begun. A copy of this approval must be provided to the Safety department.

Loose Rock or Soil

Adequate protection must be provided to protect employees from loose rock or soil that could pose a hazard by falling or rolling from an excavation face.

Such protection will consist of:

- Scaling to remove loose material;
- Installation of protective barricades, such as wire mesh or timber, at appropriate intervals on the face of the slope to stop and contain falling material; or
- Benching sufficient to contain falling material.

Excavation personnel will not be permitted to work above one another where the danger of falling rock or earth exists.

Employees must be protected from excavated materials, equipment or other materials that could pose a hazard by falling or rolling into excavations.

Protection will be provided by keeping such materials or equipment at least two feet from the edge of excavations, by the use of restraining devices that are sufficient to prevent materials or equipment from falling or rolling into excavations, or by a combination of both if necessary.

Materials and equipment may, as determined by the Project Manager, need to be stored further than two feet from the edge of the excavation if a hazardous loading condition is created on the face of the excavation.

Materials piled, grouped, or stacked near the edge of an excavation must be stable and self-supporting.

Fall Protection

Barricades, walkways, lighting and posting must be provided as necessary prior to the start of excavation operations.

Guardrails, fences, or barricades must be provided on excavations adjacent to walkways, driveways, and other pedestrian or vehicle thoroughfares. Warning lights or other illumination must be maintained as necessary for the safety of the public and employees from sunset to sunrise.

Wells, holes, pits, shafts, and all similar excavations must be effectively barricaded or covered and posted as necessary to prevent unauthorized access. All temporary excavations of this type will be backfilled as soon as possible.

Walkways or bridges protected by standard guardrails must be provided where employees and the general public are permitted to cross over excavations. Where workers in the excavation may pass under these walkways or bridges, a standard guardrail and toe board must be used. Information on the requirements for guardrails and toe boards may be obtained by reviewing the Fall Protection Program.

Daily Inspections

The Competent person will conduct daily inspections of excavations, adjacent areas, and protective systems for evidence of a situation that could result in possible cave-ins, failure of protective systems, hazardous atmospheres, or other hazardous conditions. An inspection will be conducted by the Competent person prior to the start of work and as needed throughout the shift. Inspections shall also be made after each hazard-changing event (e.g., rainstorm). These inspections are required when the excavation will be or is occupied by employees.

Where the Competent person finds evidence of a situation that could result in a possible cave-in, failure of protective systems, hazardous atmosphere, or other hazardous conditions, exposed employees shall be removed from the hazardous area until precautions have been taken to ensure their safety.

The Competent person shall maintain a written log of all inspections conducted. This log shall include the date, worksite location, results of the inspection, and a summary of any action taken to correct existing hazards.

Protection from Cave-ins

Each employee in an excavation shall be protected from cave-ins by using either an adequate sloping and benching system or an adequate support or protective system.

Exceptions to this are limited to

- Excavations made in stable rock; or
- Excavations less than four feet in depth where examination of the ground by a Competent person provides no indication of a potential cave-in.

Protective systems shall be capable of resisting all loads that could reasonably be expected to be applied to the system.

Support Systems

The design of support systems, shield systems, and other protective systems shall be selected and constructed by the Competent person in accordance with one of the following options.

Option A—Designs using OSHA Criteria

Timber shoring and aluminum hydraulic shoring must be utilized in accordance with OSHA criteria. If this option is selected, contact _____ to coordinate design and implementation of these systems.

Option B—Designs using manufacturer's tabulated data

Support systems, shield systems, or other protective systems (e.g. trench boxes) drawn from manufacturer's tabulated data shall be constructed and used in accordance with all specifications, recommendations, and limitations issued or made by the manufacturer.

Deviation from the specifications, recommendations, and limitations issued or made by the manufacturer shall only be allowed after the manufacturer issues specific written approval.

Manufacturer's specifications, recommendations, and limitations, and manufacturer's approval to deviate from the specifications, recommendations, and limitations shall be kept in written form at the jobsite during construction of the protective system. After that time this data may be stored off the jobsite, but a copy shall be made available to the safety department upon request.

Option C—Designs using other tabulated data

Designs of support systems, shield systems, or other protective systems shall be selected from and be constructed in accordance with tabulated data, such as tables and charts.

The tabulated data shall be in written form and include all of the following:

- Identification of the factors that affect the selection of a protective system drawn from such data;
- Identification of the limits of use of the data;
- Information needed by the user to make a correct selection of a protective system from the data.
- At least one copy of the tabulated data, identifying the registered professional engineer who approved the data, shall be maintained at the jobsite during construction of the system. After that time the data may be stored off the jobsite, but a copy of the data shall be made available to the safety department upon request.

Option D—Design by a registered professional engineer

Support systems, shield systems, and other protective systems not using the options detailed in options A, B, or C above shall be approved by a registered professional engineer.

Designs shall be in written form and shall include the following:

- A plan indicating the sizes, types, and configurations of the materials to be used in the protective system; and the identity of the registered professional engineer approving the design.
- At least one copy of the design shall be maintained at the jobsite during construction of the protective system. After that time, the design may be stored off the jobsite, but a copy of the design shall be made available to the safety department upon request.

Protection Systems Decision Tree

Use the following guideline to determine whether protection systems are needed.

1. If the excavation is less than four feet deep, a protection system is needed that meets the requirements under Sloping and Benching or Support Systems, unless the site excavation Competent person determines that there is no risk for cave-in.
2. If the excavation is between four and 20 feet deep, a protection system that meets the requirements under Sloping and Benching or Support Systems must be installed and utilized in all occupied areas of the excavation.
3. If the excavation is greater than 20 feet, a protection system designed by a registered engineer must be installed and utilized in all occupied areas of the excavation.

Sloping and Benching

The slope and configuration of sloping and benching systems shall be selected and constructed by the Competent person.

Employees shall not be permitted to work above other employees on the faces of sloped or benched systems except when employees at the lower levels are protected from the hazard of falling, rolling, or sliding material or equipment.

Sloping and Benching Systems

The slope and configuration of sloping and benching systems must be selected and constructed by the Competent person in accordance with one of the following:

Option A—Allowable Configurations and Slopes

Excavations must be sloped at an angle not steeper than one and one-half horizontal to one vertical (34 degrees measured from the horizontal), unless the Competent person uses one of the other options listed below.

The slopes used must be excavated in accordance with the slopes shown for Type C soil in Sloping and Benching.

Option B—Determination of Slopes and Configurations

Maximum allowable slopes and allowable configurations for sloping and benching systems must meet the requirements of Sloping and Benching and Soil Classification.

Option C—Designs Using Other Tabulated Data

The design of sloping or benching systems may be selected from, and must be constructed in accordance with, other tabulated data, such as tables and charts. The tabulated data used must be in written form and include all of the following:

- Identification of the factors that affect the selection of a sloping or benching system
- Identification of the limits of use of the data, including the maximum height and the angle of the slopes determined to be safe
- Other information needed by the user to make correct selection of a protective system
- One copy of the tabulated data that identifies the registered professional engineer who approved the data must be maintained at the jobsite during construction of the protective system. After that time, the data may be stored off the jobsite, but a copy of the data must be made available to the safety department upon request.

Option D—Design by a Registered Professional Engineer

Sloping and benching systems not utilizing Option (A), Option (B), or Option (C) above must be approved by a registered professional engineer.

Designs must be in written form and must include at least the following:

- The maximum height and angle of the slopes that were determined to be safe for the particular project
- The identity of the registered professional engineer approving the design
- At least one copy of the design must be maintained at the jobsite while the slope is being constructed. After that time the design need not be at the jobsite, but a copy must be made available to upon request.

Requirements

Soil Classification

Soil and rock deposits must be classified in accordance with Appendix A (29 CFR 1926.650 Subpart P) Soil Classification.

Maximum Allowable Slope

The maximum allowable slope for a soil or rock deposit must be determined from the table below.

Table 11.1 Maximum Allowable Slopes for Excavations Less Than 20 Feet

Soil or Rock Type	Maximum Slope (H:V)	Maximum Slope (Degrees)
Stable Rock	Vertical	90
Type A	.75:1	53
Type B	1:1	45
Type C	1.5:1	34

Actual Slope

The actual slope must not be steeper than the maximum allowable slope.

The actual slope must be less steep than the maximum allowable slope when there are signs of distress. If that situation occurs, the slope must be cut back to an actual slope which is at least one-half horizontal to one vertical (1/2H:1V) less steep than the maximum allowable slope.

When surcharge loads from stored material or equipment, operating equipment, or traffic are present, a Competent person must

- Determine the degree to which the actual slope must be reduced below the maximum allowable slope
- Ensure that such reduction is achieved
- Surcharge loads from adjacent structures must be evaluated in accordance with this program.

Configurations

Configurations of sloping and benching systems must be in accordance with the table and figures below.

Figures: Slope Configurations

All slopes stated below are in the horizontal to vertical ratio.

Excavations Made in Type A Soil

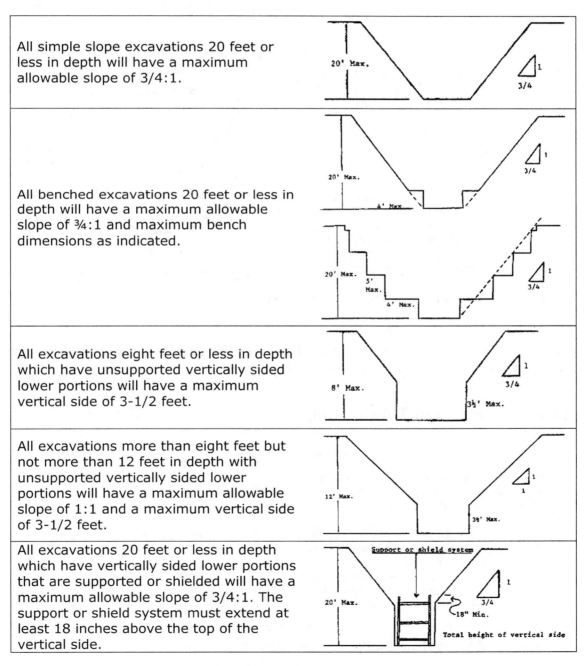

All simple slope excavations 20 feet or less in depth will have a maximum allowable slope of 3/4:1.	
All benched excavations 20 feet or less in depth will have a maximum allowable slope of ¾:1 and maximum bench dimensions as indicated.	
All excavations eight feet or less in depth which have unsupported vertically sided lower portions will have a maximum vertical side of 3-1/2 feet.	
All excavations more than eight feet but not more than 12 feet in depth with unsupported vertically sided lower portions will have a maximum allowable slope of 1:1 and a maximum vertical side of 3-1/2 feet.	
All excavations 20 feet or less in depth which have vertically sided lower portions that are supported or shielded will have a maximum allowable slope of 3/4:1. The support or shield system must extend at least 18 inches above the top of the vertical side.	

Figure 11.1 Excavations made in Type A soil

All other simple slope, compound slope, and vertically sided lower portion excavations will be in accordance with the other options described in Sloping and Benching Systems.

Excavations Made in Type B Soil

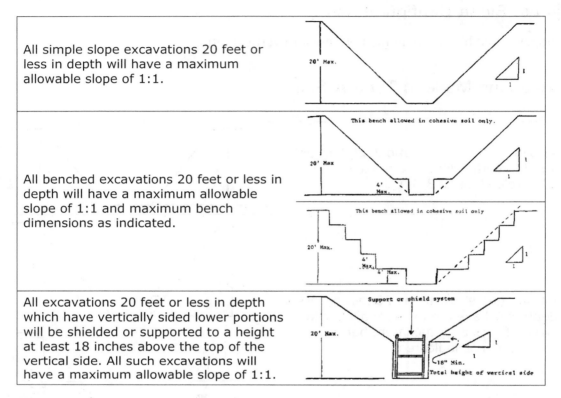

All simple slope excavations 20 feet or less in depth will have a maximum allowable slope of 1:1.	
All benched excavations 20 feet or less in depth will have a maximum allowable slope of 1:1 and maximum bench dimensions as indicated.	
All excavations 20 feet or less in depth which have vertically sided lower portions will be shielded or supported to a height at least 18 inches above the top of the vertical side. All such excavations will have a maximum allowable slope of 1:1.	

Figure 11.2 Excavations Made in Type B Soil

All other sloped excavations must be in accordance with the other options permitted in Appendix B (29 CFR 1926.650 Subpart P) Sloping and Benching.

Excavations Made in Type C Soil

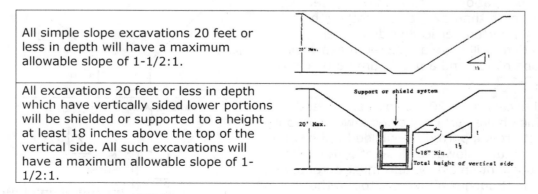

All simple slope excavations 20 feet or less in depth will have a maximum allowable slope of 1-1/2:1.	
All excavations 20 feet or less in depth which have vertically sided lower portions will be shielded or supported to a height at least 18 inches above the top of the vertical side. All such excavations will have a maximum allowable slope of 1-1/2:1.	

Figure 11.3 Excavations Made in Type C Soil

All other sloped excavations must be in accordance with the other options described in Appendix B (29 CFR 1926.650 Subpart P) Sloping and Benching.

Excavations Made in Layered Soils

All excavations 20 feet or less in depth made in layered soils will have a maximum allowable slope for each layer as set forth below.

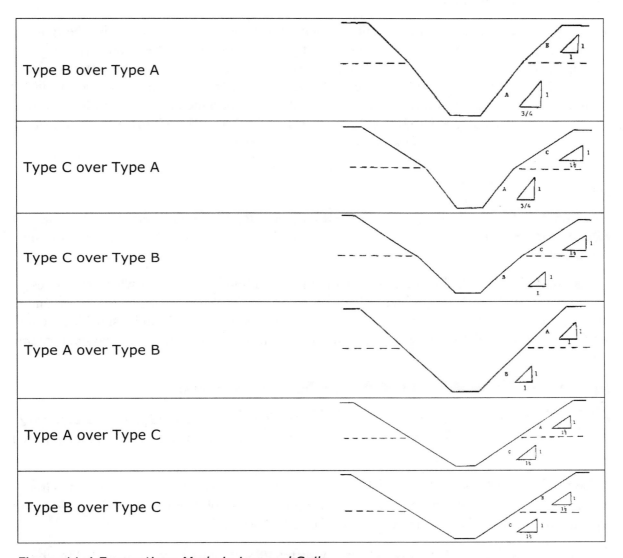

Figure 11.4 Excavations Made in Layered Soils

All other sloped excavations must be in accordance with the other options described in Sloping and Benching Systems.

SOIL CLASSIFICATION

Scope

The following describes a method of classifying soil and rock deposits based on site and environmental conditions, and on the structure and composition of the earth deposits. This section contains definitions, sets forth requirements, and describes acceptable visual and manual tests for use in classifying soils.

This applies when a sloping or benching system is designed in accordance with the requirements set forth in 1926.652(b)(2) as a method of protection for employees from cave-ins. This also applies when timber shoring for excavations is designed as a method of protection from cave-ins in accordance with subpart P of part 1926, and when aluminum hydraulic shoring is designed. This also applies if other protective systems are designed and selected for use from data prepared in accordance with the requirements set forth in 1926.652(c), and the use of the data is predicated on the use of the soil classification system.

Definitions

Cemented soil is a soil in which the particles are held together by a chemical agent, such as calcium carbonate, such that a hand-size sample cannot be crushed into powder or individual soil particles by finger pressure.

Cohesive soil is a clay (fine grained soil), or soil with a high clay content which has cohesive strength. Cohesive soil does not crumble, can be excavated with vertical side slopes, and is plastic when moist. Cohesive soil is hard to break up when dry and exhibits significant cohesion when submerged. Cohesive soils include clayey silt, sandy clay, silty clay, clay, and organic clay.

Dry soil is a soil that does not exhibit visible signs of moisture content.

Fissured soil is a soil material that has a tendency to break along definite plane of fracture with little resistance or a material that exhibits open cracks, such as tension cracks, in an exposed surface.

Granular soil is a gravel, sand, or silt (coarse grained soil) with little or no clay content. Granular soil has no cohesive strength. Some moist granular soils exhibit apparent cohesion. Granular soil cannot be molded when moist and crumbles easily when dry.

A Layered system is two or more distinctly different soil or rock types arranged in layers. Micaceous seams or weakened planes in rock or shale are considered layered.

Moist soil is a condition in which a soil looks and feels damp. Moist cohesive soil can easily be shaped into a ball and rolled into small diameter threads before crumbling. Moist granular soil that contains some cohesive material will exhibit signs of cohesion between particles.

Plastic is a property of a soil which allows the soil to be deformed or molded without cracking or appreciable volume change.

Saturated soil is a soil in which the voids are filled with water. Saturation does not require flow. Saturation or near saturation is necessary for the proper use of instruments such as a pocket penetrometer or sheer vane.

A Soil classification system is a method of categorizing soil and rock deposits in a hierarchy of Stable Rock, Type A, Type B, and Type C, in decreasing order of stability. The categories are determined based on an analysis of the properties and performance characteristics of the deposits and the characteristics of the deposits and the environmental conditions of exposure.

Stable rock is a natural solid mineral matter that can be excavated with vertical sides and remain intact while exposed.

Submerged soil is a soil which is underwater or is free seeping.

Type A soil is a cohesive soil with an unconfined, compressive strength of 1.5 ton per square foot (tsf) (144 kPa) or greater. Examples of cohesive soils are clay, silty clay, sandy clay, clay loam, and, in some cases, silty clay loam and sandy clay loam. Cemented soils such as caliche and hardpan are also considered Type A. However, no soil is Type A if

- The soil is fissured
- The soil is subject to vibration from heavy traffic, pile driving, or similar effects
- The soil has been previously disturbed
- The soil is part of a sloped, layered system where the layers dip into the excavation on a slope of four horizontal to one vertical (4H:1V) or greater
- The material is subject to other factors that would require it to be classified as a less stable material

Type B soil is a Cohesive soil with an unconfined compressive strength greater than 0.5 tsf (48 kPa) but less than 1.5 tsf (144 kPa) or

- Granular cohesionless soils including angular gravel (similar to crushed rock), silt, silt loam, sandy loam and, in some cases, silty clay loam and sandy clay loam
- Previously disturbed soils except those which would otherwise be classed as Type C soil
- Soil that meets the unconfined compressive strength or cementation requirements for Type A, but is fissured or subject to vibration, or Dry rock that is not stable
- Material that is part of a sloped, layered system where the layers dip into the excavation on a slope less steep than four horizontal to one vertical (4H:1V), but only if the material would otherwise be classified as Type B.

Type C soil is a Cohesive soil with an unconfined compressive strength of 0.5 tsf (48 kPa) or less or

- Granular soils including gravel, sand, and loamy sand
- Submerged soil or soil from which water is freely seeping

- Submerged rock that is not stable
- Material in a sloped, layered system where the layers dip into the excavation or a slope of four horizontal to one vertical (4H:1V) or steeper

Unconfined compressive strength is the load per unit area at which a soil will fail in compression. It can be determined by laboratory testing or estimated in the field using a pocket penetrometer, by thumb penetration tests, and other methods.

Wet soil is the soil that contains significantly more moisture than moist soil but in such a range of values that cohesive material will slump or begin to flow when vibrated. Granular material that would exhibit cohesive properties when moist will lose those cohesive properties when wet.

CLASSIFICATION OF SOIL AND ROCK DEPOSITS

Each soil and rock deposit shall be classified by a Competent person as Stable Rock, Type A, Type B, or Type C, in accordance with the definitions set forth below.

Basis of Classification

The classification of the deposits shall be made based on the results of at least one visual and at least one manual analysis. Such analyses shall be conducted by a Competent person using tests described below, or in other recognized methods of soil classification and testing such as those adopted by the American Society for Testing Materials.

Visual and Manual Analyses

The visual and manual analyses shall be designed and conducted to provide sufficient quantitative and qualitative information as may be necessary to identify properly the properties, factors, and conditions affecting the classification of the deposits.

Layered systems

In a layered system, the system shall be classified in accordance with its weakest layer. However, each layer may be classified individually where a more stable layer lies under a less stable layer.

Reclassification

If, after classifying a deposit, the properties, factors, or conditions affecting its classification change in any way, the changes shall be evaluated by a Competent person. The deposit shall be reclassified as necessary to reflect the changed circumstances.

Visual tests

Visual analysis is conducted to determine qualitative information regarding the excavation site in general, the soil adjacent to the excavation, the soil forming the sides of the open excavation, and the soil taken as samples from excavated material.

Observe samples of soil that are excavated and soil in the sides of the excavation. Estimate the range of particle sizes and the relative amounts of the particle sizes. Soil that is primarily composed of fine-grained material is cohesive material. Soil composed primarily of coarse-grained sand or gravel is granular material.

Observe soil as it is excavated. Soil that remains in clumps when excavated is cohesive. Soil that breaks up easily and does not stay in clumps is granular.

Observe the side of the opened excavation and the surface area adjacent to the excavation. Crack-like openings such as tension cracks could indicate fissured material. If chunks of soil fall off a vertical side, the soil could be fissured. Small spalls are evidence of moving ground and are indications of potentially hazardous situations.

Observe the area adjacent to the excavation and the excavation itself for evidence of existing utility and other underground structures, and to identify previously disturbed soil.

Observed the opened side of the excavation to identify layered systems. Examine layered systems to identify if the layers slope toward the excavation. Estimate the degree of slope of the layers.

Observe the area adjacent to the excavation and the sides of the opened excavation for evidence of surface water, water seeping from the sides of the excavation, or the location of the level of the water table.

Observe the area adjacent to the excavation and the area within the excavation for sources of vibration that may affect the stability of the excavation face.

Manual tests

Manual analysis of soil samples is conducted to determine quantitative as well as qualitative properties of soil and to provide more information in order to classify soil properly.

Plasticity—Mold a moist or wet sample of soil into a ball and attempt to roll it into threads as thin as 1/8-inch in diameter. Cohesive material can be successfully rolled into threads without crumbling. For example, if at least a two-inch (50 mm) length of 1/8-inch thread can be held on one end without tearing, the soil is cohesive.

Dry strength—If the soil is dry and crumbles on its own or with moderate pressure into individual grains or fine powder, it is granular (any combination of gravel, sand, or silt). If the soil is dry and falls into clumps which break up into smaller clumps, but the smaller clumps can only be broken up with difficulty, it may be clay in any combination with gravel, sand, or silt. If the dry soil breaks into clumps which do not break up into small clumps and which can only be broken with difficulty, and there is no visual indication the soil is fissured, the soil may be considered unfissured.

Thumb penetration—The thumb penetration test can be used to estimate the unconfined compressive strength of cohesive soils. Type A soils with an unconfined compressive strength of 1.5 tsf can be readily indented by the thumb; however, they can be penetrated by the thumb only with very great effort. Type C soils with an unconfined compressive strength of 0.5 tsf can be easily penetrated several inches by the thumb, and can be molded by light finger pressure. This test should be conducted on an undisturbed soil sample, such as a large clump of spoil, as soon as practicable after excavation to keep to a minimum the effects of exposure to drying influences. If the excavation is later exposed to wetting influences (rain, flooding), the classification of the soil must be changed accordingly.

Other strength tests—Estimates of unconfined compressive strength of soils can also be obtained by use of a pocket electrometer or by using a hand-operated shear vane.

Drying test—The basic purpose of the drying test is to differentiate between cohesive material with fissures, unfissured cohesive material, and granular material. The procedure for the drying test involves drying a sample of soil that is approximately one-inch thick (2.54 cm) and six inches (15.24 cm) in diameter until it is thoroughly dry.

If the sample develops cracks as it dries, significant fissures are indicated.

Samples that dry without cracking are to be broken by hand. If considerable force is necessary to break a sample, the soil has significant cohesive material content. The soil can be classified as an unfissured cohesive material and the unconfined compressive strength should be determined.

If a sample breaks easily by hand, it is either a fissured cohesive material or a granular material. To distinguish between the two, pulverize the dried clumps of the sample by hand or by stepping on them. If the clumps do not pulverize easily, the material is cohesive with fissures. If they pulverize easily into very small fragments, the material is granular.

Materials and Equipment

Materials and equipment used for protective systems shall be free from damage or defects that might affect their proper function.

Manufactured materials and equipment used for protective systems shall be used and maintained in accordance with the recommendations of the manufacturer and in a manner that will prevent employee exposure to hazards.

When material or equipment used for protective systems are damaged, the Competent person shall ensure that these systems are examined by a Competent person to evaluate its suitability for continued use. If the Competent person can not ensure that the material or equipment is able to support the intended loads or is otherwise suitable for safe use, then such material or equipment shall be removed from service. These materials or equipment shall be evaluated and approved by a registered professional engineer before being returned to service.

Installation and Removal of Support

Members of support systems shall be securely connected together to prevent sliding, falling, kick outs, or other potential hazards.

Support systems shall be installed and removed in a manner that protects employees from cave-ins, structural collapses, or from being struck by members of the support system.

Individual members of support systems shall not be subjected to loads exceeding those which those members were designed to support.

Before temporary removal of individual support members begins, additional precautions shall be taken as directed by the Competent person to ensure the safety of employees. These precautions could include, for example, the installation other structural members to carry the loads imposed on the support system.

Removal of support systems shall begin at, and progress from, the bottom of the excavation. Members shall be released slowly. If there is any indication of possible failure of the remaining members of the structure or possible cave-in of the sides of the excavation, the work shall be halted until it can be examined by the Project Manager.

Backfilling shall progress together with the removal of support systems from excavations.

Additional requirements for support systems for trench excavations include:

- Excavation of material to a level no greater than two feet below the bottom of the members of a support system is allowed, but only if the system is designed to resist the forces calculated for the full depth of the trench. There shall be no indications of a possible loss of soil from behind or below the bottom of the support system while the trench is open.
- Installation of a support system shall be closely coordinated with the excavation of trenches.

Shield Systems

Shield systems shall not be subjected to loads that are greater than those they were designed to withstand.

Shields shall be installed in a manner that will restrict lateral or other hazardous movement of the shield that could occur during cave-in or unexpected soil movement.

Employees shall be protected from the hazard of cave-ins when entering or exiting the areas protected by shields.

Employees shall not be allowed in shields when shields are being installed, removed, or moved vertically.

Endnote:

[1]http://www.osha.gov/dts/osta/otm_v/otm_v_2.html#8

12

✗ 2·8

Demolition Operations

Using Safe Blasting Procedures at Your Worksite

INTRODUCTION

Before the start of every demolition job, the demolition contractor should take a number of steps to safeguard the health and safety of workers at the job site. These preparatory operations involve the overall planning of the demolition job, including the methods to be used to bring the structure down, the equipment necessary to do the job, and the measures to be taken to perform the work safely. Planning for a demolition job is as important as actually doing the work. Therefore all planning work should be performed by a competent person experienced in all phases of the demolition work to be performed.

A sign should be posted where the demolition work is to be performed that says the following:

> "No employee shall be permitted in any area that can be adversely affected when demolition operations are being performed. Only those employees necessary for the performance of the operations shall be permitted in these areas."

ENGINEERING SURVEY

Prior to starting all demolition operations, 29 CFR 1926.850(a) requires that an engineering survey of the structure must be conducted by a competent person. The purpose of this survey is to determine the condition of the framing, floors, and walls so that measures can be taken, if necessary, to prevent the premature collapse of any portion of the structure. When indicated as advisable, any adjacent structure(s) or improvements should be similarly checked.

The demolition contractor must maintain a written copy of this survey. Photographing existing damage in neighboring structures is also advisable.

The engineering survey provides the demolition contractor with the opportunity to evaluate the job in its entirety. The contractor should plan for the wrecking of the structure, equipment to do the work, manpower requirements, and protection of the public. The safety of all workers on the job site should be a prime consideration. During the preparation of the engineering survey, the contractor should plan for potential hazards such as fires, cave-ins, and injuries.

If the structure to be demolished has been damaged by fire, flood, explosion, or some other cause, appropriate measures, including bracing and shoring of walls and floors, shall be taken

293

to protect workers and any adjacent structures. It shall also be determined if any type of hazardous chemicals, gases, explosives, flammable material, or similar dangerous substances have been used or stored on the site. If the nature of a substance cannot be easily determined, samples should be taken and analyzed by a qualified person prior to demolition.

During the planning stage of the job, all safety equipment needs should be determined. The required number and type of respirators, lifelines, warning signs, safety nets, special face and eye protection, hearing protection, and other worker protection devices outlined in this manual should be determined during the preparation of the engineering survey.

UTILITY LOCATION

One of the most important elements of the pre-job planning is the location of all utility services. All electric, gas, water, steam, sewer, and other services lines should be shut off, capped, or otherwise controlled, at or outside the building before demolition work is started. In each case, any utility company which is involved should be notified in advance, and its approval or services, if necessary, shall be obtained.

If it is necessary to maintain any power, water, or other utilities during demolition, such lines shall be temporarily relocated as necessary and/or protected. The location of all overhead power sources should also be determined, as they can prove especially hazardous during any machine demolition. All workers should be informed of the location of any existing or relocated utility service.

MEDICAL SERVICES AND FIRST AID

Prior to starting work, provisions should be made for prompt medical attention in case of serious injury. The nearest hospital, infirmary, clinic, or physician shall be located as part of the engineering survey. The job supervisor should be provided with instructions for the most direct route to these facilities. Proper equipment for prompt transportation of an injured worker, as well as a communication system to contact any necessary ambulance service, must be available at the job site. The telephone numbers of the hospitals, physicians, or ambulances shall be conspicuously posted.

In the absence of these of an infirmary, clinic, hospital, or physician that is reasonably accessible in terms of time and distance to the work site, a person who has a valid certificate in first aid training from the U.S. Bureau of Mines, the American Red Cross, or equivalent training should be available at the work site to render first aid.

A properly stocked first aid kit, as determined by an occupational physician, must be available at the job site. The first aid kit should contain approved supplies in a weatherproof container with individual sealed packages for each type of item. It should also include rubber gloves to prevent the transfer of infectious diseases. Provisions should also be made to provide for quick drenching or flushing of the eyes should any person be working around corrosive materials. Eye flushing must be done with water containing no additives. The contents of the kit shall be checked before being sent out on each job and at least weekly to ensure the expended items are replaced.

POLICE AND FIRE CONTACT

The telephone numbers of the local police, ambulance, and fire departments should be available at each job site. This information can prove useful to the job supervisor in the event of any traffic problems, such as the movement of equipment to the job, uncontrolled fires, or other police/fire matters. The police number may also be used to report any vandalism, unlawful entry to the job site, or accidents requiring police assistance.

A fire plan should be set up prior to beginning a demolition job. This plan should outline the assignments of key personnel in the event of a fire and provide an evacuation plan for workers on the site.

The following rules should be observed:

- All potential sources of ignition should be evaluated and the necessary corrective measures taken.
- Electrical wiring and equipment for providing light, heat, or power should be installed by a competent person and inspected regularly.
- Equipment powered by an internal combustion engine should be located so that the exhausts discharge well away from combustible materials and away from workers.
- When the exhausts are piped outside the building, a clearance of at least six inches should be maintained between such piping and combustible material.
- All internal combustion equipment should be shut down prior to refueling. Fuel for this equipment should be stored in a safe location.
- Sufficient fire fighting equipment should be located near any flammable or combustible liquid storage area.
- Only approved containers and portable tanks should be used for the storage and handling of flammable and combustible liquids.
- Roadways between and around combustible storage piles should be at least 15 feet wide and maintained free from accumulation of rubbish, equipment, or other materials.
- When storing debris or combustible material inside a structure, such storage shall not obstruct or adversely affect the means of exit.
- A suitable location at the job site should be designated and provided with plans, emergency information, and equipment, as needed. Access for heavy fire-fighting equipment should be provided on the immediate job site at the start of the job and maintained until the job is completed.
- Free access from the street to fire hydrants and to outside connections for standpipes, sprinklers, or other fire extinguishing equipment, whether permanent or temporary, should be provided and maintained at all times.
- Pedestrian walkways should not be so constructed as to impede access to hydrants.
- No material or construction should interfere with access to hydrants, Siamese connections, or fire-extinguishing equipment.
- A temporary or permanent water supply of volume, duration, and pressure sufficient to operate the fire-fighting equipment properly should be made available.

- Standpipes with outlets should be provided on large multistory buildings to provide for fire protection on upper levels. If the water pressure is insufficient, a pump should also be provided.

- An ample number of fully charged portable fire extinguishers should be provided throughout the operation. All motor driven mobile equipment should be equipped with an approved fire extinguisher.

- An alarm system (e.g., telephone system, siren, two-way radio) shall be established in such a way that employees on the site and the local fire department can be alerted in case of an emergency. The alarm code and reporting instructions shall be conspicuously posted and the alarm system should be serviceable at the job site during the demolition. Fire cutoffs shall be retained in the buildings undergoing alterations or demolition until operations necessitate their removal.

SAFE BLASTING PROCEDURES

Prior to the blasting of any structure or portion thereof, a complete written survey must be made by a qualified person of all adjacent improvements and underground utilities. When there is a possibility of excessive vibration due to blasting operations, seismic or vibration tests should be taken to determine proper safety limits to prevent damage to adjacent or nearby buildings, utilities, or other property.

The preparation of a structure for demolition by explosives may require the removal of structural columns, beams, or other building components. This work should be directed by a structural engineer or a competent person qualified to direct the removal of these structural elements. Extreme caution must be taken during this preparatory work to prevent the weakening and premature collapse of the structure.

FIRE PRECAUTIONS

The presence of fire near explosives presents a severe danger. Every effort should be made to ensure that fires or sparks do not occur near explosive materials. Smoking, matches, firearms, open flame lamps, and other fires, flame, or heat-producing devices must be prohibited in or near explosive magazines or in areas where explosives are being handled, transported, or used. In fact, persons working near explosives should not even carry matches, lighters, or other sources of sparks or flame. Open fires or flames should be prohibited within 100 feet of any explosive materials. In the event of a fire that is in imminent danger of contact with explosives, all employees must be removed to a safe area.

Electrical detonators can be inadvertently triggered by stray radio frequency (RF) signals from two-way radios. RF signal sources should be restricted from or near to the demolition site if electrical detonators are used.

PERSONNEL SELECTION

A blaster is a competent person who uses explosives. A blaster must be qualified by reason of training, knowledge, or experience in the field of transporting, storing, handling, and using explosives. In addition, the blaster should have a working knowledge of state and local regulations which pertain to explosives. Training courses are often available from manufacturers of explosives and blasting safety manuals are offered by the Institute of Makers of Explosives (IME) as well as other organizations.

Blasters shall be required to furnish satisfactory evidence of competency in handling explosives and in safety performing the type of blasting required. A competent person should always be in charge of explosives and should be held responsible for enforcing all recommended safety precautions in connection with them.

PROPER USE OF EXPLOSIVES

Blasting operations shall be conducted between sunup and sundown, whenever possible. Adequate signs should be sounded to alert to the hazard presented by blasting. Blasting mats or other containment should be used where there is danger of rocks or other debris being thrown into the air or where there are buildings or transportation systems nearby. Care should be taken to make sure mats and other protection do not disturb the connections to electrical blasting caps.

Radio, television, and radar transmitters create fields of electrical energy that can, under exceptional circumstances, detonate electric blasting caps. Certain precautions must be taken to prevent accidental discharge of electric blasting caps from current induced by radar, radio transmitters, lightning, adjacent power lines, dust storms, or other sources of extraneous or static electricity. These precautions shall include

- Ensuring that mobile radio transmitters on the job site which are less than 100 feet away from electric blasting caps, in other than original containers, shall be de-energized and effectively locked
- The prominent display of adequate signs, warning against the use of mobile radio transmitters, on all roads within 1,000 feet of the blasting operations
- Maintaining the minimum distances recommended by the IME between the nearest transmitter and electric blasting caps
- The suspension of all blasting operations and removal of persons from the blasting area during the approach and progress of an electric storm

After loading is completed, there should be as little delay as possible before firing. Each blast should be fired under the direct supervision of the blaster, who should inspect all connections before firing and who should personally see that all persons are in the clear before giving the order to fire. Standard signals, which indicate that a blast is about to be fired and a later all clear signal, have been adopted. It is important that everyone working in the area be familiar with these signals and that they be strictly obeyed.

PROCEDURES AFTER BLASTING

Immediately after the blast has been fired, the firing line shall be disconnected from the blasting machine and short-circuited. Where power switches are used, they shall be locked open or in the off position. Sufficient time shall be allowed for dust, smoke, and fumes to leave the blasted area before returning to the spot. An inspection of the area and the surrounding rubble shall be made by the blaster to determine if all charges have been exploded before employees are allowed to return to the operation. All wires should be traced and the search for unexploded cartridges made by the blaster.

Explosives, blasting agents, and blasting supplies that are obviously deteriorated or damaged should not be used, they should be properly disposed of. Explosives distributors will usually take back old stock. Local fire marshals or representatives of the United States Bureau of Mines may also arrange for its disposal. Under no circumstances should any explosives be abandoned.

Wood, paper, fiber or other materials that have contained high explosives should not be used again for any purpose, but should be destroyed by burning. These materials should not be burned in a stove, fireplace, or other confined space. Rather, they should be burned at an isolated outdoor location, at a safe distance from thoroughfares, magazines, and other structures. It is important to check that the containers are entirely empty before burning. During burning, the area should be adequately protected from intruders and all persons kept at least 100 feet from the fire.

13

Cranes and Crane Operation

Avoiding Crane Hazards

INTRODUCTION

OSHA regulations covering the use and safety of cranes and derricks are found in 29 CFR 1926.550. Cranes and their operations must always be approached with caution. A crane is designed to lift loads that are heavier than itself.

Cranes include mobile truck cranes, derricks, tower cranes, rough terrain cranes, boom trucks, railroad track mounted cranes, and speedswings. The specific characteristics of individual cranes may vary, but the principles of crane operations apply to all types of cranes.

In our highly mechanized world, cranes are the workhorses that have increased productivity in construction, and in the maintenance of production and service facilities. Statistics show, however, that because of inherent hazards that occur during normal working circumstances, a crane can be a very dangerous piece of equipment.

Serious accidents should be examined to identify hazards so that appropriate hazard prevention measures can be initiated. Unfortunately, at the worksite, a crane accident is often labeled as a freak accident because operating personnel are usually unaware of similar repetitive occurrences. Employers generally respond to safety requirements when shown that a specific hazard has caused repeated accidents and must be eliminated to avoid future occurrences from the same source.

Many cranes are inadvertently or unintentionally designed with built-in hazards. Although design engineers are very competent when it comes to crane design, they do not always have sufficient training in the most complex component of design-the people who are going to use the crane. People make errors in judgment when working with cranes, which can be used in so many different ways. Too much reliance has been placed upon the ability of those using cranes to make last-minute adjustments to avoid accidents. When safety is included in design, the errors of those who use the equipment can be overcome by the same technology that space engineers use when designing spacecraft.

It is not an insurmountable task to make the workplace safe when using cranes if the crane's limitations and the limitations of those using them are known. Analysis of crane accidents has shown that usually there was no planning of crane use in particular work circumstances. Because the man/machine aspect of design has not received the attention it should have had,

299

cranes arrive at the workplace with inherent hazards, and it becomes a challenging task to identify these hazards and apply appropriate OSHA safety requirements for their elimination.

SUPERVISOR OF EQUIPMENT

The responsibilities of the supervisor of equipment include

- Making sure that a suitable crane and competent personnel are available for the work
- Making sure that all cranes have a current certification certificate
- Making sure that cranes are equipped with the following:
    ~ Manufacturer-approved load chart mounted in the cab which is correct for the counterweight, boom, and jib supplied
    ~ Counterweight as specified by the manufacturer
    ~ A serviceable, approved fire extinguisher which is readily accessible to the operator
    ~ A serviceable seat, equipped with a seatbelt
    ~ Guards for personal protection over all exposed moving parts which are considered hazardous under normal operating conditions
    ~ Boom angle indicator
    ~ A device to determine that the crane is level
    ~ A 100 foot (30.5 meters) non-conducting tape measure
    ~ An audible horn or bell signaling device
    ~ A log book

- Making sure that a maintenance and inspection program for the crane and boom is established and followed
- Making sure that the crane is supplied with wire ropes equivalent in size, grade, and construction to those recommended by the crane manufacturer
- Making sure that the weight and capacity are permanently and legibly marked on the blocks, equalizer beams, dragline, clamshell, concrete buckets, and any other accessories which contribute to the load handled by the crane
- Making sure that the length and serial number are permanently and legibly marked on all boom and jib sections
- Making sure that the crane, and jib when used, is properly assembled and mounted
- Providing a qualified operator and making sure that the operator understands all aspects of the load chart as it applies to his or her machine
- Making sure that the crane is properly prepared for long distance moving when required

SUPERVISOR OF LIFT

The supervisor of lift may be a foreman or leadman of the trade group involved and has overall responsibility for the lift. Duties of this position are

- Supervise the work, including the machine location and making the lift.
- Determine the correct weight of load and operating radius and informs the operator.

- Supervise the rigging crew.
- Make sure that the load is properly rigged.
- Designate signalers and informs the operator. All signalers must know the standard hand signals.
- Keep all workers, except the operator and apprentice (when one is employed), clear of the crane during operation.
- Control the movements of all personnel required to work within the area affected by the lift to insure their safety.
- Make sure that all required safety precautions are taken when the lift is in the vicinity of overhead powerlines or any other energized power sources.

OPERATOR REQUIREMENTS

The operator, who is in charge of operating the crane, must meet all local requirements for certification. The operator must demonstrate adequate knowledge of

- The crane controls.
- The crane operating characteristics.
- The crane load chart and manufacturer's instructions.
- The standard hand signals.
- Starting up and shutting down the crane using proper procedures.
- The operator, or the operator's experienced oiler, moves the crane.
- Setting up the crane.
- Calculating the radius and boom angle from the charts and intended load.
- Measuring the radius from the center of the load to the center of the crane.
- Placing the crane in position to pick up and place the load within calculated limits.
- Placing the crane so that heavy loads are lifted and moved only in the strongest, most stable configuration.
- Placing the crane away from overhead powerlines and other entanglements. The minimum distance is ten feet for lines carrying up to 50 kv (50,000 volts). Higher voltages require longer distances. Wet or windy conditions also require longer distances.
- Extending outriggers to their fullest.
- Making sure that outriggers are seated on adequate support material. Anything less stable than reinforced concrete or engineered asphalt requires the use of adequate strength support under each outrigger.
- Appropriate pad construction to support the heaviest anticipated load at each outrigger for the least stable soil conditions.
- Making sure that the wheels are raised and only the outriggers support the load.
- Making sure the crane is level. Even two percent off level will seriously decrease the crane's lifting capacity.
- Anticipating the effects of windy conditions on both the crane and the load. Never lift when the wind exceeds 25 miles per hour.
- Making sure that the crane is properly rigged for the particular lifts, including correct gantry height and parts of line.

- Making sure that the lift is within the capacity of the crane. This requires a full knowledge of the crane capacity as shown on the load chart, on the site conditions, and full consideration of all other relevant factors.
- Making sure the brakes and clutch are in proper working order and properly adjusted.
- Verifying that wire ropes are in good condition and capable of lifting the maximum load shown on the load chart.
- Performing periodic operating checks of hydraulic circuits and controls.
- Following the manufacturer's operating instructions in accordance with the load chart.
- Controlling the load swing-out to avoid loading the crane beyond its capacity.
- Using tag lines to control the load.
- Operating in accordance with safe practices and applicable codes and standards when working in the vicinity of overhead powerlines, or other electrically energized apparatus.
- Considering the effect of wind conditions at the time of lift.
- Making sure the apprentice (when employed) is in a safe place during crane operation.
- Maintaining visual communication with signalers.
- Supervising and training the apprentice when one is employed.
- Making sure that the crane is kept clean and all oil and grease spills are cleaned up immediately.
- Faithfully maintaining the crane's log book.

CRANE INSPECTIONS

1. Crane inspections will be conducted on all cranes owned, rented, leased, or otherwise in the possession of the company.
2. The daily inspection will be performed at the beginning of each shift by the operator who will not begin operations until the inspection is completed.
3. Operator will adjust his schedule accordingly.
4. Operator will use this time to train oiler (apprentice) when one is assigned to the operation.
5. The inspection will be recorded on the Daily Crane Inspection Record.
6. Operator will make sure that a supply of Inspection Records is with the crane at all times.
7. If the crane requires repair or maintenance, the operator will fill out a work order and submit it to the Maintenance Department.
8. If crane operators move from crane to crane during the month, the operator for each crane on the last day of the month is responsible for submitting a copy of the inspection record to the Maintenance Department for filing.
9. If a crane is not used on a particular day, or series of days, this must be noted in the log by the next operator to use the crane.
10. Monthly inspections will be conducted during the first week of each month by the supervisor of equipment or a person designated by the supervisor of equipment to conduct the inspection.
11. The inspection will be recorded on the Monthly Crane Inspection Record.

12. The record of each inspection will be filed in the Maintenance Office.
13. Each crane must be inspected and certified by a certification and inspection company. (Some states license these companies).
14. The supervisor of equipment is responsible for scheduling the certifications and making sure that the cranes pass the inspections.
15. Every crane must be inspected, tested, and certified by a certification and inspection company every four years.
16. The supervisor of equipment is responsible for scheduling these certifications and making sure that the cranes pass the inspections and tests.
17. New, recently purchased, and rented cranes must first be inspected thorough by the supervisor of equipment and an operator before use.
18. The operator will thoroughly test the controls and all operational characteristics of the crane before using it on the job.
19. The safety of all personnel at the job-site can be affected by rented cranes with their own crews. One of the contractor's operators, or other competent person, should witness the rented equipment operator make the daily inspection. He should also make sure that each crew member is experienced in the type of work to be done.
20. Cranes recently overhauled, repaired, reconfigured, or modified will be inspected thoroughly by the supervisor of equipment and an operator before being put into use.
21. An operator will thoroughly test all controls and all operational characteristics of the crane and approve its being put into use.
22. Cranes that have undergone repairs or modifications shall be reinspected and recertified before being released for service.

HAZARD PREVENTION REQUIREMENTS

There is usually little opportunity to change the design of existing cranes, but there are measures within OSHA's span of control that can be taken to prevent a hazard from becoming an accident. The most effective ways to control hazards are preconstruction planning, job hazard analysis, hand signals, signaling devices, lifting capabilities, rigging practices, controlling the load, wire rope requirements, annual inspections, and preventative maintenance.

Preconstruction Planning

Analysis shows that most crane accidents could have been easily prevented had some basic consideration been given to the safe use of cranes and had such considerations been incorporated into the preconstruction planning meeting. The best time to address hazard avoidance is at a preconstruction planning meeting. There, with rational planning, hazards inherent with powerlines, blind lifts requiring communication, necessary lifting capacity, use of cranes and derricks on barges, and special circumstances requiring two or more cranes to lift a single load can be discussed and preventive measures taken. Pre-job planning before actual crane operations begin can eliminate major craning hazards from the jobsite.

Job Hazard Analysis

Before actual craning operations are begun at the job site, a specific job hazard analysis should be made to ensure that preconstruction planning is adequate. When pre-job planning has been neglected, this on-site job hazard analysis is necessary to ensure that craning operations can be done safely.

Hand Signals

Before any lifts are commenced, all parties, including the crane operator, signalers, riggers, and others involved, must refamiliarize themselves on appropriate hand signals. Often signals vary from job to job and region to region. It is best to ensure that everyone is familiar with the hand signals outlined in ANSI/ASME B30.5, Mobile and Locomotive Cranes. OSHA 1926.550(a)(4) states: "Hand signals to crane and derrick operators shall be those prescribed by the applicable ANSI standard for the type of crane in use. An illustration of the signals shall be posted at the job site."

Signaling Devices

On lifts where the signalers are outside the direct view of the operator due to elevation or in blind areas, either a telephone or radio is a necessity. Communication needs to be agreed upon in preconstruction planning and the prejob hazard analysis.

Lifting Capabilities

During preconstruction planning, lifting requirements should be analyzed by an engineer competent in this field to establish if the crane to be used has adequate lifting capability. The job hazard analysis should also determine and verify that the crane to be used has sufficient boom length for the lift.

Rigging Practices

OSHA 1926.251, "Rigging Equipment for Material Handling," defines requirements for rigging equipment.

Controlling the Load

The use of tag lines to control movement of the load is very important. Normally when hoisting a load, the lay or twist in wire rope causes rotation when the load becomes suspended. OSHA 1910.180(h)(3)(xvi) states: "A tag or restraint line shall be used when rotation of the load is hazardous."

Wire Rope Requirements

It is very important to comply with the crane manufacturer's recommendations for type of wire rope to be used for various hoist lines of pendants.

Annual Inspections

OSHA 1926.550(a)(6) requires an annual inspection. There are a number of firms certified to perform these inspections.

Preventive Maintenance

Cranes require continual servicing and preventive maintenance. Programs should be documented consistent with the crane manufacturer's recommendations.

COMPETENT PERSON REQUIREMENTS

Operators

As cranes become more sophisticated and are able to lift heavier loads higher, further, and faster, more and more electronic systems to monitor every aspect of operator performance are being included in design. The day of total reliance upon seat-of-the-pants operator skills is gone. Human response is not fast enough to cope with many of the rapidly changing circumstances involved in crane operation. Today's crane can be compared to an airplane, not just in terms of cost, but in its complexity of operation that requires well-qualified professional operators. Effective licensing programs for crane operators by some employers include minimum requirements for

- Education level
- Apprenticeship training and work experience
- Classroom training on crane safety
- Thorough knowledge of crane safety references
- Physical qualification
- Age (mature, intelligent)
- Vision
- Hearing
- Physical stamina
- Good coordination, reaction, and tested skill level
- No history of heart problem or other ailments that produce seizures
- Emotional stability
- Absence of addictions

Riggers, Signalers, and Others

Riggers, signalers, and others who work with cranes should have qualifications similar to those of the operator. Just as an unqualified operator can make a life-threatening error during lifting operations, the inappropriate actions of an inexperienced rigger, signaler, or anyone else involved in lifting operations can cause an accident.

HAZARDS COMMON TO MOST CRANES

The hazards which are found around movable cranes are presented below. These are the most common hazards found around crane and derrick incidents:

- Powerline contact
- Overloading
- Outrigger failure (soft ground and structural)
- Two-blocking
- Pinchpoint
- Unguarded moving parts
- Unsafe hooks
- Obstruction of vision
- Sheave-caused cable damage
- Cable kinking
- Side pull
- Boom buckling (from striking objects)
- Access to cabs, bridges, and/or runways
- Control confusion (nonuniform location)
- Turntable failure
- Removable or extendible counterweight systems

POWERLINE CONTACT

Definition

Inadvertent contact of a bare, uninsulated, high-voltage powerline by any metal part of a crane.

Description

Most powerline contacts occur when a crane is moving materials adjacent to or under energized powerlines, and the hoist line or boom touches a powerline. Sometimes the person who is electrocuted, who may be incidental to the operation or use of the crane, has been touching the crane or getting on or off of it when the hoist line or boom is in contact with an energized powerline.

In some circumstances, when sufficient ground fault is created, the electric distribution circuit is automatically deenergized by a reclosure system to avoid the blowing of intervening fuses.

The faulted circuit is automatically reenergized several seconds later. When warning is not posted on the crane or printed in the operator's manual about the functioning of such reclosure systems, victims may be reshocked and rescuers endangered if it appears that the powerline is permanently deenergized.

Risks Presented by Hazard

Powerline contact presents the highest risk in craning operations. Each year, approximately 150 to 160 people are killed by powerline contact and about three times that number are seriously injured. On an average, eight out of ten of the victims were guiding the load at the time of contact.

Preventive Measures

The key to avoiding powerline contact is pre-job safety planning, the greatest accident deterrent available in the workplace. Because of the large number of individuals and employers at a work site, a single individual should have overall supervision, coordination, and authority for the project and project safety.

Cranes and powerlines should not occupy the same work area. Contractors should not rely upon the electric utility to erect either temporary or permanent powerlines at the worksite with clearances for crane. Some work areas with existing powerlines have clearances acceptable for normal roadway traffic but not for cranes. The crane operator, those guiding the load, and those closely involved in the particular craning operation need some visual guidance on the ground so they are made aware of the danger zone so they can conduct all of their work outside of this dangerous area. The area within a radius of ten feet in any direction from powerlines is an unsafe work area and must be clearly marked off on the ground by marker tape, fences, barriers. Everyone at the worksite must ensure for themselves that the crane is positioned so that the boom and hoist line cannot intrude into the danger zone created by the powerlines.

If the danger zone can be penetrated by a crane boom, the electric utility must be notified to deenergize, relocate, bury, or insulate the lines while the crane is operating in that location.

Where cranes are working in the vicinity of powerlines, a good practice is to have the utility company lock out the automatic reclosures.

OSHA Requirements

1. OSHA 1926.550(a)(15) states: "when electrical distribution and transmission lines have been deenergized and grounded at point of work, not a part of an attachment to the equipment will be erected to prevent contact with the lines."
2. OSHA 1926.955(e)(5) states: "The automatic reclosing feature of circuit interrupting devices shall be made inoperative where practical before working on any energized line or equipment."

OVERLOADING

Definition

Overloading occurs when the rated capacity is exceeded or the load of a crane is tipped while attempting to lift or maneuver a load, resulting in upset or structural failure.

Description

Cranes can easily upset from overloading. The margin of safety between the actual tipping load and rated capacity often varies from 15 to 25 percent, based on the type of crane.

In some types of cranes, the location of the boom in relation to where the carrier is mounted determines the margin of safety between tipping load and rated capacity.

For a crane mounted on a flatbed truck with the hydraulic boom located directly behind the truck cab, the tipping load over the rear of the flatbed or over the cab can often be as much as twice the rated capacity, but the crane can easily upset when the load is slewed (rotated) to either side.

The mere weight of a boom without a load can create an imbalance and cause some high-reach hydraulic cranes to upset when the boom is positioned at a low angle. This has occurred even with outriggers extended.

Today's crane operator is confronted with a number of variables that affect lifting capacity:

- The ability to extend a hydraulic boom while raising or lowering it increases the radius very swiftly and reduces lifting capacity quickly.
- Whether the operator chooses to extend or retract outriggers is a variable that affects crane stability.
- The crane's tipping capacity can also vary when the boom is positioned at the various points of the compass or clock in relation to its particular carrier frame.
- In many instances, the operator may not know the actual weight of the load. Often the operator relies upon his perception, instinct, or experience to determine whether the load is too heavy, and the operator may not respond fast enough when the crane begins to feel light.

All of these variables create circumstances that can lead to a crane operator exceeding the rated capacity, tipping the load, and upsetting the crane or structural failure of the crane.

Risks Presented by Hazard

Estimates on crane upsets state that they occur about every 10,000 hours of crane use. Approximately three percent of upsets result in death, eight percent in lost time, and 20 percent in damage to property other than the crane. Nearly 75 percent is due to predictable human error when the operator inadvertently exceeds the crane's lifting capacity.

Preventive Measures

During the last 30 years, the advent of solid-state micro-processing electronics, for load measuring systems have evolved. Crane operation is no longer a seat-of-the-pants skill but requires both planning and training in the use of the latest technology for load-measuring.

These load-measuring systems can sense the actual load as related to boom angle and length, warn the operator as rated capacity is approached, and stop further movement. Load-measuring systems automatically prevent exceeding the rated capacity at any boom angle, length, or radius. Today most U.S. crane manufacturers are promoting the sale of user-friendly load-measuring systems to avoid this hazard.

Before computerized load-measuring systems, the only control to avoid upset from overload has been an operator's performance using difficult-to-read load charts. Formal training should be provided for all crane operators to ensure competency in the use of crane load charts.

OSHA Requirements

Generally, OSHA requirements are silent on load-measuring and/or load-moment devices but require compliance with the manufacturer's load rating charts.

ANSI requirements also address compliance with manufacturer's rating charts and are currently silent on load-measuring and load-moment devices.

FAILURE TO USE OUTRIGGERS, SOFT GROUND, AND STRUCTURAL FAILURE

Definition

Crane upset can occur when an operator does not extend the outriggers, when a crane is positioned on soft ground, and when structural defects exist.

Description

Many cranes upset because use of outriggers is voluntary and usually left to the discretion of the operator who might not perceive a potential hazard. Some of the hazards include the following:

- An operator cannot extend the outriggers because of insufficient space or a work circumstance that arises when preplanning is not done.
- Outrigger pads are too small to support the crane even on hard ground, let alone on soft ground, which poses a problem in itself.
- A mobile hydraulic rough-terrain crane on "rubber" with the outriggers retracted is completely unstable on a side lift.
- Whether a crane is on rubber or truck-mounted, if outriggers are not extended, the lifting capacity at side angles drops dramatically. As the crane boom rotates on the turntable, upset occurs so quickly that the operator cannot perceive the loss of stabil-

ity until too late. Sometimes the weight of the boom at a low angle will upset the crane even with the outriggers properly extended.

- In a few instances, outriggers have collapsed because they were not strong enough or had been damaged.
- The rear outrigger on some cranes disengages from the float or pad when the load is first lifted. When the crane cab/boom is swung around 180 degrees, the outrigger can slip from such an unfixed pad connection and cause the crane to upset or the boom to buckle as a result of loss of stability.

Risks Presented by Hazard

An analysis of some one thousand crane accidents has shown that half of the incidents involving this hazard occurred when the crane operator was either swinging the cab or extending or lowering a telescoping boom without outriggers extended. These actions rapidly increase the lifting radius, so upset occurs quickly.

Preventive Measures

Since a high proportion of accidents occur when outriggers are not extended, design changes to overcome this hazard would be desirable. Some aerial basket designs include limit switches to prevent boom movement until outriggers are extended and in place.

Outriggers/soil failures occur because either the ground is too soft or the outrigger pads are not broad enough for the soil type. Wet sand can only support 2,000 pounds per square foot; dry hard clay can support 4,000 pounds per square foot compared to well-cemented hardpan that can support as much as 10,000 pounds per square foot. When poor supporting soil is encountered, or the outriggers have inadequate floats or pads, well-designed blocking or cribbing is needed under the outriggers to extend the base of the outrigger support. ANSI requires that where floats are used on outriggers, they shall be securely attached. It also requires that blocking used to support outriggers shall be strong enough to prevent crushing, be free of defects, and be of sufficient width and length to prevent shifting or toppling under load.

OSHA Requirements

OSHA 1910.180(h)(3)(ix), Marine Terminal Standard 1917.45(f)(7) and Longshoring Standard 1918.74(a)(2) discusses outriggers.

TWO-BLOCKING

Definition

Two-blocking is the contact of the hoist block or hook assembly with the boom tip which causes parting of hoist line and loss of the load, hook, etc., endangering workers below.

Description

Both latticework and hydraulic boom cranes are prone to two-blocking. When two-blocking occurs on latticework booms, the hoist line picks up the weight of the boom and lets the pendant guys go slack.

Often a great deal of whip is created when walking a crawler crane with a long boom without a load, and the headache ball and empty chokers can drift up to the boom tip. Ordinarily, while watching the pathway of travel to avoid any rough ground that can violently jerk the crane, the crane operator does not watch the boom tip. When a hoist line two-blocks, it assumes the weight of the boom and relieves the pin-up guys of thee load. Then, if the crane crawler breaks over a rock or bump, the flypole action of a long boom is sufficient to break the hoist line.

The weight of the load plus the weight of the boom on a latticework boom, when combined with a little extra stress when lifting a load, can cause the hoist line to break if two-blocking occurs.

The power of the hydraulic rams that extend hydraulic booms is often sufficient to break the hoist line if the line two-blocks. In many circumstances, both latticework and hydraulic boom cranes will two-block when the hook is near the tip and the boom is lowered.

An operator can forget to release (payout) the load line when extending the boom. When this occurs, the hoist line can be inadvertently broken. If the load line breaks and it happens to be supporting a worker on a bosun's chair or several workers on a floating scaffold or a load above people, a catastrophe can result. Human factors logic tells us that when an operator must use two controls, one for the hoist and one for the hydraulic boom extension, the chance of error is increased.

Risks Presented by Hazard

There have been hundreds of deaths and crippling injuries resulting from two-blocking occurrences. Over the years there have probably been thousands of two-blocking occurrences that have broken the hoist line, with most of them going unrecorded because no one is injured when the hoist line fails and drops the hook and/or load.

Preventive Measures

Anti-two-blocking devices have been available for years, but industry acceptance has lagged behind when it comes to adopting this device as a preventive measure.

There are several ways to prevent two-blocking:

- Employ an electrical sensing device. Attach a weighted ring around the hoist line that is suspended on a chain from a limit switch attached to the boom tip. When the hoist block or headache ball touches the suspended, weighted ring, the limit switch opens and an alarm warns the operator. It can also be wired to intercede and stop the hoisting. The circuitry is no more complex than an electric door bell.

- On hydraulic cranes, the hydraulic valving can be sequenced to play out the hoist line when the boom is being extended, thus avoiding two-blocking.
- When making a lift, sufficient boom length is necessary to accommodate both the boom angle and sufficient space for rigging, such as slings, spreader bars, and straps. To avoid bringing the hook and headache ball into contact with the boom tip, a boom length of 150 percent of the intended lift is required for a boom angle of 45 degrees or more.
- Anti-two blocking devices should be standard equipment on all cranes. Currently most new mobile hydraulic cranes are being equipped with anti-two-blocking systems.

OSHA Requirements

OSHA 1926.550(g)(3)(ii)(c) only requires an anti-two blocking device on cranes being used to suspend a personnel platform.

PINCHPOINT

Definition

The pinchpoint is the accessible area within the swinging radius of the rotating superstructure of a crane in which people may be working and could be crushed or squeezed between the carrier frame and the crane cab, or the crane cab and an adjacent wall or other structure.

Description

A pinchpoint is created by the narrow clearance between the rotating superstructure (cab) of a crane and the stationary carrier frame. Also, when a crane must be used in a confined space, another dangerous pinchpoint is the close clearance between the rotating cab/counterweight and a wall, post, or other stationary object. This hazard is inherent in rough terrain, truck-mounted, crawler cranes, and other mobile cranes. Many people, especially oilers, have been inadvertently crushed when caught in such pinchpoints.

Analysis of these occurrences shows that the victims were usually "invited" into the danger zone for reasons such as these:

- Access to the water jug
- Access to the tool box
- Access to the outrigger controls
- Access to perform maintenance
- Access to storage of rigging materials

In all of the known cases where someone entered the danger zone and was caught in a pinchpoint, the danger zone was outside the crane operator's vision. The survivors stated they believe that the crane operator was not going to rotate or slew the boom at that particular moment.

Risks Presented by Hazard

This hazard can result in loss of life or serious crushing injuries. Over forty deaths or serious injuries have been recorded due to this hazard.

Preventive Measures

The swing area of the crane cab and counterweight shall be barricaded against entry into the danger zone.

Crane design that provides 12 to 14 inches of clearance between the cab and the carrier frame or crawler tracks would eliminate the hazard and avoid serious crushing injury.

Removal of water jugs, tool boxes, and rigging materials from this dangerous area would reduce the incentive to enter the danger zone.

Installation of rear view mirrors for the crane operator would provide a redundant safeguard so the operator could see into the turning area of the cab and counterweight.

Installation of an audible alarm that would automatically go off when movement was begun would alert those working in adjacent areas.

Installation of a panic bar or cord that could be activated by anyone entering this blind danger zone would advise the crane operator that someone was in the danger zone so the crane could stop movement until the area was clear.

OSHA Requirements

Subpart N, 1926.550(a)(9), requires the following: "Accessible areas within the swing radius of the rear of the rotating superstructure of the crane, either permanently or temporarily mounted, shall be barricaded in such a manner as to prevent an employee from being struck by the crane."

UNGUARDED MOVING PARTS

Definition

All unguarded moving or rotating parts within the crane cab, engine compartment, or service area that are accessible to people that must enter or reach into the crane's mechanical housing or compartment.

Description

Oilers, maintenance personnel, and crane operators are often injured within the crane cab when caught by one of the many unguarded moving parts. The crane cab is a convenient place to store items essential to crane operations. The cramped machinery space inherent in crane cabs makes it easy to become entangled with unguarded moving parts.

Risks Presented by Hazard

Loss of fingers, hands, or feet is very common when moving parts are unguarded. Such parts insidiously wait for the unwary to become entrapped.

Preventive Measures

It is commonly assumed by designers that access doors and panel covers serve as guards to keep people away from moving parts, and that cranes are always shut down when adjustments or servicing is to be done. These assumptions are entirely wrong. Because it is so handy, the cab is commonly used as a convenient storage space for oil, spare parts, tools, lunch boxes, clothing, and other items by maintenance people and others. For this reason, any moving parts that are accessible to these people need their own guards, even though inside the cab or a compartment.

OSHA Requirements

OSHA 1926.550(a)(8) covers "guarding" of moving parts.

UNSAFE HOOKS

Definition

Lifting hooks that do not have latches or have damaged and/or defective latches to safely retain load straps, cables, or chain are often called killer hooks.

Description

A safe lifting hook is a critical component when lifting a load. The most unsafe hook is one that has no latch to secure the load straps or chain within the throat of the hook. Many hooks have thin sheet metal latches that are easily damaged or bent and are totally worthless when it comes to securing the straps safely within the hook. Load hoist blocks found on some cranes fit flat to a boom. When a boom is in a lowered position, the straps rest deep in the throat of the hook. When the boom is raised in a nearly vertical position, the hook often rotates towards the boom so the opening is positioned downward so the straps can easily slide out of the hook. Another problem inherent with hooks is that as normal stresses gradually wear away and enlarge the throat, the possibility of the straps slipping from the hook becomes very real.

Often a lifting hook is attached to the back of the bucket of a backhoe so slings can be attached and the machine used as a lifting device. When the bucket is turned down underneath the boom, the throat of the hook can turn downward, allowing the straps to slip out of the hook. To avoid this hazard, a ring instead of a hook should be used, with a shackle to connect the lifting straps.

When two-blocking occurs, the load hook can also rotate up on the boom tip sheave, allowing the slings to fall from the throat of the hook.

When there is no latch on a hook, loose straps without a load can easily be displaced and fall.

Risks Presented by Hazard

There are countless occurrences of the straps slipping from the hook and dropping the load.

Preventive Measures

Over a hundred patented, positive-type safety latches that prevent the hook from opening are available in today's market. Most require the depression of a latch before the hook will open.

Use of a vertical swivel allows the hook to remain vertical and prevents dumping the straps from an upended hook. Wire mousing, or wrapping the hook with wire to secure the load in a place of a latch, may secure straps within the throat of a hook. Experience has shown that the use of wire mousing in place of a hook latch is inadequate, as the wire is only good for one use because it weakens with each use.

OSHA Requirements

OSHA 1926.550(g)(4)(iv)(B) discusses personnel hoisting in construction and the positive closure of hooks.

OBSTRUCTION OF VISION

Definition

Safe use of a crane is compromised when the vision of an operator, rigger, or signaler is blocked and they cannot see each other to know what the other is doing.

Description

There are two general categories off obstruction of operator's vision:

- Obstruction by the crane's own bulk
- Obstruction by the work environment

Crane size alone limits the operator's range of vision and creates many blind spots, preventing the rigger, signaler, oiler, and others who are affected by the crane's movement from having direct eye contact with the crane operator. When a cab-controlled mobile crane is moved or travels back and forth, the operator must contend with many blind spots on the right side of the crane.

Many situations arise in craning activities that can almost instantaneously turn what was first considered a simple lift into a life-taking catastrophe:

- Many people are affected by a crane's movement. Welders with their hoods on, carpenters, ironworkers, or those engaged in other crafts may be working in the immediate vicinity of a crane, may be preoccupied with their tasks, and may be momentarily oblivious of the activity of the crane. They also may be out of the range of vision of the crane operator. This lack of awareness on the part of others and the obstructed outlook of the crane operator both contribute to craning accidents.
- In many instances the work environment requires that loads be lifted to or from an area that is outside of the view of the operator.
- Often a load is lifted several stories high, and the crane operator must rely upon others to ensure safe movement of the load being handled.

Risks Presented by Hazard

When operators, riggers, signalers, oilers, and others cannot see each other or the suspended load, the risk of accident becomes very high.

Preventive Measures

The key to a safe craning operation is preplanning of all activities, starting with pre-job conferences and continuing with daily planning, to address any changes that need to be made.

The use of automatic travel alarms is an effective way to give warning to those in the immediate vicinity of crane travel movement.

To overcome the hazard of lifting loads from blind spots, the use of radio and telephones is much more effective than relying upon several signalers to relay messages by line of sight.

OSHA Requirements

OSHA 1926.16, "Rules of Construction," gives some guidance for preplanning, and OSHA 1926.201, under a section entitled "Signaling," addresses signals for cranes and refers to appropriate ANSI standards.

SHEAVE-CAUSED CABLE DAMAGE

Definition

Cable damage is accelerated by the use of sheaves of inadequate diameter or inappropriate groove configurations. This damage usually occurs because the throats or grooves of the sheaves and drums do not meet the minimum diameters recommended by wire rope manufacturers.

Description

Wire rope makes today's craning possible. Without it, we would not have our modern cranes. As indispensable as it is, wire rope is very fragile as it is very vulnerable to bending or pinching as it passes over a sheave or pulley. Sheave and drum diameters must be as large as possible. As a wire rope cable passes over a sheave, the outer strands are stretched and compressive forces are exerted on the inner strands. Wire rope has the ability to stretch in two ways, constructionally and elastically. Constructional stretch is the lay or twist that causes it to rotate. Its elasticity of each separate wire strand that comprises the rope is infinitesimal when compared to that of a rubber band. When a wire strand is stretched beyond its elasticity, it can never return to its former shape and is permanently weakened.

Crane cable is often damaged because the angle and width of the throat or groove of the sheave is insufficient, resulting in excessive wear. If a throat is too wide to support the wire rope, the rope can flatten; if it is too narrow, the rope will bind.

It must be considered that most cranes do not meet the wire rope manufacturers' design requirements, so excessive wear of crane cables will occur.

The actual number of lifts and lowerings must also be considered. Just as a car will run out of gas after a certain number of miles are driven, wire rope will wear out after a certain amount of travel over the sheave.

Preventive Measures

Inspect the sheave diameter and the groove to determine if they match the cable size specified by the crane manufacturer.

Cable must be inspected daily. Wire rope should be taken out of service when six randomly distributed broken wires occur in one lay, or three broken wires in one strand in one lay.

A properly lubricated hoist cable will reduce wear.

OSHA Requirements

OSHA 1926.550(a)(6) and (7), define and address wire rope requirements.

CABLE KINKING

Definition

Cable is sometimes damaged during work circumstances or may be otherwise abused by improper handling, creating kinks and bends.

Description

Allowing a wire rope to become slack can cause a crane cable to kink or "birdcage." Slackening of a rope can occur when there is insufficient weight on the load hook, causing the rope to ride up on the boom tip. The slack is then picked up by the heavier weight of the rope extending down from the boom tip to the hoist drum causing the rope to fall in loops in front of the hoist drum. A kink starts when a loop is formed, pulled tight, and the natural lay of the rope is lost.

On cranes with latticework booms, the wire rope can become fouled and kinked as the boom is lowered and the boom hoist assembly is slackened.

Risks Presented by Hazard

Kinked wire rope cable must not be used because it is unreliable and can pull apart, causing the suspended load to fall.

Preventive Measures

Fair leads and sheave guides should be used to prevent a slack line from rolling out of a sheave. Wire rope with kinks or birdcages must be removed from service and destroyed.

OSHA Requirements

OSHA 1926.181(g)(1) covers inspection of wire rope.

SIDE PULL

Definition

The lateral forces imposed upon a boom when the load line is tensioned to either the right or left of the boom causes side pull.

Description

When a crane sits on a slope and picks up or suspends a load, the boom is subjected to side pull. Side pulls sometimes arise when two cranes work together to lift a single load.

Risks Presented by Hazard

Side pulls can easily collapse either latticework or hydraulic cantilevered booms. The longer the boom, the more serious the hazard.

Preventive Measures

Cranes must be made level before use. Leveling bubbles or other sensing systems are available. Sensing devices can also be installed that inform the crane operator that a side pull is developing.

OSHA Requirements

OSHA 1910.179(n)(3)(iv) states: "Cranes shall not be used for side pulls except when specifically authorized by a responsible person who has determined that the stability of the crane is not thereby endangered and that various parts of the crane will not be overstressed."

OSHA 1910.180(h)(3)(iv) states: "Side loading of booms shall be limited to freely suspended loads. Cranes shall not be used for dragging loads sideways."

BOOM BUCKLING

Definition

When a boom is lowered onto or strikes a structure during slewing, the boom cannot sustain side forces, particularly when supporting a suspended load, and will easily collapse.

Description

When cranes are used to make lifts in very confining locations, the mid section of the boom sometimes strikes a nearby structure, especially when the load is being lifted over such a structure.

Risks Presented by Hazard

It does not take tremendous force to cause a boom to buckle when it strikes against a structure. Those below are in immediate danger of being killed or injured by the load or the buckling boom.

Preventive Measures

It is better to use tower cranes when the lift has to extend over and into other parts of a building.

OSHA Requirements

OSHA is silent on this specific hazard. However, OSHA "General Safety and Health Provisions," apply.

ACCESS TO CABS, BRIDGES, AND/OR RUNWAYS

Definition

Cabs, bridges and/or runways are the means by which crane operators, oilers, maintenance personnel, and others have access onto, into, or about the crane to perform their duties.

Description

The design of many cranes does not take into consideration the needs of crane operators, oilers, maintenance personnel, and others for safe entry and exit to and from their particular work areas on the crane. Often access is inconvenient, difficult, and unsafe.

Risks Presented by Hazard

When those who must work in and about a crane are provided with only marginally safe access due to inadequate ladders, steps, walkways, and platforms, the risk of falls is great.

Also, smooth surfaces over which people must gain access may be coated with oil, grease, or mud, making access slippery. Ladders, steps, walkways, and platforms may also be without sufficient handholds.

Preventive Measures

The availability of nonslip steel treads, surfacing materials, along with a wide variety of handholds, leave little reason to have unsafe access on and off of cranes. There is also available easily assembled and placed scaffold planks of adequate capacity and standard railing attachments, and/or safety belts and lifelines with safe attachment points.

OSHA Requirements

OSHA 1926.550(a)(13) covers location and other access requirements for crane cabs, bridges, and/or runways.

CONTROL CONFUSION

Definition

Non uniform control placement or location of controls for the operation of cranes contributes to operator error and inadvertent activation.

Description

The many different makes and models of cranes, like automobiles, are operated by many different people, but there is a difference. On automobiles, brakes, accelerators, and other basic controls are always in the same place, so people who rent different cars while on business travel do not have a problem with the basic controls. On cranes, the location of controls often varies, so crane operators who often change from crane to crane in normal work circumstances are continually faced with having to operate controls in different locations. Additionally, cranes do many more things than automobiles, and crane operation is much more complex. A crane operator often has to handle multiple controls concurrently.

Risks Presented by Hazard

The absence of uniformity and the many variables of operation create many opportunities for operator error. A crane operator may be simultaneously slewing (rotating the crane cab), raising the boom, extending the boom, and hoisting a load to a point eight or ten stories high. When the load nears its destination and just prior to lowering it into place, the operator must stop slewing, raising, extending, and hoisting. Such multiple concurrent actions are a test of operator skill and coordination. Even with the best of operators, there is much room for error.

Preventive Measures

Since work circumstances often require operators to work with a variety of cranes, controls need to be made uniform and consistent from make to make, model to model, to reduce operator error. When possible, crane operators should be assigned to the same crane or the same make and model to avoid having to work with dissimilar controls.

OSHA Requirements

OSHA is silent on this specific hazard.

TURNTABLE FAILURE

Definition

Turntable failure is a failure of the rotating mechanism of the crane, which causes it to fall from its turntable.

Description

During a heavy lift, the bolts holding the rotating crane hoist structure to the turntable ring or securing the pedestal have been known to fail. An overload usually triggers such failure. Investigation usually shows that the connecting bolts or pins had become loose and/or were fatigued from previous overstressing.

Crane accident investigations of turntable failure reveal that crane operators are generally unaware of this hazard.

Risks Presented by Hazard

Because cranes are designed to operate under such close tolerances, a lift in excess of rated capacity is possible, particularly when a lift over the cab of a truck cab is undertaken (see Overloading).

Preventive Measures

Inspection of the turntable assembly should be part of the annual inspection to determine whether damage or dangerous overstressing has occurred. Also, inspections should be made for a positive program to prevent overloading of the crane.

OSHA Requirements

OSHA 1910.180(h)(3)(vii) states: "On truck-mounted cranes, no loads shall be lifted over the front area except as approved by the crane manufacturer."

REMOVABLE OR EXTENDIBLE COUNTERWEIGHT SYSTEMS

Definition

Many cranes utilize removable or adjustable counterweights to make the crane transportable. However, methods off removing or adjusting counterweights are often error-provocative and dangerous.

Description

The need for more counterweights increases as cranes are designed to lift greater loads higher and further. Because many of our streets, highways, and bridges have safe load limits, many cranes arrive at the work site without counterweights attached. The design of some counterweights includes a self-lifting system for loading and unloading the counterweights.

Some self-lifting systems are designed only to lift or lower the counterweight from a trailer. If a trailer is unavailable, a second crane is need to lift the counterweight if the crane does not have an alternate lifting system.

A hydraulic ram is used on a few cranes to increase the capacity of the counterweight. On one model the control panel for the ram is located so that the person operating the controls cannot ascertain whether the area surrounding the counterweight is clear before extension or retraction is begun. On this particular model, the storage bin for outrigger floats is located under the area of extension or retraction, which creates an invitation for workers to enter this danger zone. In one instance, a worker's head was crushed against the cab by the retracting counterweight. Other means are used on other cranes to extend counterweights further out.

Risks Presented by Hazard

Counterweights can unintentionally be dropped during attachment/detachment due to the close tolerances required. Serious crippling injuries have resulted from error-provocative counterweight systems.

Preventive Measures

Removable or extendible counterweight systems need to be examined to identify the inherent hazards.

OSHA Requirements

OSHA Subpart C, "General Safety and Health Provisions," 1926.20(a) and (b), require a safe workplace.

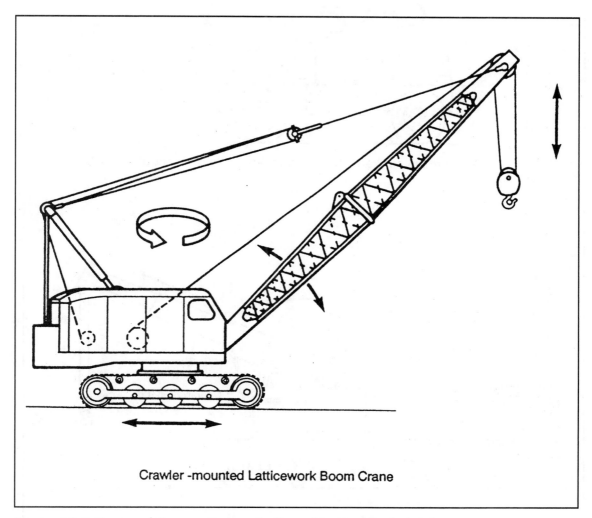

Crawler-mounted Latticework Boom Crane

Figure 13.1 Types of Cranes Generally Found in the Workplace[1]

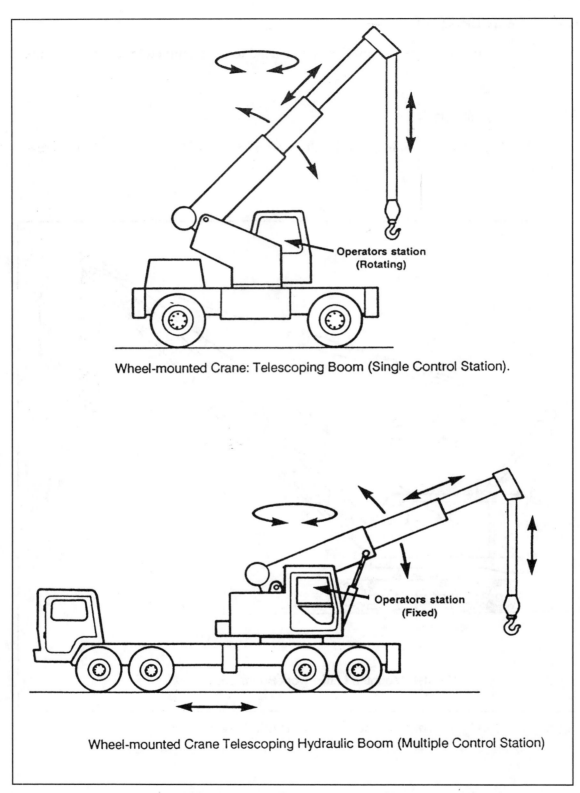

Wheel-mounted Crane: Telescoping Boom (Single Control Station).

Wheel-mounted Crane Telescoping Hydraulic Boom (Multiple Control Station)

Figure 13.1 Types of Cranes Generally Found in the Workplace, continued

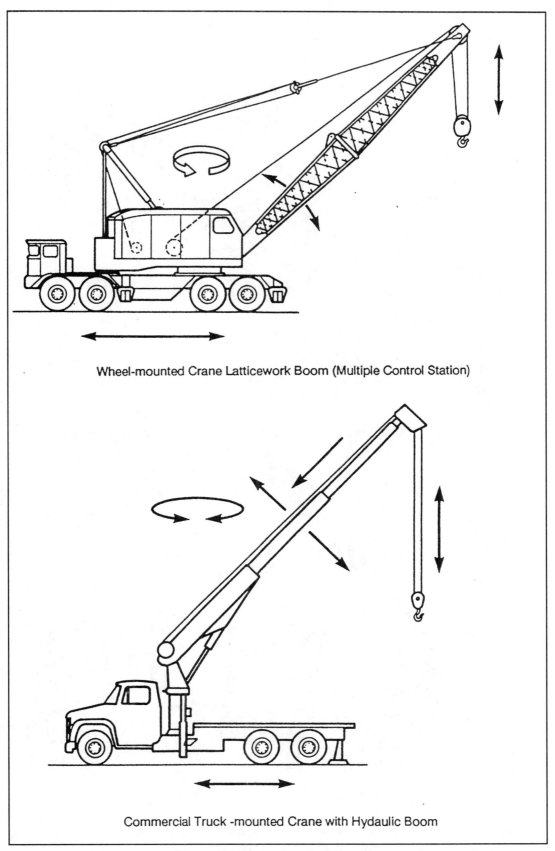

Wheel-mounted Crane Latticework Boom (Multiple Control Station)

Commercial Truck -mounted Crane with Hydaulic Boom

Figure 13.1 Types of Cranes Generally Found in the Workplace, continued

325

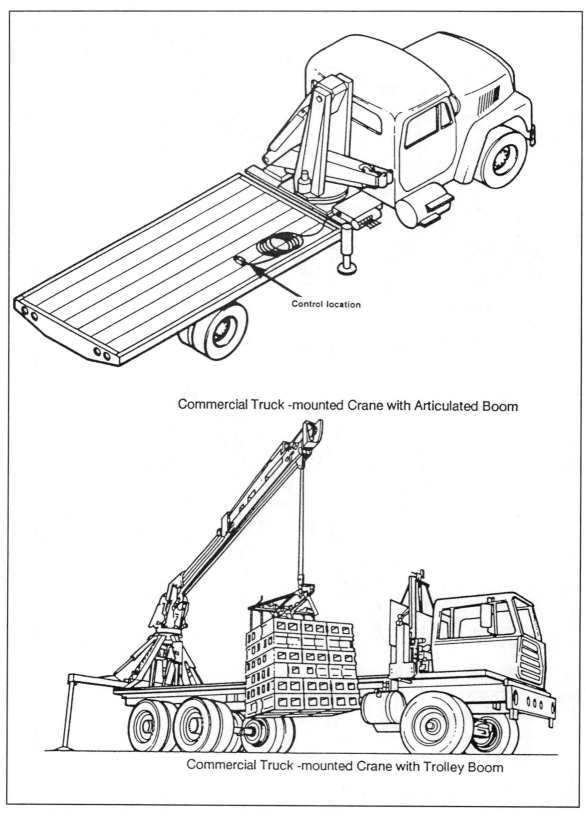

Control location

Commercial Truck -mounted Crane with Articulated Boom

Commercial Truck -mounted Crane with Trolley Boom

Figure 13.1 Types of Cranes Generally Found in the Workplace, continued

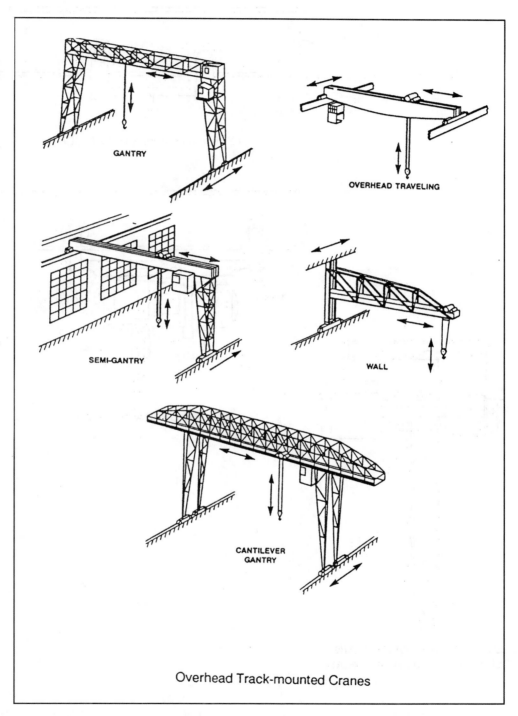

GANTRY

OVERHEAD TRAVELING

SEMI-GANTRY

WALL

CANTILEVER GANTRY

Overhead Track-mounted Cranes

Figure 13.1 Types of Cranes Generally Found in the Workplace, continued

327

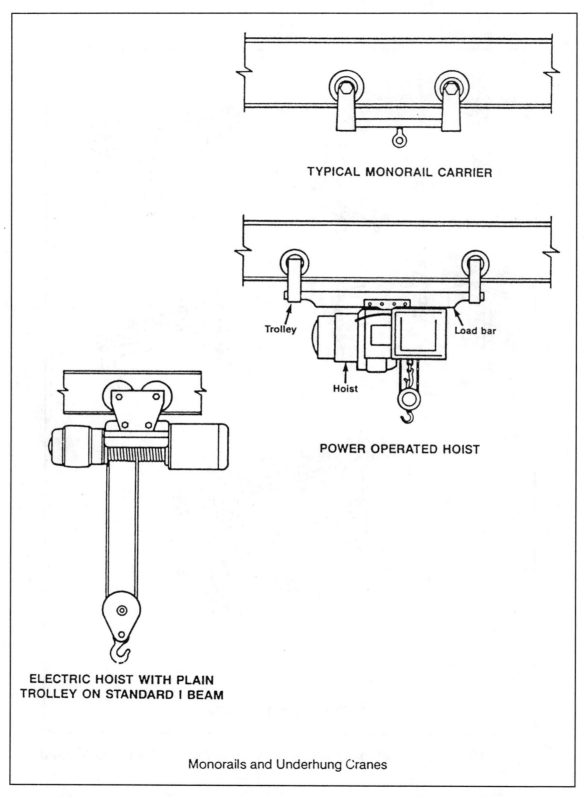

TYPICAL MONORAIL CARRIER

Trolley

Load bar

Hoist

POWER OPERATED HOIST

ELECTRIC HOIST WITH PLAIN
TROLLEY ON STANDARD I BEAM

Monorails and Underhung Cranes

Figure 13.1 Types of Cranes Generally Found in the Workplace, continued

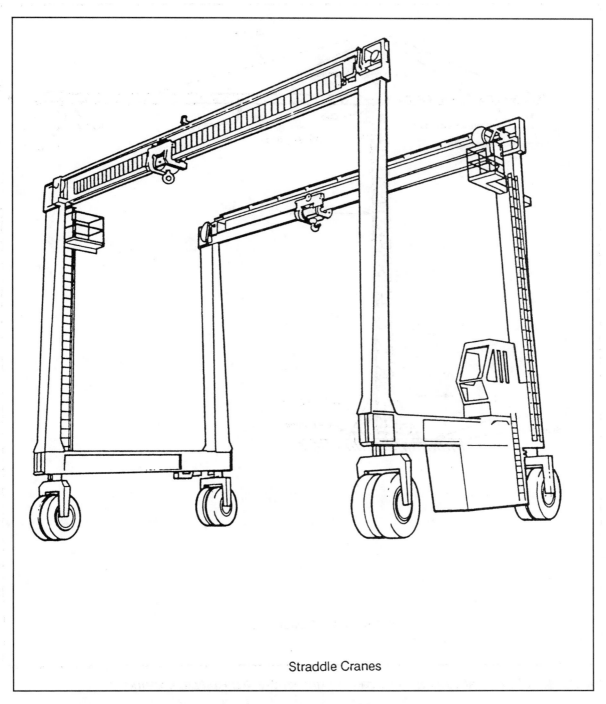

Straddle Cranes

Figure 13.1 Types of Cranes Generally Found in the Workplace, continued

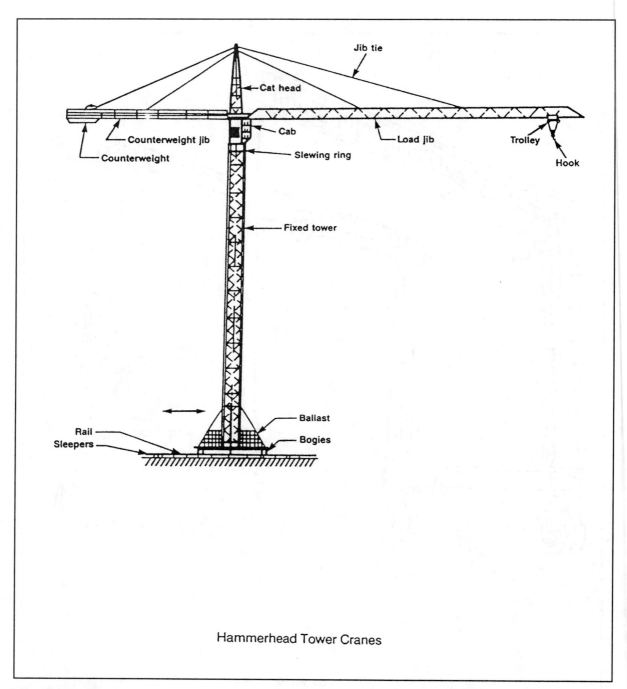

Hammerhead Tower Cranes

Figure 13.1 Types of Cranes Generally Found in the Workplace, continued

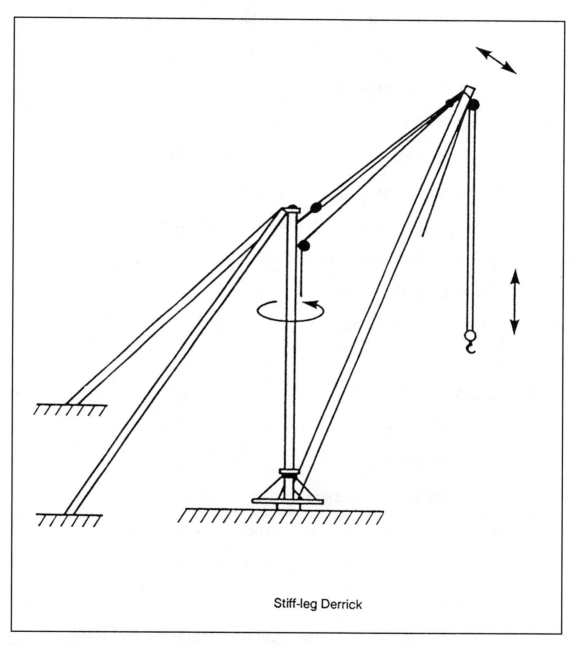

Stiff-leg Derrick

Figure 13.1 Types of Cranes Generally Found in the Workplace, continued

331

FOR USE WITH DAILY CRANE INSPECTION RECORDS

In the following table, please place Y for Yes, or N for No, or NA for Not Applicable to the right of each item.

1. Is item properly installed on machine?

2. Check for cracked or broken glass.

3. Check for proper working order.

4. Check for proper service level.

5. Check for proper reading while operating machine.

6. Check for excess dirt, grease, or foreign matter.

7. Check for proper operation while operating machine.

8. Check for proper adjustment.

9. Check for cracks or leaks.

10. Check for excessive wear.

11. Check torque.

12. Clean or replace.

13. Drain moisture.

14. Change oil filters.

15. Check for bent or cracked cord or lacings.

16. Check for cracked or broken structures and welds.

17. Service as required by engine manufacturer.

18. Lubricate.

19. Check connectors & wiring for proper alignment and insulation.

20. Check wire rope for damaged, frayed, or broken strands.

Figure 13.2 Sample Inspection Checklist

[1] OSHA Instruction CPC 2-2.20B, Chapter 1. 13 November 1990.

14

Forklift Safety

Training Your Forklift Operators in Safe Driving Techniques

The Occupational Safety and Health Administration (OSHA) has revised its existing requirements for powered industrial truck operator training at 29 CFR 1926.602 and has issued new requirements to improve the training of these operators. The new requirements are intended to reduce the number of injuries and deaths that occur as a result of inadequate operator training. They apply to all industries (general industry, construction, shipyards, marine terminals, and longshoring operations) in which the trucks are being used, except agricultural operations.

These provisions mandate a training program that bases the amount and type of training required on

- The operator's prior knowledge and skill
- The types of powered industrial trucks the operator will operate in the workplace
- The hazards present in the workplace
- The operator's demonstrated ability to operate a powered industrial truck safely

Refresher training is required if

- The operator is involved in an accident or a near-miss incident
- The operator has been observed operating the vehicle in an unsafe manner
- The operator has been determined during an evaluation to need additional training
- There are changes in the workplace that could affect safe operation of the truck
- The operator is assigned to operate a different type of truck

Evaluations of each operator's performance are required as part of the initial and refresher training and at least once every three years.

"Only trained and authorized operators shall be permitted to operate a powered industrial truck. Methods shall be devised to train operators in the safe operation of powered industrial trucks."[1]

POWERED INDUSTRIAL TRUCK HAZARDS

Powered industrial trucks are used in almost all industries. They can be used to move, raise, lower, or remove large objects or a number of smaller objects on pallets or in boxes, crates, or other containers. Because powered industrial truck movement is controlled by the operator and is not restricted by the frame of the machine or other impediments, virtually unrestricted movement of the vehicle about the workplace is possible.

The hazards commonly associated with powered industrial trucks vary for different vehicle types, makes, and models. Each type of truck presents different operating hazards. For example, a sit-down, counterbalanced, high-lift rider truck is more likely than a motorized hand truck to be involved in a falling load accident, because the sit- down rider trucks can lift a load much higher than can a hand truck.

The method or means to prevent an accident and to protect employees from injury varies for different types of trucks. For example, operators of sit-down rider trucks are often injured in tip over accidents when they attempt to jump clear of the vehicle as it tips over. Because the operator's natural tendency is to jump downward, he or she lands on the floor or ground and is then crushed by the vehicle's overhead guard. Therefore, operators of sit-down trucks need to be trained to remain in the operator's position in a tip over accident and to lean away from the direction of fall to minimize the potential for injury.

On the other hand, when a stand-up rider truck tips over, the truck operator can exit the vehicle by simply stepping backward, perpendicular to the direction of the vehicle's fall, to avoid being crushed. In this situation, the operator usually should attempt to jump clear of the vehicle, and should be trained accordingly.

Driving a powered industrial truck at excessive speed can result in loss of control, causing the vehicle to skid, tip over, or fall off a loading dock or other elevated walking or working surface. This condition can be made more dangerous because the load being carried sometimes partially obscures the operator's vision. A vehicle that is out of control or being operated by a driver whose view in the direction of travel is restricted can strike an employee, run into a column or other part of the building, or strike stored material, causing the material to topple and injure employees in the area. Effective driver training teaches operators to act properly to minimize these hazards to themselves and other employees.

Other characteristics of a powered industrial truck that affect safe truck operation are the truck's tendency to become unstable; its ability to carry loads high off the ground; and its characteristic mode of steering, i.e., with the rear wheels while being powered by the front wheels. Moving loads upward, downward, forward, and backward causes a shift of the center of gravity and can adversely affect the vehicle's stability. When a load is raised or moved away from the vehicle, the vehicle's longitudinal stability is decreased. When the load is lowered or moved closer to the vehicle, its longitudinal stability is increased. Training also is needed to avoid accidents that can be caused by these characteristics.

Qualifications of Machinery Operators

Only those employees who are determined by the employer to be competent by reason of training or experience; who understand the signs, notices, and operating instructions; and who are familiar with the signal code in use shall be permitted to operate a crane, winch, or other power-operated cargo-handling apparatus, or any power-operated vehicle, or give signals to the operator of any hoisting apparatus.

Exception: Employees being trained and supervised by a designated person may operate such machinery and give signals to operators during training.

Therefore, by incorporating by reference the same ANSI standard that was the source document for 29 CFR 1910.178, this provision imposes the identical truck operator training requirements on the construction industry as they apply to general industry.

Operator Training

Personnel who have not been trained to operate powered industrial trucks may operate a truck for the purposes of training only, and only under the direct supervision of the trainer. This training should be conducted in an area away from other trucks, obstacles, and pedestrians.

The operator training program should include the user's policies for the site where the trainee will operate the truck, the operating conditions for that location, and the specific truck the trainee will operate. The training program shall be presented to all new operators regardless of previous experience.

The training program shall inform the trainee that

- The primary responsibility of the operator is to use the powered industrial truck safely following the instructions given in the training program.
- Unsafe or improper operation of a powered industrial truck can result in death or serious injury to the operator or others; or damage to the powered industrial truck or other property.
- The training program shall emphasize safe and proper operation to avoid injury to the operator and others and prevent property damage, and shall cover the following areas:

  Fundamentals of the powered industrial truck(s) the trainee will operate, including
    ~ Characteristics of the powered industrial truck(s), including variations between trucks in the workplace
    ~ Similarities to and differences from automobiles
    ~ Significance of nameplate data, including rated capacity, warnings, and instructions affixed to the truck
    ~ Operating instructions and warnings in the operating manual for the truck, and instructions for inspection and maintenance to be performed by the operator
    ~ Type of motive power and its characteristics
    ~ Method of steering
    ~ Braking method and characteristics, with and without load
    ~ Visibility, with and without load, forward and reverse
    ~ Load handling capacity, weight and load center
    ~ Stability characteristics with and without load, and with and without attachments
    ~ Controls—location, function, method of operation, identification of symbols

~ Load handling capabilities, forks, attachments
~ Fueling and battery charging
~ Guards and protective devices for the specific type of truck
~ Other characteristics of the specific industrial truck

Operating environment and its effect on truck operation:
~ Floor or ground conditions including temporary conditions
~ Ramps and inclines, with and without load
~ Trailers, railcars, and dockboards (including the use of wheel chocks, jacks, and other securing devices)
~ Fueling and battery charging facilities
~ The use of "classified" trucks in areas classified as hazardous due to risk of fire or explosion, as defined in ANSI/NFPA 505
~ Narrow aisles, doorways, overhead wires and piping, and other areas of limited clearance
~ Areas where the truck may be operated near other powered industrial trucks, other vehicles, or pedestrians
~ Use and capacity of elevators
~ Operation near edge of dock or edge of improved surface
~ Other special operating conditions and hazards which may be encountered

Operation of the powered industrial truck:
~ Proper preshift inspection and approved method for removing from service a truck which is in need of repair
~ Load handling techniques, lifting, lowering, picking up, placing, tilting
~ Traveling, with and without loads; turning corners
~ Parking and shutdown procedures
~ Other special operating conditions for the specific application

Operational training practice:
~ If feasible, practice in the operation of powered industrial trucks shall be conducted in an area separate from other workplace activities and personnel
~ Training practice shall be conducted under the supervision of the trainer
~ Training practice shall include the actual operation or simulated performance of all operating tasks such as load handling, maneuvering, traveling, stopping, starting, and other activities under the conditions which will be encountered in the use of the truck

Testing, Retraining, and Enforcement

(a) During training, performance and oral and/or written tests shall be given by the employer to measure the skill and knowledge of the operator in meeting the requirements of the Standard. Employers shall establish a pass/fail requirement for such tests. Employers may delegate such testing to others but shall remain responsible for the testing. Appropriate records shall be kept.

(b) Operators shall be retrained when new equipment is introduced, existing equipment is modified, operating conditions change, or an operator's performance is unsatisfactory.

(c) The user shall be responsible for enforcing the safe use of the powered industrial truck according to the provisions of this Standard.

Powered Industrial Truck Characteristics

The term "powered industrial truck" is defined as a "mobile, power-propelled truck used to carry, push, pull, lift, stack, or tier material." Vehicles that are used for earth moving and over-the-road hauling are excluded.

Powered industrial trucks are classified by their manufacturers according to their individual characteristics. There are seven classes of powered industrial trucks:

Class 1—Electric Motor, Sit-down Rider, Counter-Balanced Trucks (Solid and Pneumatic Tires)

Class 2—Electric Motor Narrow Aisle Trucks (Solid Tires)

Class 3—Electric Motor Hand Trucks or Hand/Rider Trucks (Solid Tires)

Class 4—Internal Combustion Engine Trucks (Solid Tires)

Class 5—Internal Combustion Engine Trucks (Pneumatic Tires)

Class 6—Electric and Internal Combustion Engine Tractors (Solid and Pneumatic Tires)

Class 7—Rough Terrain Forklift Trucks (Pneumatic Tires)

Each of the different types of powered industrial trucks has its own unique characteristics and some inherent hazards. To be most effective, training must address the unique characteristics of the type of vehicle(s) the employee is being trained to operate.

Powered industrial trucks may operate on almost any type of surface, from smooth and level floors to rocky, uneven ground, provided they were manufactured to operate on that type of floor or ground and the surface does not have an excessive slope. For example, construction forklift trucks (most commonly, those that are classified as Class 7, rough terrain forklifts) are more often operated on uneven, ungraded terrain than is the case for trucks in other industries.

Different trucks are designed and manufactured to operate in different work environments. Some powered industrial trucks are used for moving material in a particular type of workplace. For example, high lift trucks can be used to raise loads up to 30 or 40 feet above the ground, deposit the material on a rack, mezzanine, roof under construction, scaffold, or another elevated location, and subsequently retrieve and lower the material. Some vehicles are used to raise a palletized load just a few inches above the floor and move that load to another location in a warehouse or other indoor workplace.

Powered industrial trucks can be equipped with, or can be modified to accept, attachments that permit the truck to move odd-shaped material or carry out tasks that may not have been envisioned when the truck was designed and manufactured. Many of these attachments are added to or installed on the vehicle by the dealer or the employer. For example, there are powered industrial truck attachments for grasping barrels or drums of material. Some of these attachments not only grasp a barrel or drum but allow the vehicle operator to rotate the barrel or drum to empty it or lay it on its side.

Modifications and additions which affect capacity and safe operation shall not be performed by the customer or user without manufacturer's prior written approval. Capacity, operation, and maintenance instruction plates, tags, or decals shall be changed accordingly.

This is *(Enter Company Name Here)* Forklift Program. It meets all OSHA requirements and applies to all our work operations.

(Enter Name Here) will be responsible for overall direction of the Safety Program.

Introduction

This program has been prepared to assist employers in their efforts to prevent worker injury and to comply with the OSHA's Powered Industrial Truck Standard. It will also make recommendations for general safe practices for powered industrial truck use.

We have adopted this program in order to implement 29 CFR 1926.602, the OSHA standard regulating Powered Industrial Trucks.

The OSHA standard contains safety requirements relating to fire protection, design, maintenance, use, and operator training for industrial powered trucks. Industrial powered trucks include forklifts, tractors, platform lift trucks, motorized hand trucks, and other specialized industrial trucks. Powered industrial trucks are used to carry, push, pull, lift, stack or tier material.

This standard does not cover vehicles used for earth moving or over-the-road hauling.

The standard requires a training program based on the trainee's prior knowledge and skill, types of powered industrial trucks used in the workplace, and the operator's demonstrated ability to handle a powered industrial truck safely. Evaluation of each operator's performance is required as part of the initial and refresher training and at least once every three years.

The objective of this program is to make employers and operators of powered industrial trucks aware of their responsibility in the safe use of the equipment. It will provide a guideline safety program for powered industrial truck use.

Figure 14.1 A Sample Forklift Safety Program[2]

General Requirements

1. Approved trucks must bear a label or some other identifying mark indicating approval by the testing laboratory.
2. All name plates and markings on the truck must be in place and in a legible condition. Name plates should show the weight of the truck and its rated capacity.
3. Modifications and additions which affect capacity and safe operation must not be performed without manufacturers prior written approval.
4. If the truck is equipped with front-end attachments other than a factory installed attachments, the user shall request that the truck be marked to identify the attachments and show the approximate weight of the truck and attachment combination at maximum elevation with load laterally centered.

Designations

Listed below are powered industrial truck designations as prescribed by the OSHA standard:

D—diesel powered with minimum acceptable safeguards against inherent hazards.

DS—diesel powered with additional safeguards to the exhaust, fuel, and electrical systems.

DY—diesel powered with all of the safeguards as DS; does not have any electrical equipment including ignition; equipped with temperature limitation features.

E—electrical powered with minimum acceptable safeguards against inherent hazards.

ES—electrical powered, with all the requirements of E; have additional safeguards to the electrical system to prevent emission of hazardous sparks and to limit surface temperatures.

EE—electrical powered, with all the requirements of E and ES; electrical motors; all other electrical equipment completely enclosed.

EX—electrically powered units that differ from E, ES, and EE units in that the electrical fittings and equipment are designed, constructed, and assembled so that the units may be used in certain atmospheres containing flammable vapors and dusts.

G—gasoline powered with minimum acceptable safeguards against inherent hazards.

GS—gasoline powered with additional safeguards to the exhaust, fuel, and electrical systems.

LP—liquefied petroleum gas powered; minimum acceptable safeguards against inherent hazards.

LPS—liquefied petroleum with additional safeguards to the exhaust, fuel, and electrical systems.

Figure 14.1 A Sample Forklift Safety Program, continued

Designated Locations

The atmosphere or location must be classified as hazardous or non-hazardous before consideration of powered industrial truck to be used at that location. Use only trucks that meet the designation requirements for hazardous locations and atmospheres.

Safe Guards

1. High lift rider trucks must be fitted with an overhead guard when the truck has the capability of lifting loads overhead or where falling object hazards exist.
2. Overhead guards should not interfere with the drivers visibility and operation of the truck. Openings on the guard should be small enough to protect the operator from being struck by falling material from overhead.
3. If the type of load presents a hazard, the user shall equip fork trucks with a vertical load backrest extension.
4. Other safeguards that can be used to increase safety and reduce driver fatigue include convex mirrors at blind spots; backup alarms; headlights; turn signals; enhanced front and rear vision; noise-reducing insulation; fail-safe brakes; and comfortable wraparound seats.
5. Employers should develop a company policy to encourage seat belt use on forklifts when it will be driven over rough ground, uneven ground, or for long distances. Seat belts are required in the construction standard except for standup operations.
6. Guards should be installed over exposed tires, hazardous moving parts, such as chain-and-socket drives, and exposed gears.
7. Horns and flashing overhead lights can be used to warn employees of approaching trucks.
8. Specifications of steering, braking, and other controls should also conform to ANSI B56.1.

Fuel Handling and Storage

1. The storage and handling of liquid fuels such as gasoline and diesel fuel shall be in accordance with NFPA Flammable and Combustible Liquids Code (NFPA No. 30-1969).
2. Fill fuel tanks on trucks at designated locations, preferably in the open air, with the filling hose and equipment properly bonded and grounded.
3. Engines must be stopped and operators must be off trucks before they are refilled.
4. Do not permit smoking during refueling.
5. The storage and handling of liquefied petroleum gas fuel shall be in accordance with NFPA Storage and Handling of Liquefied Petroleum Gases (NFPA No. 58-1969).
6. Fire extinguishers should be located at the refueling site and must comply with OSHA's Forklift Standard.

Figure 14.1 A Sample Forklift Safety Program, continued

Batteries

1. Battery charging installations shall be located in a designated area.
2. Facilities must be provided for flushing and neutralizing spilled electrolyte, for fire protection, for protecting charging apparatus from damage by trucks, and for adequate ventilation for dispersal of fumes from gassing batteries.
3. A conveyor, overhead hoist, or equivalent material handling equipment must be provided for handling batteries.
4. All reinstalled batteries must be properly positioned and secured in the truck.
5. When charging batteries, acid must be poured into water instead of water poured into acid.
6. A carboy tilter or siphon must be provided for handling electrolyte.
7. Make sure trucks are properly positioned and brakes applied before attempting to change or charge batteries.
8. Ensure that battery vent caps are functioning and the battery covers are open to dissipate heat.
9. Eliminate all possible ignition sources in the area.
10. Smoking must not be permitted in the battery handling area.
11. Keep tools and other metal objects away from the top of uncovered batteries.
12. Workers handling batteries should wear goggles, rubber gloves, aprons, and rubber boots, to protect against acid burns.
13. Rubber mats, wood-slat mats, or clean floorboards can be used to help prevent slips and falls and will protect against electric shock.
14. If acid is spilled on workers' clothing it should be washed off immediately with water.
15. Refer to manufacturers recommendations before performing charging and maintenance procedures.
16. Make sure battery chargers are off when batteries are connected and disconnected.

Control of Noxious Gases and Fumes

1. Concentration levels of carbon monoxide gas created by powered industrial truck operations must not exceed the levels specified in 29 CFR 1926.1000.
2. Do not leave trucks idling unnecessarily inside buildings or outside near windows or ventilation in-take ducts.

Trucks and Railroad Cars

1. The brakes of highway trucks must be set and chock blocks placed under the rear wheels to prevent the truck from rolling.
2. Wheel stops or other recognized positive protection shall be provided to prevent railroad cars from moving during loading or unloading.

Figure 14.1 A Sample Forklift Safety Program, continued

341

3. Fixed jacks may be necessary to support a semi-trailer and prevent up-ending during the loading or unloading when the trailer is not coupled to a tractor.

4. Flags should be used during unloading and loading, to warn the driver not to move the trailer.

Truck Operation

1. Trucks must not be driven up to anyone standing in front of a bench or other fixed objects.
2. Workers must not be allowed to pass under the elevated portion of any truck, whether loaded or empty.
3. Unauthorized personnel must not be allowed to ride on powered industrial trucks. A safe place to ride must be provided where riding of trucks is authorized.
4. Operator's arms or legs must be prohibited from being placed between the uprights of the mast or outside the running lines of the truck.
5. When a powered industrial truck is left unattended, load engaging means must be fully lowered, controls shall be in neutral, power must be shut off, and brakes set. Wheels need to be blocked if the truck is parked on an incline.
6. A safe distance must be maintained from the edge of ramps or platforms while on any elevated dock, platform, or freight car. Trucks shall not be used for opening or closing freight doors.
7. Looking out for pedestrians is also the truck operator's responsibility. They should sound the horn when approaching pedestrians.

Traveling

1. Operators must observe all traffic regulations, including authorized plant speed limits.
2. A safe distance must be maintained approximately three truck lengths from the truck ahead, and the truck shall be kept under control at all times.
3. Do not pass other trucks traveling in the same direction at intersections, blind spots, or other dangerous locations.
4. Divers must be required to slow down and sound the horn at cross aisles and other locations where vision is obstructed. If the load being carried obstructs forward view, the driver must travel with the load trailing.
5. Ascend or descend grades slowly.
6. When ascending or descending grades in excess of ten percent, loaded trucks must be driven with the load upgrade.
7. Dockboard or bridge plates, must be properly secured before they are driven over and their rated capacity never exceeded.
8. Elevators must be approached slowly, and then entered squarely after the elevator car is properly leveled. Once on the elevator, the controls shall be neutralized, power shut off, and the brakes set.
9. When transporting material, keep the load no more than six to ten inches off the floor and use extreme caution when turning.

Figure 14.1 A Sample Forklift Safety Program, continued

Loading

1. Only stable or safely arranged loads shall be handled. Caution must be taken when handling off-center loads which cannot be centered.
2. Place heavy, odd-shaped objects with the weight as low as possible.
3. Loads should never exceed the trucks rated capacity.
4. Use extreme caution when tilting the load forward or backward, particularly when high tiering.
5. Tilting forward with the load engaging means elevated shall be prohibited except to pick up a load.
6. An elevated load must not be tilted forward except when the load is in deposit position over a rack or stack.

Maintenance

1. Any power-operated industrial truck not in safe operating condition shall be removed from service and all repairs made by authorized personnel.
2. No truck should be operated with a leak in the fuel system until the leak has been corrected.
3. Repairs to the fuel and ignition system which involve fire hazards must be conducted only in designated repair locations.
4. When working on the electrical system, the battery must be disconnected.
5. All parts that are replaced must be replaced with parts equivalent with regard to safety with those used in the original design.
6. Before and end-of-shift maintenance and safety checks need to be preformed before the forklift is used. Defects need to be reported and corrected before the fork lift is used.
7. Water mufflers shall be filled daily or as frequently as necessary to prevent depletion of the water supply below 75 percent of the filled capacity.
8. Vehicles with mufflers having screens or other parts that may become clogged must be cleaned of any debris or unclogged before operation.
9. Any vehicle that emits hazardous sparks or flames from the exhaust system shall immediately be removed from service and repaired.

Required Operator Training

Listed below are the requirements for operator training:

Safe Operation

1. The employer shall ensure that each powered industrial truck operator is competent to operate a powered industrial truck safely, as demonstrated by the successful completion of the training evaluation. Employees will not be allowed to operate forklifts, except for training, unless they have met all training requirements.
2. Trainees may operate a powered industrial truck only under direct supervision of persons who have knowledge, training, and experience to train and evaluate operators competence; and where operation does not endanger the trainee or other employees.

Figure 14.1 A Sample Forklift Safety Program, continued

3. Training must consist of a combination of formal instruction, practical training, and evaluation of the operator's performance in the workplace.
4. All training and evaluation must be conducted by persons who have the knowledge, training, and experience to train and evaluate powered industrial truck operators.

Training Program

Forklift Operators must receive training in the following topics:

1. Operating instructions, warnings, and precautions for the type of truck the operator will be authorized to operate
2. Differences between the truck and the automobile
3. Truck controls and instrumentation: where they are located, what they do, and how they work
4. Engine and motor operation; steering and maneuvering
5. Visibility (including restrictions due to loading)
6. Fork and attachment adaptation, operation, and use limitations
7. Vehicle capacity and stability
8. Any vehicle inspection and maintenance that the operator will be required to perform
9. Refueling and/or charging and recharging batteries
10. Operating limitations; any other operating instructions, warnings, or precautions listed in the operator's manual for the types of vehicle that the employee is being trained to operate

Workplace-Related Topics

- Surface conditions where the vehicle will be operated
- Composition of loads to be carried and load stability
- Load manipulation, stacking, and unstacking
- Pedestrian traffic in areas where the vehicle will be operated
- Narrow aisles and other restricted places where the vehicle will be operated
- Hazardous (classified) locations where the vehicle will be operated
- Ramps and other sloped surfaces that could affect the vehicle's stability
- Closed environments and other areas where insufficient ventilation or poor vehicle maintenance could cause a buildup of carbon monoxide or diesel exhaust
- Determining whether the load is safe to handle
- Other unique or potentially hazardous environmental conditions in the workplace that could affect safe operation; and the training requirements

Figure 14.1 A Sample Forklift Safety Program, continued

Refresher Training and Evaluation

Refresher training will be provided upon the following circumstances:

- The operator has been observed to operate the vehicle in an unsafe manner
- The operator has been involved in an accident or near-miss incident
- The operator has received an evaluation that reveals that the operator is not operating the truck safely
- The operator is assigned to drive a different type of truck; or a condition in the workplace changes in a manner that could affect safe operation of the truck
- An evaluation of each powered industrial truck operator's performance shall be conducted at least once every three years

Avoidance of Duplicate Training

If an operator has previously received training in a topic and such training is appropriate to the truck and working conditions encountered, additional training in that topic is not required if the operator has been evaluated and found competent to operate the truck safely.

Training Certification

The employer must certify each operator has been trained and evaluated as required by this standard. The certification must include the name of the operator, the date of the training, the date of the evaluation, and the identity of the person(s) performing the training and evaluation.

All employees who operate a forklift, if only occasionally, need training first. New operators with previous experience or training need at least a test of operating proficiency and instruction in the hazards of the particular job and in the truck maintenance program.

An operator with extensive experience on one brand or model of truck still needs time to get used to controls placed in different positions or to different attachments or truck types.

Figure 14.1 A Sample Forklift Safety Program[2], continued

POWERED INDUSTRIAL TRUCK TRAINING

OSHA 29 CFR 1926.602 describes methods to use for compliance with powered industrial truck operations, maintenance, inspection, and training.

This is to certify that I have attended the above training program which has informed me of the following:

_____ Common forklift accidents

_____ Safety procedures for picking up loads

_____ Safety procedures for traveling with loads

_____ Safety procedures for stacking and dropping loads

_____ Discussion on lift truck capacity, load center, and center of gravity

_____ Safe loading/unloading procedure

_____ General safe operating procedures

_____ OSHA training requirements for forklift operators

_____   _____

DATE   Employee's Signature

_____   _____

DATE   Trainer's Signature

Figure 14.2   Sample Certificate of Training

Endnotes:

[1] ANSI Standard B56.1-1969

[2] Adapted from *Forklift Safety Guide*, 29 CFR 1910.178, Occupational Safety & Health Bureau, Montana Department of Labor & Industry. Available at http://erd.dli.state.mt.us/Safety/SB_brochures/Forklift_Safety_Guide.pdf.

15

Controlling Lead Exposures

Working with Lead in the Construction Industry

INTRODUCTION

The current OSHA standard (29 CFR 1926.62) for lead exposure in construction has a permissible exposure limit (PEL) of 50 micrograms per cubic meter of air (50 µg/m3), measured as an eight-hour time-weighted average (TWA). As with all OSHA health standards, when the PEL is exceeded, the hierarchy of controls requires employers to institute feasible engineering and work practice controls as the primary means to reduce and maintain employee exposures to levels at or below the PEL. When all feasible engineering and work practice controls have been implemented but have proven inadequate to meet the PEL, employers must nonetheless implement these controls and must supplement them with appropriate respiratory protection. The employer also must ensure that employees wear the respiratory protection provided when it is required. Lead has been poisoning workers for thousands of years. In the construction industry, traditionally most overexposures to lead are found in the trades, such as plumbing, welding, and painting.

In building construction, lead is frequently used for roofs, cornices, tank linings, and electrical conduits. In plumbing, soft solder is used chiefly for soldering tinplate and copper pipe joints (an alloy of lead and tin). Soft solder has, in fact, been banned for many uses in the United States. The use of lead-containing paint in residential applications has also been banned by the Consumer Product Safety Commission. However, since lead-containing paint inhibits the rusting and corrosion of iron and steel, it is still used on bridges, railways, ships, lighthouses, and other iron and steel structures, although substitute coatings are available.

Significant lead exposures can also arise from removing paint from surfaces previously coated with lead-containing paint, such as in bridge repair, residential renovation, and demolition. With the increase in highway work, including bridge repair, residential lead abatement, and residential remodeling, the potential for exposure to lead-containing paint has become more common. The trades potentially exposed to lead include iron work; demolition work; painting; lead-based paint abatement work; plumbing; heating/air-conditioning; electrical work; and carpentry, renovation, and remodeling.

Operations that generate lead dust and fumes include

- Flame-torch cutting, welding, grinding of lead painted surfaces in repair, reconstruction, dismantling, and demolition work

- Abrasive blasting of bridges and other steel structures containing lead-based paints
- Abatement of residential lead-based paint using torches, heat guns, and sanding machines
- Maintaining process equipment or exhaust ductwork

Operations that involve exposure to lead-containing products include

- Spray painting bridges and other structures with lead-based paints and primers
- Using solder in plumbing and electrical work

SAMPLE LEAD EXPOSURE CONTROL PROGRAM

The company has established this program in order to (1) protect our employees and other people who use our facilities from the hazardous effects of lead; (2) implement 29 CFR §1926.62, the OSHA standard regulating occupational exposure to lead; (3) limit occupational exposure to lead to or below the OSHA permissible exposure limit (PEL) of 50 micrograms per cubic meter of air (50 µg/m3) when averaged over an eight-hour work day; and (4) be prepared to deal with any emergency situations involving substantial releases of airborne lead that might occur.

The hazard of exposure can be minimized or eliminated by the use of a combination of engineering and work practice controls, personal protective clothing and equipment, training, medical surveillance, housekeeping, signs and labels, and other provisions. That is exactly what this program provides. The program also solicits the cooperation of every employee in our effort to minimize or eliminate lead exposure and makes it a requirement that employees immediately notify their supervisor of any actual or suspected instance of noncompliance with any provision of this program.

This program will be reviewed and updated at least annually, or more often if necessary, in order to reflect changes that may occur in our operations, our compliance status, new or modified tasks and procedures which affect occupational exposure to lead, and to reflect new or revised employee positions with occupational exposure. A copy will be accessible to our employees and their representatives. It will also be made available to OSHA and NIOSH representatives in accordance with applicable legal and constitutional provisions.

Health Effects of Lead

There are several ways in which lead enters your body:

1. When absorbed into your body in certain doses, lead is a toxic substance. The object of the OSHA lead standard is to prevent absorption of harmful quantities of lead. The standard is intended to protect you not only from the immediate toxic effects of lead, but also from the serious toxic effects that may not become apparent until years of exposure have passed.
2. Lead can be absorbed into your body by inhalation (breathing) and ingestion (eating). Lead (except for certain organic lead compounds not covered by the standard, such as tetraethyl lead) is not absorbed through your skin. When lead is scattered in the air as

a dust, fume, or mist, it can be inhaled and absorbed through your lungs and upper respiratory tract. Inhalation of airborne lead is generally the most important source of occupational lead absorption. You can also absorb lead through your digestive system if lead gets into your mouth and is swallowed. If you handle food, cigarettes, chewing tobacco, or cosmetics which have lead on them or handle them with hands contaminated with lead, this will contribute to ingestion.

3. A significant portion of the lead that you inhale or ingest gets into your blood stream. Once in your blood stream, lead is circulated throughout your body and stored in various organs and body tissues. Some of this lead is quickly filtered out of your body and excreted, but some remains in the blood and other tissues. As exposure to lead continues, the amount stored in your body will increase if you are absorbing more lead than your body is excreting. Even though you may not be aware of any immediate symptoms of disease, this lead stored in your tissues can be slowly causing irreversible damage, first to individual cells, then to your organs and whole body systems.

4. Cumulative exposure to lead, which is typical in construction settings, may result in damage to the blood, nervous system, kidneys, bones, heart, and reproductive system and contributes to high blood pressure.

The symptoms of lead poisoning include the following:

- Headache
- Poor appetite
- Dizziness
- Irritability/anxiety
- Constipation
- Pallor
- Excessive tiredness
- Numbness
- Metallic taste in the mouth
- Muscle and joint pain or soreness
- Sleeplessness
- Hyperactivity
- Weakness
- Reproductive difficulties
- Nausea
- Fine tremors
- Insomnia
- "Lead line" on the gums
- Hyperactivity
- "Wrist drop" (weakness of extensor muscles)

DEFINITIONS

As used in this Program and in the OSHA Lead Standard, the terms listed below have the following meanings:

Action level means employee exposure, without regard to the use of respirators, to an airborne concentration of lead of 30 micrograms per cubic meter of air (30 µg/m3), averaged over an eight-hour period.

Employee exposure (and similar language referring to the air lead level to which an employee is exposed) means the exposure to airborne lead that would occur if the employee were not using respiratory protective equipment.

Lead means metallic lead, all inorganic lead compounds, and organic lead soaps. Excluded from this definition are all other organic lead compounds.

Permissible exposure limit (PEL) means the concentration of airborne lead to which an average person may be exposed without harmful effects. OSHA has established a PEL of fifty micrograms per cubic meter of air (50 µg/m3) averaged over an eight-hour period. If an employee is exposed to lead for more than eight hours in any work day, the permissible exposure limit, a time weighted average (TWA) for that day, shall be reduced according to the following formula:

Maximum permissible limit (in µg/m3) = 400 X hours worked in the day.

When respirators are used to supplement engineering and work practice controls to comply with the PEL and all the requirements of the lead standard's respiratory protection rules have been met, employee exposure, for the purpose of determining whether the employer has complied with the PEL, may be considered to be at the level provided by the protection factor of the respirator for those periods the respirator is worn. Those periods may be averaged with exposure levels during periods when respirators are not worn to determine the employee's daily TWA exposure.

Time Weighted Average (TWA) means a worker's cumulative airborne exposure to lead during his work shift. It permits excursions above the permissible exposure limit so long as they are compensated for by corresponding periods below the limit.

µg/m3 means micrograms per cubic meter of air. A microgram is one millionth of a gram. There are 454 grams in a pound.

THE OSHA PERMISSIBLE EXPOSURE LIMIT (PEL)

The OSHA PEL for airborne concentrations of lead is 50 micrograms per cubic meter of air (50 µg/m3), calculated as an eight-hour time-weighted average exposure (TWA). [29 CFR §1926.62(c)]

The measures we have taken to date together with those set forth in this Program will ensure that employee exposure is within that PEL. See also the definition above of "Permissible exposure limit (PEL)."

EXPOSURE MONITORING

1. On or before the date on which this program is adopted, we will make an initial determination of the airborne concentrations of lead to which our employees may be exposed. That determination will be made by one of the following methods permitted by the OSHA standard:
   ~ Any information, observations, or calculations which would indicate employee exposure to lead
   ~ Any previous measurements of airborne lead and/or analytical methods used that meet the accuracy and confidence level of 29 CFR §1926.62(d)(9)
   ~ Any employee complaints of symptoms which may be attributed to exposure to lead

2. Where a determination based upon the above shows the possibility of any employee exposure at or above the action level, we will conduct monitoring which is representative of the exposure for each employee in the workplace who is exposed to lead. We will also do the following:
   ~ If the initial determination or subsequent monitoring reveals employee exposure to be at or above the action level but below the permissible exposure limit, we will repeat the monitoring at least every six months. We will continue monitoring at that frequency until at least two consecutive measurements, taken at least seven days apart, are below the action level at which time we will discontinue monitoring unless there is a change.
   ~ If the initial monitoring reveals that employee exposure is above the permissible exposure limit, we will repeat monitoring quarterly. We will continue monitoring at that frequency until at least two consecutive measurements taken at least seven days apart are below the PEL but at or above the action level, at which time we will repeat monitoring for that employee at the frequency specified above.
   ~ Whenever there has been a production, process, control or personnel change which may result in new or additional exposure to lead, or whenever we have any other reason to suspect a change which may result in new or additional exposures to lead, additional monitoring shall be conducted at the same frequencies stated above.
   ~ Within five working days after the receipt of monitoring results, we will notify each employee in writing of the results which represent that employee's exposure. If the results indicate that the representative employee exposure, without regard to respirators, exceeds the permissible exposure limit, we will include in the written notice, a statement that the permissible exposure limit was exceeded and a description of the corrective action taken or to be taken to reduce exposure to or below the permissible exposure limit.
   ~ We will use a method of monitoring and analysis which has an accuracy (to a confidence level of 95 percent) of not less than plus or minus 20 percent for airborne concentrations of lead equal to or greater than 30 μg/m3.

~ Monitoring for the initial determination may be limited to a representative sample of the exposed employees who are reasonably believed to be exposed to the greatest airborne concentrations of lead in the workplace.

~ If the initial monitoring reveals employee exposure to be below the action level, the measurements will not be repeated.

~ Where a determination, conducted as described above, is made that no employee is exposed to airborne concentrations of lead at or above the action level, we will make a written record of such determination. The record will set forth the basis for the determination and shall include the date of determination, location within the worksite, and the name and social security number of each employee monitored.

3. We will provide affected employees or their designated representatives an opportunity to observe any monitoring of employee exposure to lead conducted pursuant to this Program in the manner stated below:

~ Whenever observation of the monitoring of employee exposure to lead requires entry into an area where the use of respirators, protective clothing, or equipment is required, we will provide the observer with and ensure the use of such respirators, clothing, and equipment, and shall require the observer to comply with all other applicable safety and health procedures.

~ Without interfering with the monitoring, observers shall be entitled to receive an explanation of the measurement procedures, observe all steps related to the monitoring of lead performed at the place of exposure, and record the results obtained or receive copies of the results when returned by the laboratory.

METHODS OF COMPLIANCE

We will implement all feasible engineering and work practice controls that are necessary in order to reduce and maintain employee exposure to lead at or below the PEL. However, as stated in 29 CFR §1926.62(e), controls of that kind will not be necessary when

- The employee is only intermittently exposed
- The employee is not exposed above the PEL on 30 or more days per year (12 consecutive months)

Wherever feasible engineering and work practice controls are required and are not sufficient to reduce employee exposure to or below the PEL, we will nonetheless implement such controls to reduce exposures to the lowest levels achievable.

In the event that the OSHA PEL is exceeded, we will adopt as a supplement to this Program whichever of the following are applicable:

- A description of each operation in which lead is emitted; e.g., machinery used, material processed, controls in place, crew size, employee job responsibilities, operating procedures, and maintenance practices

- A description of the specific means that will be employed to achieve compliance, including engineering plans and studies used to determine the methods we have selected for controlling exposure to lead, as well as, where necessary, the use of appropriate respiratory protection to achieve the PEL
- A report of the technology considered in meeting the PEL
- Air monitoring data that document the sources of lead emissions
- A detailed schedule for implementation of the program to reduce employee exposure to or below the PEL, including documentation such as copies of purchase orders for equipment, construction contracts, etc.
- A work practice program that includes items required under 29 CFR §§1926.62(g),(h) and (i)
- An administrative job rotation schedule which includes (i) name or identification number of each affected employee; (ii) duration and exposure levels at each job or work station where each affected employee is located; and (iii) any other information which may be useful in assessing the reliability of administrative controls to reduce exposure to lead

In the event that a supplement to this Program is necessary as stated above, it will be submitted upon request to OSHA and NIOSH and shall be available at the worksite for examination and copying by any affected employee or authorized employee representatives. The supplement to his Program shall be revised and updated at least every six months to reflect the current status of the program.

Whenever ventilation is used to control exposure, measurements that demonstrate the effectiveness of the system in controlling exposure, such as capture velocity, duct velocity, or static pressure, will be made as necessary to maintain its effectiveness-or at least every three months. That process will include the following:

- Measurements of the system's effectiveness in controlling exposure will be made as necessary within five working days of any change in production, process, or control that might result in a significant increase in employee exposure to lead.
- If air from exhaust ventilation is recirculated into the workplace, the system shall have a high efficiency filter with a reliable back-up filter and be monitored to ensure effectiveness.
- Special procedures will be developed and implemented to minimize employee exposure to lead during the time when maintenance of ventilation systems and changing of filters is being conducted.
- Controls to monitor the concentration of lead in the return air and to bypass the recirculation system automatically if it fails will be installed, operating, and maintained.

RESPIRATORY PROTECTION

Respirators are required and must be used in the following circumstances:

- Where exposure levels exceed the PEL

- In those maintenance and repair activities and during those brief or intermittent operations where exposures exceed the PEL and engineering and work practice controls are not feasible or are not required
- During the time period necessary to install or implement feasible engineering and work practice controls
- Where the feasible engineering and work practice controls that have been implemented are not sufficient to reduce exposures to or below the PEL
- In emergencies
- Wherever an employee requests a respirator
- Wherever an employee is exposed to lead above the PEL and engineering controls are not required by the OSHA lead standard

No employee shall be required to wear a negative pressure respirator longer than 4.4 hours per day.

Whenever respirators are required, we will select and provide (at no cost to employees) only respirators that are appropriate, as specified in "Table II-Respiratory Protection for Lead Aerosols" of 29 CFR §1926.62(f), and Approved for protection against lead dust, fume and mist by the Mine Safety and Health Administration (MSHA) and the National Institute for Occupational Safety and Health (NIOSH) under the provisions of 30 CFR part 11. Table II provides as follows:

> In order to determine an employee's fitness for using a respirator, we will provide a limited medical examination in accordance with the provisions of 29 CFR §1926.62(b)(10) on or before the time when this Program is adopted or prior to the employee's assignment to a job that requires respirator use, whichever date is later.

We will provide a powered, air-purifying respirator (PAPR) in lieu of the respirator specified in Table II wherever:

- An employee who is entitled to a respirator chooses to use that type of respirator
- The PAPR respirator will provide adequate protection to the employee

Each employee who is required to use an air purifying respirator will be permitted to leave the work area to change the filter elements or replace the respirator whenever an increase in breathing resistance is detected. We will maintain an adequate supply of filter elements for that purpose.

Each employee who is required to wear a respirator will be permitted to leave the work area to wash his or her face and the respirator facepiece whenever necessary in order to prevent skin irritation associated with respirator use.

No employee shall be assigned a task requiring the use of a respirator if, based upon his or her most recent examination, an examining physician determines that the employee will be unable to continue to function normally while wearing a respirator. If the physician determines that the employee must be limited in, or removed from his or her current job because of the employee's inability to wear a respirator, the limitation or removal will be taken in accordance with the provisions of the medical removal provisions of the OSHA lead standard of this Program.

Any employee who has difficulty in breathing while wearing a respirator must immediately notify his or her supervisor.

If an employee exhibits difficulty in breathing while wearing a respirator during a fit test or during use, we will make available to the employee a medical examination in order to determine if the employee can wear a respirator while performing the required duties.

In the event that respiratory protection is required, we will institute a respiratory protection program in accordance with the OSHA respiratory protection standard, 29 CFR §1926.103(b), (d), (e) and (f).

Respirator fit testing (either quantitative or qualitative) will be required and conducted at the time of initial fitting and at least every six months thereafter for each employee wearing negative pressure respirators. The qualitative fit tests may be used only for testing the fit of half-mask respirators where they are permitted to be worn, and shall be conducted in accordance with Appendix D of the OSHA lead standard. The tests shall be used to select facepieces that provide the required protection as prescribed in Table II.

It shall be the responsibility of the employee's immediate supervisor to ensure that the respirator issued to the employee exhibits minimum facepiece leakage and that the respirator is fitted properly.

PROTECTIVE WORK CLOTHING AND EQUIPMENT

If an employee is exposed to airborne lead above the PEL, or where the possibility of skin or eye irritation exists, we will provide at no cost to the employee, appropriate protective work clothing and equipment that prevents contamination of the employee and the employee's garments.

We will also provide for the cleaning, laundering, or disposal of such protective clothing and equipment. Each employee who is so provided will be required to use that clothing and equipment in accordance with instructions. The clothing and equipment includes, but is not limited to

- Coveralls or similar full-body work clothing
- Gloves, head coverings, and shoes or disposable shoe coverings
- Face shields, vented goggles, or other appropriate protective equipment that complies with 29 CFR §1910.132-110.138, the OSHA standard on personal protective equipment

The clothing will be provided in a clean and dry condition at least weekly. That frequency will be changed to daily for those employees whose exposure levels without regard to a respirator are over 200 ug/m3 of lead as an eight-hour TWA.

All protective clothing and equipment that is contaminated with lead must be removed at the completion of the work shift. Removal must take place only in change rooms provided in accordance with 29 CFR §1926.62(i)(2).

No employee may take lead-contaminated protective clothing or equipment from the workplace, except for those employees who have been authorized to do so for purposes of laundering, cleaning, maintaining, or disposing of lead-contaminated protective clothing and equipment at an appropriate location or facility away from the workplace.

Contaminated protective clothing and equipment, when removed for laundering, cleaning, maintenance, or disposal, must be placed and stored in a closed container that is designed to prevent dispersion of lead outside the container.

Containers of contaminated protective clothing and equipment that are to be taken out of the change rooms or the workplace for laundering, cleaning, maintenance or disposal shall bear labels with the following information:

CAUTION:

CLOTHING CONTAMINATED WITH LEAD. DO NOT REMOVE DUST BY BLOWING OR SHAKING. DISPOSE OF LEAD CONTAMINATED WASH WATER IN ACCORDANCE WITH APPLICABLE LOCAL, STATE, OR FEDERAL REGULATIONS.

Any person who cleans or launders protective clothing or equipment shall be informed in writing of the potentially harmful effects of exposure to lead.

The required protective clothing and equipment must be maintained in as clean and as dry a condition as is necessary to maintain its effectiveness and must be discarded in accordance with applicable disposal requirements whenever it is no longer suitable or fit for its intended purpose. Cleaning and laundering must occur at least in accordance with the schedule set forth above. In addition, the following must be observed:

- The repairing or replacing of required protective clothing and equipment must occur as needed in order to maintain its effectiveness. When rips or tears are detected while an employee is working, the rips or tears shall be immediately mended or the clothing or equipment shall be immediately replaced.
- Removal of lead from protective clothing and equipment by blowing, shaking, or any other means that disperses lead into the air is prohibited.
- Any laundering of contaminated clothing or cleaning of contaminated equipment in the workplace must be done in a manner that prevents the release of airborne lead in excess of the PEL.
- Any person who launders or cleans protective clothing or equipment contaminated with lead must be informed of the potentially harmful effects of exposure to lead and that the clothing and equipment should be laundered or cleaned in a manner to effectively prevent the release of airborne lead in excess of the PEL.

Each employee must immediately notify his supervisor whenever (a) he or she experiences eye or skin irritation that is (or may be) associated with lead exposure, and (b) whenever there are rips or tears in his or her protective clothing or equipment or it is no longer suitable or fit for its intended purpose.

HYGIENE AREAS AND PRACTICES

Employees whose airborne exposure to lead is above the PEL, will be provided with clean change rooms, lavatories, handwashing facilities, showers, and lunchroom facilities that comply with 29 CFR §1926.51(f) & (i), the OSHA standard on sanitation. The change rooms will be equipped with separate storage facilities for street clothes and for protective work clothing and equipment, that are designed to prevent dispersion of lead and contamination of the employee's street clothes.

Employees who work in areas where exposure to lead is above the PEL shall be required to

- Shower at the end of the work shift.
- Wash their hands and faces prior to eating, drinking, smoking, chewing tobacco or gum, or applying cosmetics. In any case where lead exposure exceeds the PEL, no food or beverage can be present or consumed, no tobacco products can be present or used, and no cosmetics can be applied.
- Refrain from leaving the workplace wearing any clothing or equipment worn during the work shift.

The lunchroom facilities will be readily accessible to employees and will have a temperature-controlled, positive pressure, filtered air supply. Tables for eating will be maintained free of lead.

No employee in a lunchroom facility may be exposed at any time to lead at or above the action level and no employee may enter lunchroom facilities with protective work clothing or equipment unless surface lead has been removed from the clothing and equipment by vacuuming downdraft booth or some other method that removes lead dust without dispersing it.

HOUSEKEEPING

All workplace surfaces must be maintained as free as practicable of accumulations of lead at all times.

All spills and sudden releases of material containing lead shall be cleaned up as soon as possible.

Surfaces contaminated with lead shall, wherever possible, be cleaned by vacuuming or other methods that minimize the likelihood of lead becoming airborne.

HEPA-filtered vacuuming equipment or equally effective filtration methods shall be used for vacuuming. The equipment shall be used and emptied in a manner that minimizes the reentry of lead into the workplace.

Shoveling, dry or wet sweeping, and brushing may be used only where vacuuming or other methods that minimize the likelihood of lead becoming airborne have been tried and found not to be effective.

Compressed air shall not be used to remove lead from floors or any other surface.

MEDICAL SURVEILLANCE

We will provide a medical surveillance program that includes all the provisions of 29 CFR §1926.62(j) for employees who are or may be exposed to lead at or above the action level unless the employee is not, and will not be, exposed at or above the action level on 30 or more days per year (12 consecutive months).

All medical examinations and procedures will be performed by or under the supervision of a licensed physician. The required medical surveillance shall include multiple physician review under the provisions of §1926.62(j)(3)(iii) without cost to employees and at a reasonable time and place. Medical examinations and consultations will be provided on the following schedule:

- At least annually, for each employee for whom a blood sampling test conducted at any time during the preceding twelve months indicated a blood lead level at or above 40 µg/100g
- Prior to assignment, for each employee being assigned for the first time to an area in which airborne concentrations of lead are at or above the action level
- As soon as possible, upon notification by an employee either that the employee has developed signs or symptoms commonly associated with lead intoxication, that the employee desires medical advice concerning the effects of current or past exposure to lead on the employee's ability to procreate a healthy child, or that the employee has demonstrated difficulty in breathing during a respirator fitting test or during use
- As medically appropriate, for each employee either removed from exposure to lead due to a risk of sustaining material impairment to health, or otherwise limited pursuant to a final medical determination

We will provide biological monitoring in the form of blood sampling and analysis for lead and zinc protoporphyrin levels to each such employee on the following schedule:

- At least every six months.
- At least every two months for each employee whose last blood sample and analysis indicated a blood lead level at or above 40 µg/100 g of whole blood. This frequency shall continue until two consecutive blood samples and analyses indicate a blood lead level below 40 µg/100 g of whole blood.
- At least monthly during the removal period of each employee removed from exposure to lead due to an elevated blood lead level.

Blood lead level sampling and analysis shall have an accuracy (to a confidence level of 95 percent) within plus or minus 15 percent or 6 µg/100ml, whichever is greater, and shall be conducted by a laboratory licensed by the Center for Disease Control, United States Department of Health and Human Services (CDC) or which has received a satisfactory grade in blood lead proficiency testing from CDC in the prior twelve months.

Within five working days after the receipt of biological monitoring results, each employee whose blood lead level exceeds 40 µg/100 g shall be notified in writing (a) of that employee's blood lead level, and (b) that the standard requires temporary medical removal with Medical Removal

Protection benefits when an employee's blood lead level exceeds the numerical level set forth in 29 CFR §1926.62(k)(l)(i).

Whenever the results of a blood lead level test indicate that an employee's blood lead level exceeds the numerical criterion for medical removal under §1926.62(k)(l)(i), we will provide a second (follow-up) blood sampling test within two weeks after we receive the results of the first blood sampling test.

Each medical examination shall include the following:

- A detailed work history and a medical history, with particular attention to past lead exposure (occupational and non-occupational), personal habits (smoking, hygiene), and past gastrointestinal, hematologic, renal, cardiovascular, reproductive and neurological problems.
- A thorough physical examination, with particular attention to teeth, gums, hematologic, gastrointestinal, renal, cardiovascular, and neurological systems. Pulmonary status should be evaluated if respiratory protection will be used.
- A blood pressure measurement.
- A blood sample and analysis which determines:
    - ~ Blood lead level
    - ~ Hemoglobin and hematocrit determinations, red cell indices, and examination of peripheral smear morphology
    - ~ Zinc Protoporphyrin
    - ~ Blood urea nitrogen
    - ~ Serum creatinine

- A routine urinalysis with microscopic examination.
- Any laboratory or other test which the examining physician deems necessary by sound medical practice.

If we select the initial physician who conducts any medical examination or consultation provided to an employee under this Program, the employee may designate a second physician:

- To review any findings, determinations or recommendations of the initial physician
- To conduct such examinations, consultations, and laboratory tests as the second physician deems necessary to facilitate this review

In all such cases, the multiple physician review system included in §1926.62(j)(3)(iii) and (vi) will be applied.

We will provide any physician conducting a medical examination or consultation under this Program with the following information:

- A copy of the OSHA lead standard including all Appendices
- A description of the affected employee's duties as they relate to the employee's exposure
- The employee's exposure level or anticipated exposure level to lead and to any other toxic substance (if applicable)

- A description of any personal protective equipment used or to be used
- Prior blood lead determinations
- All prior written medical opinions concerning the employee in the employer's possession or control

We will obtain and furnish the employee with a copy of the written medical opinion from each examining or consulting physician which will contain the following information:

- The physician's opinion as to whether the employee has any detected medical condition which would place the employee at increased risk of material impairment of the employee's health from exposure to lead
- Any recommended special protective measures to be provided to the employee, or limitations to be placed upon the employee's exposure to lead
- Any recommended limitation upon the employee's use of respirators, including a determination of whether the employee can wear a powered air purifying respirator if a physician determines that the employee cannot wear a negative pressure respirator
- The results of the blood lead determinations

We will instruct each examining and consulting physician to

- Not reveal either in the written opinion or in any other means of communication with us findings, including laboratory results or diagnoses, unrelated to the employee's occupational exposure to lead
- Advise the employee of any medical condition, occupational or nonoccupational, which dictates further medical examination or treatment

MEDICAL REMOVAL PROTECTION

We will remove any employee from work having an exposure to airborne lead at or above the action level on each of the following occasions:

- When required by 29 CFR §1926.62(k)(l)(i)
- Whenever a final medical determination results in a medical finding, determination, or opinion that the employee has a detected medical condition which places the employee at increased risk of material impairment to health from exposure to lead

For such purposes, the phrase "final medical determination" shall mean the outcome of the multiple physician review mechanism or alternate medical determination mechanism used pursuant to the medical surveillance provisions of the OSHA lead standard.

Where a final medical determination results in any recommended special protective measures for an employee or limitations on an employee's exposure to lead, we will implement the same and act consistent with the recommendations.

Any employee temporarily removed from his job pursuant to the above provisions will be returned to his or her former job status as follows:

- For an employee removed due to a blood lead level at or above 80 µg/100g, when two consecutive blood sampling tests indicate that the employee's blood lead level is at or below 60 µg/100 g of whole blood.
- For an employee removed due to a blood level at or above 70 µg/100 g, when two consecutive blood sampling tests indicate that the employee's blood lead level is at or below 50 ug/100 g of whole blood.
- For an employee removed due to a blood lead level at or above 60 µg/100 g, or due to an average blood lead level at or above 50 µg/100 g, when two consecutive blood sampling tests indicate that the employee's blood lead level is at or below 40 µg/100 g of whole blood.
- For an employee removed due to a final medical determination, when a subsequent final medical determination results in a medical finding, determination, or opinion that the employee no longer has a detected medical condition which places the employee at increased risk of material impairment to health from exposure to lead.
- We will remove any limitations placed on an employee or end any special protective measures provided to an employee pursuant to a final medical determination when a subsequent final medical determination indicates that the limitations or special protective measures are no longer necessary.

We will provide to an employee up to eighteen (18) months of medical removal protection benefits on each occasion that an employee is removed from exposure to lead or is otherwise limited pursuant to the provisions of this Program. Medical removal protection benefits means that we will maintain the earnings, seniority, and other employment rights and benefits of an employee as though the employee had not been removed from normal exposure to lead or otherwise limited. However, the provision of medical removal protection benefits will be conditioned upon the employee's participation in follow-up medical surveillance made available pursuant to this Program and the amount of the benefits to be paid will be reduced to the extent that the employee receives compensation for earnings lost during the period of removal either from a publicly or employer-funded compensation program, or receives income from employment with another employer made possible by virtue of the employee's removal.

The following measures shall be taken with respect to any employee removed from exposure to lead due to an elevated blood lead level whose blood lead level has not sufficiently declined within the past eighteen (18) months of removal to permit return to his or her former job status:

- We will make available to the employee a medical examination in order to obtain a final medical determination with respect to the employee.
- We will ensure that the final medical determination obtained indicates whether or not the employee may be returned to his or her former job status, and if not, what steps would be taken to protect the employee's health.
- Where the final medical determination has not yet been obtained, or once obtained indicates that the employee may not yet be returned to his or former job status, we will continue to provide medical removal protection benefits to the employee until either the employee is returned to former job status, or a final medical determination is made that the employee is incapable of ever safely returning to his or her former job status.

- If we should act pursuant to a final medical determination which permits the return of the employee to his or her former job status despite what would otherwise be an unacceptable blood lead level and later questions concerning removing the employee arise, they will again be decided by a final medical determination. However, we will not automatically remove such an employee pursuant to the blood lead level removal criteria provided by this Program.

EMPLOYEE INFORMATION AND TRAINING

We will cover the hazards of lead as part of our Hazard Communication Program including but not limited to the requirements concerning warning signs and labels, material safety data sheets (MSDS), and employee information and training.

Warning signs will be provided and displayed in all work areas where the PEL is exceeded. The signs shall bear the following information:

```
LEAD WORK AREA

POISON

NO SMOKING OR EATING

WARNING
```

The signs shall be illuminated, cleaned, and maintained as necessary so that the legend is readily visible.

All employees who are potentially exposed to lead at or above the action level or for whom the possibility of skin or eye irritation exists will be required to participate in our lead training program. The training will be provided prior to, or at the time of, initial assignment to a job involving potential exposure to lead and at least annually thereafter.

The training program will be made understandable to those required to participate in it, and will inform them of the following:

- The adverse health hazards associated with lead exposure, with special attention to the information incorporated in Appendix A and B of the OSHA Lead Standard and the adverse reproductive effects of lead on both males and females.
- The quantity, location, manner of use, release, and storage of lead in the workplace and the specific nature of operations that could result in exposure to lead, especially exposures above the action level.

- The engineering controls and work practices associated with the employee's job assignment.
- The measures employees can take to protect themselves from exposure to lead, including modification of such habits as smoking and personal hygiene, and specific procedures we have implemented to protect employees from exposure to lead such as appropriate work practices and the provision of personal protective clothing and equipment.
- The purpose, proper selection, fitting, proper use, and limitations of respirators and protective clothing.
- The purpose and a description of our medical surveillance and medical removal protection program, as set forth in this Program above.
- The contents of this program as well as the OSHA Lead Standard and its appendices.
- The employee's rights of access to his medical and exposure records under 29 CFR §1926.33(g)(1) and (2).
- The employee's obligation to immediately notify supervision whenever (a) there is any actual or suspected instance of noncompliance with any provision of this Program, (b) any difficulty in breathing is experienced while wearing a respirator, (c) eye or skin irritation is experienced that is (or may be) associated with lead exposure, or (d) rips or tears are detected in protective clothing or equipment or it is no longer suitable or fit for its intended purpose.
- Chelating agents should not routinely be used to remove lead from their bodies and should not be used at all except under the direction of a licensed physician.

The following training program records shall be made at the completion of each employee training session and maintained for at least one year thereafter:

- The contents of the training program
- A record of each training session which shall include (1) the name and job assignment of each person trained, (2) the signature of the person who conducted the training or an official of this company, and (3) the calendar date the training was completed

A copy of the OSHA Lead Standard and its appendices will be kept on the premises, made readily available to all employees, and will be provided to them free of charge upon request.

All materials relating to our employee information and training program will be provided in accordance with law to authorized OSHA and NIOSH representatives upon request. In addition, we will include as part of the training program, and distribute to employees, any materials pertaining to the Occupational Safety and Health Act, the regulations issued pursuant to that Act, and the lead standard, which are made available to us by OSHA.

RECORDKEEPING

Whenever lead monitoring is conducted at this workplace, we will establish and keep (for at least 40 years or for the duration of employment plus 20 years, whichever is longer) an accurate record that will include at least the following information:

- The monitoring date(s), number, duration, location, and results of each of the samples taken including a description of the sampling procedure used to determine representative employee exposure where applicable
- The name, social security number, and job classification of the employees monitored and of all other employees whose exposure the monitoring is intended to represent
- A description of the sampling and analytical methods used and evidence of their accuracy
- The type of respiratory protective device, if any, worn by the monitored employee
- A notation of the environmental variables and any other conditions that might have affected the monitoring results

The following records will be established and maintained (for at least 40 years or for the duration of employment plus 20 years, whichever is longer) for each employee covered by the lead standard's medical surveillance provisions:

- Name, social security number, and description of his duties
- A copy of each physician's written opinions and a copy of the medical examination results including medical and work history
- Results of any airborne exposure monitoring done for that employee and the representative exposure levels supplied to the physician
- The employee's medical symptoms that might be related to exposure to lead
- A description of the laboratory procedures and a copy of any standards or guidelines used to interpret the test results or references to that information
- A copy of the results of biological monitoring

We shall also keep (for at least 40 years or for the duration of employment plus 20 years, whichever is longer) or ensure that the examining physician keeps the following medical records:

- A copy of the medical examination results including medical and work history required under this Program
- A description of the laboratory procedures and a copy of any standards or guidelines used to interpret the test results or references to that information
- A copy of the results of biological monitoring

Those employee medical and monitoring records that we are required by 29 CFR §1926.62(n) to keep will be made available (within 15 days after a request) for examination and copying by the subject employee; designated representatives; anyone having the specific written consent of the subject employee; and, after the employee's death or incapacitation, the employee's family members. In the event that we cease to do business and there is no successor employer to receive and retain records for the prescribed period or we find it appropriate to dispose of any records that are required to be preserved for at least thirty years, we will comply with the requirements concerning transfer of records set forth in 29 CFR §1926.33(h)(3).

The following records will be established and maintained for each employee removed from lead exposure pursuant to this Program. (Such records will be kept for at least the duration of such employee's employment):

- Name and social security number of the employee
- The date on each occasion that the employee was removed from current exposure to lead as well as the corresponding date on which the employee was returned to his or her former job status
- A brief explanation of how each removal was or is being accomplished
- A statement with respect to each removal indicating whether the reason for the removal was an elevated blood lead level

---

NOTIFICATION FROM: _____

TO:
_____
Employee's name

The results of airborne lead monitoring taken on _____(date) and
_____(date) have been received. They show a lead level of
_____(number) micrograms of lead per cubic meter of air, a concentration that is below the OSHA lead standard's permissible exposure level of 50 micrograms of lead per cubic meter of air based upon an eight-hour time-weighted average. If you are not familiar with the metric system of measurement, you may want to know that a microgram is one millionth of a gram and there are 454 grams in a pound.

The OSHA lead standard, 29 CFR §1926.62, requires that an employer must notify employees of the result of airborne lead monitoring. We have a copy of the text of the OSHA lead standard and its appendices available on the premises for you to consult whenever you want to do so.

Figure 15.1 Sample Airborne Lead Notification Form

NOTIFICATION FROM: _____

TO:_____
Employee's name

The results of the airborne lead monitoring taken on _____(date) and
_____(date) have been received. They show a lead level of
_____(number) micrograms of lead per cubic meter of air, a concentra-
tion that exceeds the OSHA lead standard's permissible exposure limit of 50 micrograms of lead per cubic meter of air based upon an eight-hour, time-weighted
average. If you are not familiar with the metric system of measurement, you may
want to know that a microgram is one millionth of a gram and there are 454 grams
in a pound.

The OSHA lead standard, 29 CFR §1926.62, requires that an employer must notify
employees of the result of the monitoring of airborne lead levels and a description
of the corrective action(s) being taken to reduce employee exposure to or below
the permissible exposure level. We have adopted a Written Lead Compliance Program that describe the actions being taken to reduce lead exposure. A copy of that
Program as well as a copy of the OSHA lead standard is kept on the premises for
you to consult whenever you want to do so.

Figure 15.2 Results of Airborne Lead Monitoring Form

NOTIFICATION FROM: _____

TO: _____

_____
Employee's name

The results of the blood test you took on _____(date) have been received. It shows a blood/lead level of _____(number) micrograms per
one hundred grams of whole blood. If you are not familiar with the metric system
of measurement, you may want to know that a microgram is one millionth of a
gram and there are 454 grams in a pound.

The OSHA lead standard, 29 CFR §1926.62, requires that an employer must notify
employees of the result of the of the result of their blood tests and also tell them
that, with some exceptions, employees must be removed from working in areas
where the lead in the air exceeds the standard's "action level" (30 µg/m) if the
average of three consecutive blood tests shows a blood/lead level of 50 µg/100g or
more of whole blood but, while so removed, they retain the same earnings, seniority and other rights and benefits as before. The standard contains many details on
this and other matters and should be consulted for complete information. A copy is
posted on the bulletin board.

Figure 15.3 Blood Lead Test Form

16

Asbestos

Controlling Asbestos Exposures at Your Worksite

INTRODUCTION

Asbestos is the generic term for a group of naturally occurring, fibrous minerals with high tensile strength, flexibility, and resistance to thermal, chemical, and electrical conditions.

In the construction industry, asbestos is found in installed products such as shingles, floor tiles, cement pipe and sheet, roofing felts, insulation, ceiling tiles, fire-resistant drywall, and acoustical products. Very few asbestos-containing products are currently being installed. Consequently, most worker exposures occur during the removal of asbestos and the renovation and maintenance of buildings and structures containing asbestos.

Asbestos fibers enter the body by the inhalation or ingestion of airborne particles that become embedded in the tissues of the respiratory or digestive systems. Exposure to asbestos can cause disabling or fatal diseases, such as asbestosis, an emphysema-like condition; lung cancer; mesothelioma, a cancerous tumor that spreads rapidly in the cells of membranes covering the lungs and body organs; and gastrointestinal cancer. The symptoms of these diseases generally do not appear for 20 years or more after initial exposure.

OSHA began regulating workplace asbestos exposure in 1970, adopting a permissible exposure limit (PEL) to regulate worker exposures. Over the years, more information on the adverse health effects of asbestos exposure has become available, prompting the agency to revise the asbestos standard several times to better protect workers. On August 10, 1994, OSHA issued a revised final standard regulating asbestos exposure in all industries. The newly revised standard for the construction industry lowers the PEL, cutting it in half from 0.2 fibers per cubic centimeter of air (f/cc) to 0.1 f/cc.

The Asbestos Standard, 29 CFR 1926.1101 includes

1. A reduced time-weighted average permissible exposure limit (PEL) of 0.1 fiber per cubic centimeter (f/cc) for all asbestos work in all industries
2. A new classification scheme for asbestos construction and shipyard industry work which ties mandatory work practices to work classification
3. A presumptive asbestos identification requirement for certain asbestos containing building materials
4. Limited notification requirements for employers who use unlisted compliance methods in high risk asbestos abatement work

SCOPE AND APPLICATION

The asbestos standard for the construction industry (29 CFR 1926.1101) regulates asbestos exposure for the following activities:

- Demolishing or salvaging structures where asbestos is present
- Removing or encapsulating asbestos-containing materials
- Constructing, altering, repairing, maintaining, or renovating asbestos-containing structures or substrates
- Installing asbestos-containing products
- Cleaning up asbestos spills/emergencies
- Transporting, disposing, storing, containing, and housekeeping involving asbestos or asbestos-containing products on a construction site

PERMISSIBLE EXPOSURE LIMIT (PEL)

Employers must ensure that no employee is exposed to an airborne concentration of asbestos in excess of 0.1 f/cc as an eight-hour time-weighted average (TWA).

OSHA also established a short-term exposure limit (STEL) for asbestos. Employers must ensure that no employee is exposed to an airborne concentration of asbestos in excess of 0.1 f/cc as averaged over a sampling period of thirty minutes.

WORK CLASSIFICATION

OSHA's revised standard establishes a new classification system for asbestos construction work, which clearly spells out mandatory work practices to follow to reduce worker exposures. Four classes of construction activity are matched with increasingly stringent control requirements.

1. Class I Asbestos Work-The most potentially hazardous class of asbestos jobs, Class I work involves the removal of thermal system insulation and sprayed-on or troweled-on surfacing asbestos-containing materials or presumed asbestos-containing materials. Thermal system insulation includes asbestos-containing materials applied to pipes, boilers, tanks, ducts, or other structural components to prevent heat loss or gain. Surfacing materials include decorative plaster on ceilings, acoustical asbestos-containing materials on decking, or fireproofing on structural members.
2. Class II Asbestos Work-Class II work includes the removal of other types of asbestos-containing materials that are not thermal system insulation such as resilient flooring and roofing materials containing asbestos. Examples of Class II work include removal of floor or ceiling tiles, siding, roofing, or transite panels.
3. Class III Asbestos Work-Class III work includes repair and maintenance operations where asbestos-containing or presumed asbestos-containing materials are disturbed.
4. Class IV Asbestos Work-Class IV work includes custodial activities where employees clean up asbestos-containing waste and debris. This includes dusting contaminated surfaces, vacuuming contaminated carpets, mopping floors, and cleaning up asbestos-containing or presumed asbestos-containing materials from thermal system insulation.

EXPOSURE MONITORING

Employee exposure measurements must be made from breathing zone air samples representing the eight-hour TWA and thirty-minute short-term exposures for each employee.

Employers must take one or more samples representing full-shift exposure to determine the eight-hour TWA exposure in each work area. To determine short-term employee exposures, employers must take one or more samples representing thirty-minute exposures for the operations most likely to expose employees above the STEL in each work area.

Employers must allow affected employees and their designated representatives to observe any employee exposure monitoring. When observation requires entry into a regulated area, the employer must provide and require the use of protective clothing and equipment.

Employers must keep records of all employee exposure monitoring for at least 30 years, including

- The date of measurement
- The operation involving asbestos exposure that was monitored
- Sampling and analytical methods used and evidence of their accuracy
- The number, duration, and results of samples taken
- The type of protective devices worn
- The name, social security number, and exposures of the represented employees

COMPETENT PERSON REQUIREMENTS

On all construction sites with asbestos operations, employers must name a "competent person" qualified and authorized to ensure worker safety and health, as required by Subpart C., "General Safety and Health Provisions for Construction" (29 CFR 1926.20). Under the requirements for safety and health prevention programs, the competent person must frequently inspect job sites, materials, and equipment.

In addition, for Class I jobs the competent person must inspect onsite at least once during each work shift and upon employee request. For Class II and III jobs, the competent person must inspect often enough to assess changing conditions and upon employee request.

REGULATED AREAS

A regulated area is a marked off site where employees work with asbestos, including any adjoining area(s) where debris and waste from asbestos work accumulates or where airborne concentrations of asbestos exceed or can possible exceed the PEL.

All Class I, II, and III asbestos work or any other operations where airborne asbestos exceeds the PEL must be done within regulated areas. Authorized personnel only may enter. The designated competent person supervises all asbestos work performed in the area. (See Competent Person Requirements.)

Employers must mark off the regulated area in any manner that minimizes the number of persons within the area and protects persons outside the area from exposure to airborne asbestos.

Posted warning signs demarcating the area must be easily readable and understandable. The signs must bear the following information:

> ASBESTOS
>
> CANCER AND LUNG DISEASE HAZARD
>
> AUTHORIZED PERSONNEL ONLY
>
> RESPIRATORY AND PROTECTIVE CLOTHING ARE REQUIRED
>
> IN THIS AREA
>
> DANGER

EMPLOYEE INFORMATION AND TRAINING

Employers must, at no cost to employees, provide a training program for all employees installing and handling asbestos-containing products and for employees performing Class I through IV asbestos operations. Employees must receive training prior to or at initial assignment and at least annually thereafter.

Training courses must be easily understandable for employees and must inform them of

- Ways to recognize asbestos
- The adverse health effects of asbestos exposure
- The relationship between smoking and asbestos in causing lung cancer
- Operations that could result in asbestos exposure and the importance of protective controls to minimize exposure
- The purpose, proper use, fitting instruction, and limitations of respirators
- The appropriate work practices for performing asbestos jobs
- Medical surveillance program requirements
- The contents of the standard

RESPIRATORY PROTECTION

Employers must provide respirators at no cost to employees, selecting the appropriate type from among those approved by the Mine Safety and Health Administration (MSHA) and NIOSH.

For all employees performing Class I work in regulated areas and for jobs without a negative exposure assessment, employers must provide full-facepiece supplied-air respirators operated in pressure-demand mode and equipped with an auxiliary positive-pressure, self-contained breathing apparatus.

Employers must provide half-mask purifying respirators—other than disposable respirators—equipped with high-efficiency filters for Class II and III asbestos jobs without a negative exposure assessment and for Class III jobs where work disturbs thermal system insulation or surfacing asbestos-containing or presumed asbestos-containing materials.

If a particular job is not covered above and exposures are above the PEL or STEL, the asbestos standard, Occupational Exposure to Asbestos, 29 CFR 1926.1101, contains a table specifying types of respirators to use.

Employers must institute a respiratory program in accordance with OSHA's Respiratory Protection Standard, 29 CFR 1926.103. Employers must permit employees using filter respirators to change the filter elements when breathing resistance increases and must maintain an adequate supply of filters for this purpose. Employers must permit employees wearing respirators to leave work areas to wash their faces and respirator facepieces as necessary to prevent skin irritation.

Employers must ensure that the respirators issued have the least possible facepiece leakage and that they fit properly. For employees wearing negative-pressure respirators, employers must perform either quantitative or qualitative face fit tests with the initial fitting and at least every six months thereafter. The qualitative fit tests can be used only for fit testing of half-mask respirators where they are permitted or for full-facepiece air-purifying respirators where they are worn at levels where half-facepiece air-purifying respirators are permitted. Employers must conduct qualitative and quantitative fit tests in accordance with Occupational Exposure to Asbestos (29 CFR 1926.1001, Appendix C) and use the tests to select facepieces that provide the required protection.

Respirators must be used during

- All Class I asbestos jobs
- All Class II work where an asbestos-containing material is not removed substantially intact
- All Class II and III work not using wet methods
- All Class II and III work without a negative exposure assessment
- All Class III jobs where thermal system insulation or surfacing asbestos-containing or presumed asbestos-containing material is cut, abraded, or broken
- All Class IV work within a regulated area where respirators are required
- All work where employees are exposed above the PEL or STEL
- All emergencies

PROTECTIVE CLOTHING

Employers must provide and require the use of protective clothing, such as coveralls or similar whole-body clothing, head coverings, gloves, and foot coverings for

- Any employee exposed to airborne asbestos exceeding the PEL or STEL
- Work without a negative exposure assessment

- Any employee performing Class I work involving the removal of over 25 linear or ten square feet of thermal system insulation or surfacing asbestos-containing or presumed asbestos-containing materials.

Employers must launder contaminated clothing to prevent the release of airborne asbestos in excess of the PEL or STEL. Any employer who gives contaminated clothing to another person for laundering must inform him or her of the contamination.

Employers must transport contaminated clothing in sealed, impermeable bags or closed impermeable containers bearing appropriate labels.(See the Hazard Communication Standard 29 CFR 1926.59 for label requirements.)

HOUSEKEEPING

If employees are required to perform maintenance and custodial operations around installed asbestos products, these general OSHA housekeeping requirements must be followed:

- Dust Collection—Keep all surfaces as free as possible of dust and waste that contain asbestos.
- Spills—Clean up all spills and sudden releases of ACM as soon as possible.
- Compressed Air-Never use compressed air to clean surfaces contaminated with asbestos.
- Vacuuming—Always use special asbestos vacuums equipped with high-efficiency particulate-air (HEPA) filters and empty them in a manner that minimizes the re-entry of airborne fibers into the workplace.
- Wet Methods—Never shovel, dry-sweep or use other dry clean-up methods for asbestos debris unless vacuuming or wet cleaning methods are not feasible.
- Waste Disposal—Dispose of all waste, scrap, debris, empty containers, equipment and clothing contaminated with asbestos only in sealed, impermeable bags or containers. Warning labels must be placed on airtight containers of asbestos waste before they are transported.

MEDICAL SURVEILLANCE

A licensed physician must perform or supervise all medical exams and procedures, which are to be provided at no cost to employees and at a reasonable time.

If the employee was examined within the past 12 months and that exam meets the criteria of the standard, however, another medical exam is not required.

Employers must provide a medical surveillance program for all employees

- Who, for a combined total of 30 or more days per year, engage in Class I, II, or III work or are exposed at or above the PEL or STEL
- Who wear negative-pressure respirators

- Prior to employee assignment to an area where negative-pressure respirators are worn
- Within ten working days after the 30th day of exposure for employees assigned to an area where exposure is at or above the PEL for 30 or more days per year
- At least annually after an exposure
- When the examining physician suggests them more frequently

Medical exams must include the following:

- A medical and work history
- Completion of a standardized questionnaire with the initial exam (See 29 CFR 1926.1101, Appendix D, Part 1) and an abbreviated standardized questionnaire with annual exams (See 29 CFR 1926.1101, Appendix D, Part 2.)
- A physical exam focusing on the pulmonary and gastrointestinal systems

Employers must provide the examining physician

- A copy of OSHA's asbestos standard and its appendices
- A description of the affected employee's duties relating to exposure
- The employee's representative exposure level or anticipated exposure level
- A description of any personal protective equipment and respiratory equipment used
- Information from previous medical exams not otherwise available

It is the employer's responsibility to obtain the physician's written opinion containing results of the medical exam and

- Any medical conditions of the employee that increase health risks from asbestos exposure.
- Any recommended limitations on the employee or protective equipment used.
- A statement that the employee has been informed of the results of the medical exam and any medical conditions resulting from asbestos exposure.
- A statement that the employee has been informed of the increased risk of lung cancer from the combined effect of smoking and asbestos exposure.
- The physician must not reveal in the written opinion specific findings or diagnoses unrelated to occupational exposure to asbestos..
- The employer must provide a copy of the physician's written opinion to the affected employee within 30 days after receipt.

RECORDKEEPING

Where employers use objective data to demonstrate that products made from or containing asbestos cannot release fibers in concentrations at or above the PEL or STEL, they must keep an accurate record for as long as it is relied on and include the following:

- The exempt product
- The source of the objective data
- The testing protocol, test results, and analysis of the material for release of asbestos
- A description of the exempt operation and support data

- Other data relevant to operations, materials, processes, or employee exposures

Monitoring Records

Employers must keep records of all employee exposure monitoring for at least 30 years, including

- The date of measurement
- The operation involving asbestos exposure that was monitored
- Sampling and analytical methods used and evidence of their accuracy
- The number, duration, and results of samples taken
- The type of protective devices worn
- The name, social security number, and exposures of the represented employees

Exposure records must be made available when requested by affected employees, former employees, their designated representatives, and/or OSHA's Assistant Secretary.

Medical Surveillance Records

Employers must keep all medical surveillance records for the duration of the employee's employment plus 30 years, including

- The employee's name and social security number
- The employee's medical exam results, including the medical history, questionnaires, responses, test results, and physician's recommendations
- The physician's written opinions
- Any employee medical complaints related to asbestos exposure
- A copy of the information provided to the examining physician

Employee medical surveillance records must be available to the subject employee.

Other Recordkeeping Requirements

- Employers must maintain all employee training records for one year beyond the last date of employment for each employee.
- Where data demonstrates presumed asbestos-containing materials do not contain asbestos, building owners or employers must keep the records for as long as they rely on them. Building owners must maintain written notifications on the identification, location, and quantity of any asbestos-containing or presumed asbestos-containing materials for the duration of ownership and transfer the records to successive owners.

1. NAME _____

2. SOCIAL SECURITY # ___ ___ ___ - ___ ___ - ___ ___ ___ ___

3. CLOCK NUMBER _____

4. PRESENT OCCUPATION _____

5. COMPANY _____

6. ADDRESS _____

   _____
                                                   (Zip Code)

8. TELEPHONE NUMBER _____

9. INTERVIEWER _____

10. DATE _____

11. What is your marital status? 1. Single ____   3. Separated/Divorced ____
                                 2. Married ____   4. Widowed ____

12. OCCUPATIONAL HISTORY

In the past year, did you work full time   1.   Yes ____   2. No ____
       (30 hours per week or more) for 6 months or more?

        If No, Go to Question 13.

In the past year, did you work   1.   Yes ____   2. No ____
       in a dusty job?

Was dust exposure: 1.   Mild ____   2. Moderate ____   3.   Severe ____

In the past year, were you   1.   Yes ____   2. No ____
       exposed to gas or chemical
       fumes in your work?

Was exposure: 1.   Mild ____   2. Moderate ____   3.   Severe ____

Figure 16.1 Sample Asbestos Medical Questionnaire

375

12.   In the past year, What was your:
      1. Job/Occupation? _____

      2. Position/Job Title? _____

13.   RECENT MEDICAL HISTORY

13A. Do you consider yourself to be in good health?

                                        Yes _____   No _____

      If NO, state reason _____

      In the past year, have you   Yes   No
            developed:
      Epilepsy?   ___   ___

      Rheumatic fever?   ___   ___

      Kidney disease?   ___   ___

      Bladder disease?   ___   ___

      Diabetes?   ___   ___

      Jaundice?   ___   ___

      Cancer?   ___   ___

14.   CHEST COLDS AND CHEST ILLNESSES

14A. If you get a cold, does it *usually* go to your chest?
                        (Usually means more than 1/2 the time)
                                        1. Yes ____ 2. No ____
                                        3. Don't get colds ____

15A. During the past year, have you had any
            chest illnesses that have kept you off work,
            indoors at home, or in bed?   1. Yes ____ 2. No ____
                                            3. Does Not Apply ____

            If NO, Go to Question 16.

Figure 16.1 Sample Asbestos Medical Questionnaire, continued

15B. Did you produce phlegm with any
of these chest illnesses?   1. Yes _____ 2. No _____
                                3. Does Not Apply _____

15C. In the past year, how many such illnesses
with (increased) phlegm did you have
Number of illnesses _____ which lasted a week or more?
No such illnesses _____

16. RESPIRATORY SYSTEM

In the past year have you had:

	Yes	No
(Further Comments of Positive Answers)		
Asthma	_____	_____
Bronchitis	_____	_____
Hay Fever	_____	_____
Other Allergies	_____	_____
Pneumonia	_____	_____
Tuberculosis	_____	_____
Chest Surgery	_____	_____
Other Lung Problems	_____	_____
Heart Disease	_____	_____

Do you have:

	Yes	No
(Further Comment on Positive Answers)		
Frequent Colds	_____	_____
Chronic cough	_____	_____

Figure 16.1 Sample Asbestos Medical Questionnaire, continued

Shortness of breath
when walking or
climbing one flight
of stairs   _____   _____

Do you:

Wheeze   _____   _____

Cough up phlegm   _____   _____

Smoke cigarettes   _____   _____
Packs per day _____   How many years?_____

Date _____Patient's Signature _____

Figure 16.1 Sample Asbestos Medical Questionnaire, continued

17

Process Safety Management

Conducting a Process Safety Analysis at Your Worksite

INTRODUCTION

The Process Safety Management Standard (PSM) 29 CFR §1926.64 regulates "highly hazardous chemicals." This standard's objective is to prevent fires, explosions, and toxic releases by imposing stringent management practices upon companies that handle specified quantities of any one or more of 130 specified toxic and reactive chemicals or 10,000 pounds or more of flammable and liquid gases.

Employers who wonder if the standard covers them should check the list of chemicals in the standard. If they do not handle any of those chemicals in the specified threshold quantities (TQ) listed in pounds, they will not need to concern themselves with the PSM standard. If, on the other hand, they do handle any of them at the TQ level or above, they will have to find the ways and means to cope with what may well be OSHA's single greatest intrusion to date in the management of private business enterprises.

The program detailed below will help minimize the consequences of catastrophic releases by stating polices and procedures for the management of process hazards in design, construction, start-up, operation, inspection, maintenance, and other matters addressed in the OSHA standard.

This program calls for maximum employee participation and includes all elements of the employee participation provisions of the OSHA standard. It shall, as a minimum (1) consult with employees and their representatives on the conduct and development of process hazards analyses, and on the development of the OSHA standard's other elements of process safety management, and (2) provide to our employees and their representatives access to the process hazard analyses and all other information required to be developed under that standard.

This Process Safety Management Program must be regularly reviewed and updated as provided below, and whenever necessary to reflect new or modified tasks and procedures.

A copy of this program as well as the other documents and reports that are created pursuant to this program, must be accessible to employees and made available to other affected persons in accordance with applicable OSHA requirements and the provisions of this program.

379

SAMPLE PROCESS SAFETY MANAGEMENT PROGRAM

Scope and Application

This program covers

- Each process that involves a chemical listed in Appendix A of 29 CFR §1926.64 at or above its specified threshold quantities
- Each process that involves a flammable liquid or gas as defined in 29 CFR §1926.64, on site in one location, in a quantity of 10,000 pounds (4535.9kg) or more, except for
- Hydrocarbon fuels used solely for workplace consumption as a fuel (e.g., propane used for comfort heating, gasoline for vehicle refueling), if such fuels are not a part of a process containing another highly hazardous chemical covered by the OSHA standard
- Flammable liquids stored in atmospheric tanks or transferred which are kept below their normal boiling point without benefit of chilling or refrigeration

Process Safety Information

In conjunction with the Process Hazard Analysis (part of this Program, listed below), we will complete a compilation of written process safety information in order to enable our supervisory personnel and our employees who operate the processes to identify and understand the hazards posed by processes involving highly hazardous chemicals. This safety information will be made readily available to those involved in the processes, and shall include

- Information pertaining to hazards of the highly hazardous chemicals used in the process
- Information pertaining to the technology of the process
- Information pertaining to the equipment in the process

Further details on each of those three kinds of information follows.

Information pertaining to hazards of the chemicals used in the process will consist of at least the following:

- Toxicity information
- Permissible exposure limits
- Physical data
- Reactivity data
- Corrosivity data
- Thermal and chemical stability data
- Hazardous effects of inadvertent mixing of different materials, that could foreseeably occur

Usually, the Material Safety Data Sheets (MSDS) will be used to supply such information. If any MSDS fails to include any of the above-listed matters, we will obtain that information and provide it in the same manner as the MSDS information is provided.

Information pertaining to the technology of the process will include at least the following:

- A block flow diagram or simplified process flow diagram
- Process chemistry
- Maximum intended inventory
- Safe upper and lower limits for such items as temperatures, pressures, flows and/or compositions
- An evaluation of the consequences of deviations, including those affecting the safety and health of employees

In those cases where the original technical information no longer exists, we will develop the information in conjunction with the Process Hazard Analysis in sufficient detail to support that Analysis.

Information pertaining to the equipment in the process will include:

- Materials of construction
- Piping and instrument diagrams (P&ID's)
- Electrical classification
- Relief system design and design basis
- Ventilation system design
- Design codes employed
- Material and energy balances for processes that are built after May 26, 1992
- Safety systems (such as interlocks, detection and suppression systems, etc.)

We have ascertained (and have documented) that our equipment complies with applicable codes and standards, such as those published by the American Society of Mechanical Engineers, the American Petroleum Institute, the American Institute of Chemical Engineers, the American National Standards Institute, the American Society of Testing and Materials, and the National Fire Protection Association, where they exist, or, recognized and generally accepted good engineering practices.

For those items of existing equipment that were designed and constructed in accordance with codes, standards, or practices that are no longer in general use, we have determined (and documented) that the equipment is designed, maintained, inspected, tested, and operated in such a way that safe operation is ensured.

Process Hazard Analysis

We have assembled a team with expertise in engineering and process operations to perform the process hazard analysis hazard evaluation required by the OSHA standard. At least one member of the team is knowledgeable in the specific process hazard analysis methodology being used. The team will also include at least one employee who has experience and knowledge specific to the process being evaluated.

The team will perform the initial process hazard analysis on all processes covered by the OSHA standard.

That analysis shall be appropriate to the complexity of the process and shall identify, evaluate, and control the hazards involved in the process.

The team will determine and document the priority order for conducting the process hazard analysis based upon a rationale that includes such consideration as

- The extent of the process hazards
- Number of potentially affected employees
- Age of the process
- Operating history of the process

All initial process hazards analyses shall be completed by May 26, 1997.

Once the initial process hazard analysis has been completed, it shall be updated and revalidated at least every five years by the same (or an equally qualified and representative) team in order to ensure that the process hazard analysis is consistent with the current process. In the event that a hazard analysis performed previously, but not earlier than May 26, 1987, is acceptable as an initial process hazard analysis under the OSHA standard, it shall be our process hazard analysis. It will then be updated and revalidated at least every five years after the date it was completed (but that completion date cannot be earlier than May 26, 1987 in order for it to be acceptable under the OSHA standard).

The process hazard analysis shall utilize one or more of the following methodologies that are appropriate to determine and evaluate the hazards of the processes being analyzed:

- What-If
- Checklist
- What-If/checklist
- Hazard and Operability Study (HAZOP)
- Failure Mode and effects Analysis (FMEA)
- Fault Tree Analysis
- An appropriate equivalent methodology

The process hazard analysis will address the following:

- The hazards of the process
- The identification of any previous incident which had a likely potential for catastrophic consequences in the workplace
- Engineering and administrative controls applicable to the hazards and their interrelationships such as appropriate application of detection methodologies to provide early warning of releases. (Note: Acceptable detection methods might include process monitoring and control instrumentation with alarms, and detection hardware such as hydrocarbon sensors.)
- Consequences of failure of engineering and administrative controls
- Facility sitting
- Human factors
- A qualitative evaluation of a range of the possible safety and health effects of failure of controls on employees in the workplace

When the team's findings and recommendations are made, we will establish a system to promptly address them. That system will do the following:

- Ensure that the recommendations are resolved in a timely manner and that the resolution is documented
- Document what actions are to be taken
- Complete actions as soon as possible
- Develop a written schedule of when these actions are to be completed
- Communicate the actions to operating, maintenance and other employees whose work assignments are in the process and who may be affected by the recommendations or actions

The process hazard analysis and all updates and revalidation for each covered process, including the documented resolution of the recommendations made pursuant to the system stated immediately above, shall be retained for the life of the process.

Operating Procedures

We will develop and implement written operating procedures that are consistent with the Process Safety Information. Those operating procedures will provide clear instructions for safely conducting activities involved in each process consistent with the process safety information. They will at least address the following:

The steps for each operating phase

- Initial startup
- Normal operation
- Temporary operations as the need arises
- Emergency operations, including emergency shutdowns, and who may initiate these procedures
- Normal shutdown
- Startup following a turnaround or after an emergency shutdown / MOBILIZATION.

RECOGNIZED BY THE GOV'T.

Operating limits

- Consequences of deviation
- Steps required to correct and/or avoid deviation
- Safety systems and their functions

Safety and health considerations

- Properties of, and hazards presented by, the chemicals used in the process
- Precautions necessary to prevent exposure, including administrative controls, engineering controls, and personal protective equipment
- Control measures to be taken if physical contact or airborne exposure occurs
- Safety procedures for opening process equipment (such as pipe line breaking)

- Quality control for raw materials and control of hazardous chemical inventory levels
- Any special or unique hazards

A copy of the operating procedures will be readily accessible to employees who work in or maintain a process.

The operating procedures will be reviewed as often as necessary to ensure that they reflect current operating practice, including changes that result from changes in process chemicals, technology, and equipment and changes to facilities.

We will make an annual certification that those operating procedures are both current and accurate.

We have developed and implemented additional safe work practices in order to provide for the control of hazards during the operations listed below. They apply to contractor employees as well as our own employees.

- Lockout/tagout
- Confined space entry
- Opening process equipment or piping
- Control over entrance to a facility by maintenance, contractor, laboratory, or other support personnel

Training

Initial training. Each employee presently involved in a process and each employee before working in a newly assigned process will be trained in an overview of the process and in the operating procedures of this Program and 29 CFR §1926.64(g). The training will include emphasis on the specific safety and health hazards, procedures, and safe practices applicable to the employee's job tasks.

Initial training need not be provided to those employees already involved in operating a process so long as there is written certification that the employee has the required knowledge, skills, and abilities to safely carry out the duties and responsibilities specified in the Operating Procedures of this Program and 29 CFR §1926.64(g).

Refresher and supplemental training. Refresher and supplemental training shall be provided to each employee involved in operating a process at least every three years (and more often if necessary) in order to ensure that the employee understands and adheres to the current operating procedures of the process.

Training documentation. We shall ascertain that each such employee has received, successfully completed, and understood such training. We will also prepare a record that will contain

- The identity of the employee
- The date of training
- The means and methods that were used in order to verify that the employee understood the training

Contractors

This section of our Program applies to contractors performing maintenance or repair, turn-around, major renovation, or specialty work on (or adjacent to) a covered process. It does not apply to contractors providing incidental services that do not influence process safety, such as janitorial work, food and drink services, laundry, and delivery or other supply services.

When selecting a contractor, we will obtain and evaluate information regarding the contract employer's safety performance and programs. Before the contractor begins work, we will do the following:

- Inform contract employers of the known potential fire, explosion, or toxic release hazards related to the contractor's work and the process
- Explain to contract employers, the applicable provisions of our emergency action plan
- Develop and implement additional safe work practices consistent with those listed above in order to control the entrance, presence and exit of contract employers and contract employees in covered process areas
- Periodically evaluate the performance of contract employers in fulfilling their obligations as specified in the OSHA standard, §1926.64(h)(3)
- Maintain a contract employee injury and illness log for those OSHA recordable injuries and illnesses sustained by contractor employees that are related to the contractor's work in process areas

We will make it a condition of each contract that the contract employer must do the following:

- Ensure that each contract employee is trained in the work practices necessary to safely perform his/her job
- Ensure that each contract employee is instructed in the known potential fire, explosion, or toxic release hazards related to his/her job and the process, as well as the applicable provisions of our Emergency Action Plan (see emergency planning and responses of this Program below)
- Document that each contract employee has received and understood the training required by the OSHA standard, and that the contract employer will prepare a record which contains the identity of the contract employee, the date of training, and the means used to verify that the employee understood the training
- Ensure that each contract employee follows our safety rules including the additional safe work practices listed above
- Advise us of any unique hazards presented by the contract employer's work, or of any hazards found by the contract employers' work

Pre-Startup Safety Review

Whenever we establish new facilities or modify facilities for which the modification necessitates a change in the process safety information, we will perform a pre-startup safety review.

The pre-startup safety review must confirm that, prior to the introduction of highly hazardous chemicals to a process:

- Construction is in accordance with design specifications
- Safety, operating, maintenance, and emergency procedures are in place and are adequate
- Process hazard analysis recommendations have been addressed and actions necessary for startup have been completed
- Operating procedures are in place and training of each operating employee has been completed

Mechanical Integrity

We will establish and implement written procedures in order to maintain the ongoing integrity of the following items of process equipment:

- Pressure vessels and storage tanks
- Piping systems (including piping components such as valves)
- Relief and vent systems and devices
- Emergency shutdown systems
- Controls (including monitoring devices and sensors, alarms, and interlocks)
- Pumps

Each employee involved in maintaining the ongoing integrity of those items of process equipment will be trained in an overview of the process and its hazards and in the procedures applicable to the employees' job tasks in order to ensure that the employee can perform the job tasks in a safe manner.

An inspection and testing program will be established that includes the following:

- Inspections and tests shall be performed on that process equipment.
- Inspection and testing procedures shall follow applicable codes and standards, such as those published by the American Society of Mechanical Engineers, the American Petroleum Institute, the American Institute of Chemical Engineers, the American National Standards Institute, the American Society of Testing and Materials, and the National Fire Protection Association, where they exist; or, recognized and generally accepted engineering practices.
- The frequency of inspections and tests shall be consistent with applicable codes and standards, applicable manufacturers' recommendations, and good engineering practices. They will be conducted more frequently if determined necessary by prior operating experience.

Each such inspection and test shall be documented in order to identify the following:

- The date of the inspection
- The name of the person who performed the inspection or test
- The serial number or other identifier of the equipment on which the inspection or test was performed
- A description of the inspection or test performed
- The results of the inspection or test

Deficiencies in equipment that are outside acceptable limits (defined by the process safety information mentioned in this Program and §1926.64(j)(5) of the OSHA standard) must be corrected either before further use, or in a safe and timely manner provided necessary means are taken to ensure safe operation.

Whenever a new plant or new equipment is constructed, we will make sure that

- The equipment, as it is fabricated, is suitable for the process application for which it will be used
- Equipment as fabricated meets design specifications
- Appropriate checks and inspections are performed as necessary in order to ensure that equipment is installed properly and is consistent with design specifications and manufacturer's instructions
- Maintenance materials, spare parts, and equipment meet design specifications.

Hot Work Permit

A permit will be required for all hot work operations.

- The permit must document that the fire prevention and protection requirements in 29 CFR §1926.252(k), the OSHA standard that specifies the precautions for fire prevention in welding or cutting work, have been implemented prior to beginning the hot work operations.
- Each such permit shall indicate the date(s) authorized for hot work and identify the object on which hot work is to be performed.
- The permit shall be kept on file until completion of the hot work operations.

Management of Change

All contemplated changes to a process must be thoroughly evaluated in order to assess their potential impact upon the safety and health of employees and to determine what modifications to operating procedures may be necessary. We have therefore, established written procedures to manage those changes to process chemicals, technology, equipment, procedures, and facilities that affect a covered process.

Those procedures will ensure that the following are addressed prior to any change:

- The technical basis for the proposed change
- Impact of the change on safety and health
- Modifications to operating procedures
- Necessary time period for the change
- Authorization requirements for the proposed change

Employees involved in operating a process and those maintenance and contract employees whose job tasks will be affected by a change in the process will be informed of and trained in the change prior to start-up of the process or affected part of the process.

If such a change results in a change to the process safety information covered in this Program and 29 CFR §1926.64(L), that information will be appropriately updated.

If such a change results in a change to the operating procedures or practices covered in this Program and 29 CFR §1926.64(f), those procedures and practices will be appropriately updated.

Incident Investigation

All incidents that have either resulted in a catastrophic release of a highly hazardous chemical in the workplace or could reasonably have so resulted will be investigated.

Each such incident investigation will be initiated as promptly as possible but not later than 24 hours following the incident.

We have established an incident investigation team to thoroughly investigate and analyze each such incident. The membership of the investigation team will include at least one person knowledgeable in the process involved, including a contract employee if the incident involved work of the contractor, and other persons with appropriate knowledge and experience.

The investigation team must prepare a report at the conclusion of the investigation which includes at a minimum

- Date of the incident
- Date investigation began
- A description of the incident
- The factors that contributed to the incident
- Any recommendations resulting from the investigation

We will establish a system to promptly address and resolve the incident report's findings and recommendations. All corrective actions and resolutions that are taken in response shall be documented.

The report of the investigation team will be reviewed with all affected personnel whose job tasks are relevant to the incident findings including contract employees where applicable.

All incident investigation reports shall be retained for five years.

Emergency Planning and Response

We have established an Emergency Action Plan for the entire plant in accordance with the provisions of 29 CFR §1926.35(a), the OSHA standard that provides for the adoption of an Emergency Action Plan. Our Plan also includes procedures for handling small releases. In the event that we are covered by similar OSHA requirements (such as the OSHA hazardous waste and emergency response standard), we will establish whatever plans or programs those OSHA standards may require.

Compliance Audits

Our process safety management system will be regularly evaluated in order to

- Measure its effectiveness.
- Direct our attention to potential weaknesses in the system.
- Verify that the procedures and practices that we have developed under the OSHA standard are adequate.
- Ensure that all required procedures and practices are being followed.
- The evaluation will be conducted at least every three years.
- It will be conducted by at least one person knowledgeable in the process.
- A report of the findings of the audit will be developed.
- We will promptly determine and document an appropriate response to each of the findings of the compliance audit and document that deficiencies have been corrected.
- The two most recent compliance audit reports will be retained in our records.

Trade Secrets

Without regard to the trade secret status of such information, all information that is needed in order to comply with the OSHA standard will be made available to those persons responsible for compiling the process safety information, those assisting in the development of the process hazard analysis, those responsible for developing the Operating Procedures, as well as those involved in incident investigations, Emergency Planning and Response, and Compliance Audits. Those persons with access to information that is classified as a trade secret, however, may be required to enter into confidentiality agreements against disclosure of such information as provided in the OSHA Hazard Communication standard, 29 CFR §1926.59(i).

Subject to the rules and procedures contained in that standard (see §§1926.59(i)(1) through 1926.59(i)(12)), employees and their designated representatives shall be permitted access to trade secret information contained within our Process Hazard Analysis and other documents that are required to be developed by the OSHA Process Safety Management standard, 29 CFR §1926.64.

What Are the Basic Requirements of the Standard?

Requirements	Recommended Response
1. This program calls for maximum employee participation. Employees will be consulted on the development of Process Hazard Analyses, and other elements of Process Safety Management.	Employee Participation
2. Written Compilation of Safety Information	Use the Hazard Communication Program as a basis. Needs engineering documents. Coordinate through corporate engineering department.
3. Initial Process Hazard Analysis	Corporate guidance from risk management department allows up to five years to complete.
4. Reaction to Process Hazard Analysis	Facility or site will use the system in maintenance and safety to implement recommendations.
5. Written Operating Procedures.	We have developed and implemented written procedures that will certify annually that procedures are current.
6. Initial training A)presently involved B) newly assigned	Facility or site certifies those presently involved, provides training for newly assigned employees. Done by facility with corporate coordination.
Refresher Training	Will be done at least every three years.
7. A) Employer Responsibility when selecting contractors	Use the facility or site process to involve facility or site safety and purchasing contracts groups.
B) Contractor Responsibilities	Facility requires as a part of the contract.
8. Pre Start-up Safety Review	Whenever new facilities necessitate a change, we will perform a pre-start-up safety review.

Figure 17.1 Process Safety Management Standard Requirements
and Recommended Response

Requirements	Recommended Response
9. Mechanical Integrity	Facility or site corporate engineering departments develop programs.
10. Hot Work Permits	A permit will be required for all hot work operations.
11. Management of change	All changes to a process must be thoroughly evaluated. We have established written procedures to manage changes to process chemicals.
12. Incident Investigation	All incidents will be investigated that have resulted in catastrophic release of a highly hazardous chemical. Records will be retained for five years.
13. Emergency Planning	An emergency action plan has been established in accordance with 29 C.F.R. 1926.35(a).
14. Compliance Audits	Coordinated with Facility & Corporate Safety and Engineering departments and be conducted every three years.
15. Trade Secrets	Subject to the rules contained in the Hazard Communication Program. They are containedn 1926.5(i)(1)-1926.59(i)(12).

Figure 17.1 Process Safety Management Standard Requirements and Recommended Response, continued

```
TRAINING CERTIFICATION FORM


In lieu of initial training, I certify that _____
                                                        (Name)

has the required knowledge, skills, and abilities to be involved in the operation of the


_____.

  (name of listed process)












_____          _____
       Certifying Manager                              Date
```

Figure 17.2 Process Safety Management Training Certification Document

APPENDIX A TO § 1926.64—LIST OF HIGHLY HAZARDOUS CHEMICALS, TOXICS AND REACTIVES (MANDATORY)

This Appendix contains a listing of toxic and reactive highly hazardous chemicals which present a potential for a catastrophic event at or above the threshold quantity.

Chemical Name	CAS*	TQ**
Acetaldehyde	75-07-0	2500
Acrolein (2-Propenal)	107-02-8	150
Acrylyl Chloride	814-68-6	250
Allyl Chloride	107-05-1	1000
Allylamine	107-11-9	1000
Alkylaluminums	Varies	5000
Ammonia, Anhydrous	7664-41-7	10000
Ammonia solutions (greater than 44% ammonia by weight)	7664-41-7	15000
Ammonium Perchlorate	7790-98-9	500
Ammonium Permanganate	7787-36-2	7500
Arsine (also called Arsenic Hydride)	7784-42-1	100
Bis(Chloromethyl) Ether	542-88-1	100
Boron Trichloride	10294-34-5	2500
Boron Trifluoride	7637-07-2	250
Bromine	7726-95-6	1500
Bromine Chloride	13863-41-7	1500
Bromine Pentafluoride	7789-30-2	2500
Bromine Trifluoride	7787-71-5	15000
3-Bromopropyne (also called Propargyl Bromide)	106-96-7	100
Butyl Hydroperoxide (Tertiary)	75-91-2	5000
Butyl Perbenzoate (Tertiary)	614-45-9	7500
Carbonyl Chloride (see Phosgene)	75-44-5	100
* Carbonyl Fluoride	353-50-4	2500
Cellulose Nitrate (concentration greater than 12.6% nitrogen	9004-70-0	2500
Chlorine	7782-50-5	1500
Chlorine Dioxide	10049-04-4	1000
Chlorine Pentrafluoride	13637-63-3	1000
Chlorine Trifluoride	7790-91-2	1000
Chlorodiethylaluminum (also called Diethylaluminum Chloride)	96-10-6	5000
1-Chloro-2,4-Dinitrobenzene	97-00-7	5000
Chloromethyl Methyl Ether	107-30-2	500
Chloropicrin	76-06-2	500
Chloropicrin and Methyl Bromide mixture	None	1500
Chloropicrin and Methyl Chloride mixture	None	1500
Cumene Hydroperoxide	80-15-9	5000
Cyanogen	460-19-5	2500
Cyanogen Chloride	506-77-4	500
Cyanuric Fluoride	675-14-9	100
Diacetyl Peroxide (concentration greater than 70%)	110-22-5	5000
Diazomethane	334-88-3	500
Dibenzoyl Peroxide	94-36-0	7500
Diborane	19287-45-7	100
Dibutyl Peroxide (Tertiary)	110-05-4	5000

Figure 17.3 List of Highly Hazardous Chemicals

Chemical Name	CAS*	TQ**
Dichlorosilane	4109-96-0	2500
Diethylzinc	557-20-0	10000
Diisopropyl Peroxydicarbonate	105-64-6	7500
Dilauroyl Peroxide	105-74-8	7500
Dimethyldichlorosilane	75-78-5	1000
Dimethylhydrazine, 1,1-	57-14-7	1000
Dimethylamine, Anhydrous	124-40-3	2500
2,4-Dinitroaniline	97-02-9	5000
Ethyl Methyl Ketone Peroxide (also Methyl Ethyl Ketone Peroxide; concentration greater than 60%)	1338-23-4	5000
Ethyl Nitrite	109-95-5	5000
Ethylamine	75-04-7	7500
Ethylene Fluorohydrin	371-62-0	100
Ethylene Oxide	75-21-8	5000
Ethyleneimine	151-56-4	1000
Fluorine	7782-41-4	1000
Formaldehyde (Formalin)	50-00-0	1000
Furan	110-00-9	500
Hexafluoroacetone	684-16-2	5000
Hydrochloric Acid, Anhydrous	7647-01-0	5000
Hydrofluoric Acid, Anhydrous	7664-39-3	1000
Hydrogen Bromide	10035-10-6	5000
Hydrogen Chloride	7647-01-0	5000
Hydrogen Cyanide, Anhydrous	74-90-8	1000
Hydrogen Fluoride	7664-39-3	1000
Hydrogen Peroxide (52% by weight or greater)	7722-84-1	7500
Hydrogen Selenide	7783-07-5	150
Hydrogen Sulfide	7783-06-4	1500
Hydroxylamine	7803-49-8	2500
Iron, Pentacarbonyl	13463-40-6	250
Isopropylamine	75-31-0	5000
Ketene	463-51-4	100
Methacrylaldehyde	78-85-3	1000
Methacryloyl Chloride	920-46-7	150
Methacryloyloxyethyl Isocyanate	30674-80-7	100
Methyl Acrylonitrile	126-98-7	250
Methylamine, Anhydrous	74-89-5	1000
Methyl Bromide	74-83-9	2500
Methyl Chloride	74-87-3	15000
Methyl Chloroformate	79-22-1	500
Methyl Ethyl Ketone Peroxide (concentration greater than 60%)	1338-23-4	5000
Methyl Fluoroacetate	453-18-9	100
Methyl Fluorosulfate	421-20-5	100
Methyl Hydrazine	60-34-4	100
Methyl Iodide	74-88-4	7500
Methyl Isocyanate	624-83-9	250
Methyl Mercaptan	74-93-1	5000
Methyl Vinyl Ketone	79-84-4	100
Methyltrichlorosilane	75-79-6	500

Figure 17.3 List of Highly Hazardous Chemicals, continued

Chemical Name	CAS*	TQ**
Nitric Acid (94.5% by weight or greater)	7697-37-2	500
Nitric Oxide	10102-43-9	250
Nitroaniline (para Nitroaniline	100-01-6	5000
Nitromethane	75-52-5	2500
Nitrogen Dioxide	10102-44-0	250
Nitrogen Oxides (NO; NO(2); N2O4; N2O3)	10102-44-0	250
Nitrogen Tetroxide (also called Nitrogen Peroxide)	10544-72-6	250
Nitrogen Trifluoride	7783-54-2	5000
Nitrogen Trioxide	10544-73-7	250
Oleum (65% to 80% by weight; also called Fuming Sulfuric Acid)	8014-94-7	1000
Osmium Tetroxide	20816-12-0	100
Oxygen Difluoride (Fluorine Monoxide)	7783-41-7	100
Ozone	10028-15-6	100
Pentaborane	19624-22-7	100
Peracetic Acid (concentration greater 60% Acetic Acid; also called Peroxyacetic Acid)	79-21-0	1000
Perchloric Acid (concentration greater than 60% by weight)	7601-90-3	5000
Perchloromethyl Mercaptan	594-42-3	150
Perchloryl Fluoride	7616-94-6	5000
Peroxyacetic Acid (concentration greater than 60% Acetic Acid; also called Peracetic Acid)	79-21-0	1000
Phosgene (also called Carbonyl Chloride)	75-44-5	100
Phosphine (Hydrogen Phosphide)	7803-51-2	100
Phosphorus Oxychloride (also called Phosphoryl Chloride)	10025-87-3	1000
Phosphorus Trichloride	7719-12-2	1000
Phosphoryl Chloride (also called Phosphorus Oxychloride)	10025-87-3	1000
Propargyl Bromide	106-96-7	100
Propyl Nitrate	627-3-4	2500
Sarin	107-44-8	100
Selenium Hexafluoride	7783-79-1	1000
Stibine (Antimony Hydride)	7803-52-3	500
Sulfur Dioxide (liquid)	7446-09-5	1000
Sulfur Pentafluoride	5714-22-7	250
Sulfur Tetrafluoride	7783-60-0	250
Sulfur Trioxide (also called Sulfuric Anhydride)	7446-11-9	1000
Sulfuric Anhydride (also called Sulfur Trioxide)	7446-11-9	1000
Tellurium Hexafluoride	7783-80-4	250
Tetrafluoroethylene	116-14-3	5000
Tetrafluorohydrazine	10036-47-2	5000
Tetramethyl Lead	75-74-1	1000
Thionyl Chloride	7719-09-7	250
Trichloro (chloromethyl) Silane	1558-25-4	100
Trichloro (dichlorophenyl) Silane	27137-85-5	2500
Trichlorosilane	10025-78-2	5000
Trifluorochloroethylene	79-38-9	10000
Trimethyoxysilane	2487-90-3	1500

* Chemical Abstract Service Number

** Threshold Quantity in Pounds (Amount necessary to be covered by this standard.)

Figure 17.3 List of Highly Hazardous Chemicals, continued

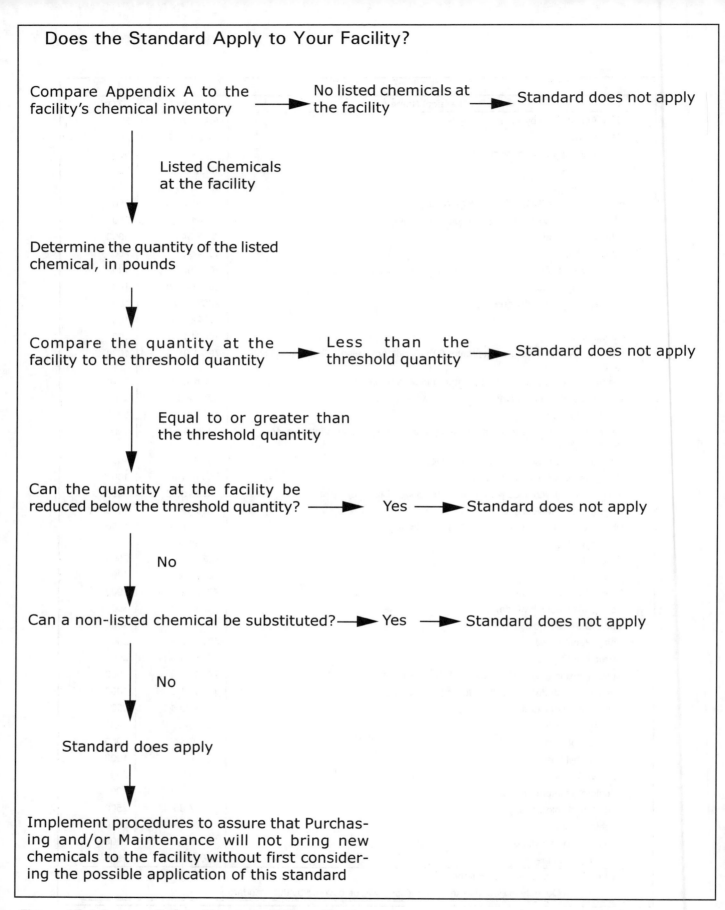

Does the Standard Apply to Your Facility?

Compare Appendix A to the facility's chemical inventory → No listed chemicals at the facility → Standard does not apply

Listed Chemicals at the facility

Determine the quantity of the listed chemical, in pounds

Compare the quantity at the facility to the threshold quantity → Less than the threshold quantity → Standard does not apply

Equal to or greater than the threshold quantity

Can the quantity at the facility be reduced below the threshold quantity? → Yes → Standard does not apply

No

Can a non-listed chemical be substituted? → Yes → Standard does not apply

No

Standard does apply

Implement procedures to assure that Purchasing and/or Maintenance will not bring new chemicals to the facility without first considering the possible application of this standard

Figure 17.4 Process Safety Managment Flow Chart

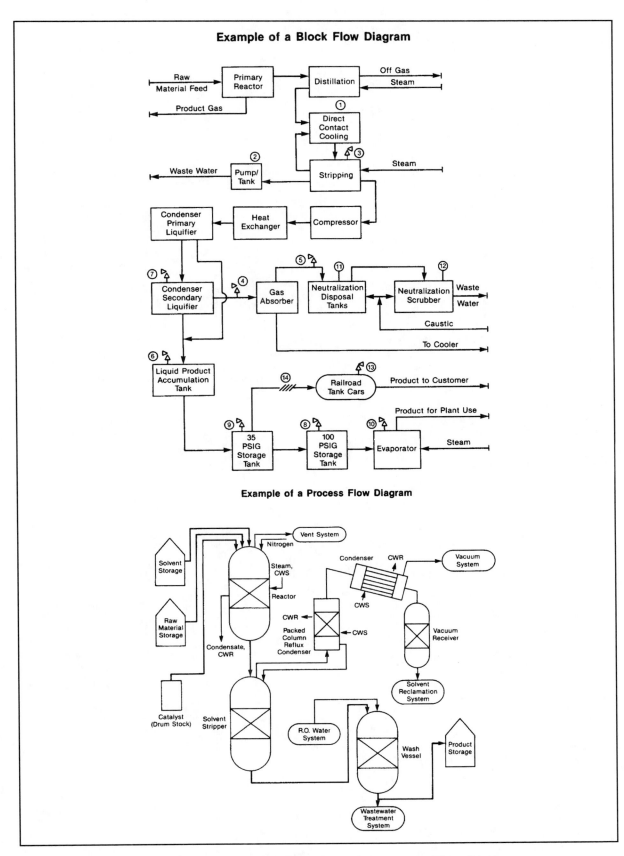

Figure 17.5 Examples of a Block Flow Diagram and a Process Flow Diagram

18

Heat Stress[1]

Setting up an Effective Program to Reduce Heat Exhaustion

INTRODUCTION

Operations involving high air temperatures, radiant heat sources, high humidity, direct physical contact with hot objects, or strenuous physical activities have a high potential for inducing heat stress in employees engaged in such operations. Such places include outdoor operations conducted in hot weather such as construction, asbestos removal, and hazardous waste site activities.

CAUSAL FACTORS

Age, weight, degree of physical fitness, degree of acclimatization, metabolism, use of alcohol or drugs, and a variety of medical conditions such as hypertension all affect a person's sensitivity to heat. However, even the type of clothing worn must be considered. Prior heat injury predisposes an individual to additional injury.

It is difficult to predict just who will be affected and when because individual susceptibility varies. In addition, environmental factors include more than the ambient air temperature. Radiant heat, air movement, conduction, and relative humidity all affect an individual's response to heat.

HEAT STROKE

Heat stroke occurs when the body's system of temperature regulation fails and body temperature rises to critical levels. This condition is caused by a combination of highly variable factors, and its occurrence is difficult to predict.

Heat stroke is a medical emergency. The primary signs and symptoms of heat stroke are confusion; irrational behavior; loss of consciousness; convulsions; a lack of sweating; hot, dry skin; and an abnormally high body temperature, e.g., a rectal temperature of 41°C (105.8°F). If body temperature is too high, it causes death. The elevated metabolic temperatures caused by a combination of work load and environmental heat load, both of which contribute to heat stroke, are also highly variable and difficult to predict.

399

If a worker shows signs of possible heat stroke, professional medical treatment should be requested and the victim placed in a shady area. The outer clothing should be removed. The worker's skin should be wetted and air movement around the worker should be increased to improve evaporative cooling until professional methods of cooling are initiated and the seriousness of the condition can be assessed. Fluids should be replaced as soon as possible. The medical outcome of an episode of heat stroke depends on the victim's physical fitness and the timing and effectiveness of first aid treatment.

Regardless of the worker's protests, no employee suspected of being ill from heat stroke should be sent home or left unattended unless a physician has specifically approved such an order.

HEAT EXHAUSTION

The signs and symptoms of heat exhaustion are headache, nausea, vertigo, weakness, thirst, and giddiness. Fortunately, this condition responds readily to prompt treatment.

Heat exhaustion should not be taken lightly. The fainting associated with heat exhaustion can be dangerous because the victim may be operating machinery or controlling an operation that should not be left unattended; moreover, the victim may be injured when he or she faints. Also, the signs and symptoms seen in heat exhaustion are similar to those of heat stroke, a medical emergency.

Workers suffering from heat exhaustion should be removed from the hot environment and given fluid replacement. They should also be encouraged to get adequate rest.

HEAT CRAMPS

Heat cramps are usually caused by performing hard physical labor in a hot environment. These cramps have been attributed to an electrolyte imbalance caused by sweating. It is important to understand that cramps can be caused by both too much and too little salt.

Cramps appear to be caused by the lack of water replenishment. Because sweat is a hypotonic solution, excess salt can build up in the body if the water lost through sweating is not replaced. Thirst cannot be relied on as a guide to the need for water. Instead, water must be taken every fifteen to twenty minutes in hot environments.

Under extreme conditions, such as working for six to eight hours in heavy protective gear, a loss of sodium may occur. Recent studies have shown that drinking commercially available carbohydrate-electrolyte replacement liquids is effective in minimizing physiological disturbances during recovery.

HEAT COLLAPSE (FAINTING)

In heat collapse, the brain does not receive enough oxygen because blood pools in the extremities. As a result, the exposed individual may lose consciousness.

This reaction is similar to that of heat exhaustion and does not affect the body's heat balance. However, the onset of heat collapse is rapid and unpredictable. To prevent heat collapse, the worker should gradually become acclimatized to the hot environment.

HEAT RASHES

Heat rashes are the most common problem in hot work environments. Prickly heat is manifested as red papules and usually appears in areas where the clothing is restrictive. As sweating increases, these papules give rise to a prickling sensation. Prickly heat occurs in skin that is persistently wetted by unevaporated sweat, and heat rash papules may become infected if they are not treated.

In most cases, heat rashes will disappear when the affected individual returns to a cool environment.

HEAT FATIGUE

A factor that predisposes workers to heat fatigue is lack of acclimatization. The use of a program of acclimatization and training for work in hot environments is advisable.

The signs and symptoms of heat fatigue include impaired performance of skilled sensorimotor, mental, or vigilance jobs. There is no treatment for heat fatigue except to remove the heat stress before a more serous heat-related condition develops.

PREVENTING HEAT STRESS

Most heat-related health problems can be prevented or the risk of developing them reduced. Following a few basic precautions should lessen heat stress.

Ventilation, air cooling, fans, shielding, and insulation are the five major types of engineering controls used to reduce heat stress in hot work environments.

Heat reduction can also be achieved by using power assists and tools that reduce the physical demands placed on a worker. However, for this approach to be successful, the metabolic effort required for the worker to use or operate these devices must be less than the effort required without them. Another method is to reduce the effort necessary to operate power assists.

The worker should be allowed to take frequent rest breaks in a cooler environment.

ACCLIMATIZATION

The human body can adapt to heat exposure to some extent. This physiological adaptation is called acclimatization. After a period of acclimatization, the same activity will produce fewer cardiovascular demands. The worker will sweat more efficiently (causing better evaporative cooling), and thus will more easily be able to maintain normal body temperatures.

A properly designed and applied acclimatization program decreases the risk of heat-related illnesses. Such a program basically involves exposing employees to work in hot environment for progressively longer periods.

FLUID REPLACEMENT

Cool water or any cool liquid (except alcoholic beverages) should be made available to workers to encourage them to drink small amounts frequently, e.g., one cup every twenty minutes. Ample supplies of liquids should be placed close to the work area. Although some commercial replacement drinks contain salt, this is not necessary for acclimatized individuals because most people add enough salt to their summer diets.

ENGINEERING CONTROLS

General ventilation is used to dilute hot air with cooler air (generally cooler air that is brought in from the outside). This technique clearly works better in cooler climates than in hot ones. A permanently installed ventilation system usually handles large areas or entire buildings. Portable or local exhaust systems may be more effective or practical in smaller areas.

Air treatment/air cooling differs from ventilation because it reduces the temperature of the air by removing heat (and sometimes humidity) from the air. Air conditioning is a method of air cooling, but it is expensive to install and operate. An alternative to air conditioning is the use of chillers to circulate cool water through heat exchangers over which air from the ventilation system is then passed; chillers are more efficient in cooler climates or in dry climates where evaporative cooling can be used.

Local air cooling can be effective in reducing air temperature in specific areas. Two methods have been used successfully in industrial settings. One type, cool rooms, can be used to enclose a specific workplace or to offer a recovery area near hot jobs. The second type is a portable blower with built-in air chiller. The main advantage of a blower, aside from portability, is minimal set-up time.

CONVECTION

Another way to reduce heat stress is to increase the air flow or convection using fans, etc., in the work area (as long as the air temperature is less than the worker's skin temperature).

Changes in air speed can help workers stay cooler by increasing both the convective heat exchange (the exchange between the skin surface and the surrounding air) and the rate of evaporation. Because this method does not actually cool the air, any increases in air speed must impact the worker directly to be effective.

If the dry bulb temperature is higher than 35°C (95°F), the hot air passing over the skin can actually make the worker hotter. When the temperature is more than 35°C and the air is dry, evaporative cooling may be improved by air movement, although this improvement will be offset by the convective heat. When the temperature exceeds 35°C and the relative humidity is

100 percent, air movement will make the worker hotter. Increases in air speed have no effect on the body temperature of workers wearing vapor-barrier clothing.

ADMINISTRATIVE CONTROLS

Training is the key to good work practices. Unless all employees understand the reasons for using new or changing old work practices, the chances of such a program succeeding are greatly reduced.

A good heat stress training program should include at least the following components:

- Knowledge of the hazards of heat stress
- Recognition of predisposing factors, danger signs, and symptoms
- Awareness of first-aid procedures for, and the potential health effects of, heat stroke
- Employee responsibilities in avoiding heat stress
- Dangers of using drugs, including therapeutic ones, and alcohol in hot work environments
- Use of protective clothing and equipment
- Purpose and coverage of environmental and medical surveillance programs and the advantages of worker participation in such programs
- Hot jobs should be scheduled for the cooler times of day, and routine maintenance and repair work in hot areas should be scheduled for the cooler seasons of the year

OTHER ADMINISTRATIVE CONTROLS

Circulating air is the most highly effective, as well as the most complicated, personal cooling system. By directing compressed air around the body from a supplied air system, both evaporative and convective cooling are improved. The greatest advantage occurs when circulating air is used with impermeable garments or double cotton overalls.

Circulating air, however, is noisy and requires a constant source of compressed air supplied through an attached air hose. One problem with this system is the limited mobility of workers whose suits are attached to an air hose. Another is that of getting air to the work area itself. These systems should therefore be used in work areas where workers are not required to move around much or to climb.

Another concern with these systems is that they can lead to dehydration. The cool, dry air feels comfortable and the worker may not realize that it is important to drink liquids frequently.

The following administrative controls can be used to reduce heat stress:

- Reduce the physical demands of work, e.g., excessive lifting or digging with heavy objects
- Provide recovery areas, e.g., air-conditioned enclosures and rooms
- Use shifts, e.g., early morning, cool part of the day, or night work
- Use intermittent rest periods with water breaks
- Use relief workers

- Use worker pacing
- Assign extra workers and limit worker occupancy, or the number of workers present, especially in confined or enclosed spaces

HEAT STRESS QUESTIONNAIRE

Listed below are sample questions that you may wish to consider when investigating heat stress in the workplace.

1. Are heat-producing equipment or processes used?
2. Are there any employees with previous history of heat-related problems?
3. Is the heat steady or intermittent?
4. How many employees are exposed?
5. Is potable water available?
6. Are supervisors trained to detect/evaluate heat stress symptoms?
7. Is ventilation in place?
8. Is ventilation operating?
9. Is air conditioning in place?
10. Is air conditioning operating?
11. Are fans operating?
12. Are fans being used in the right place?
13. Is there a training program in place?
14. Is there a liquid replacement program?

Endnotes:

[1] *OSHA Technical Manual*, Section III, Chapter 4. Available at http://www.osha-slc.gov/dts/osta/otm/otm_iii/otm_iii_4.html

19

OSHA Inspections

How to Prepare for and Respond to an Inspection

INTRODUCTION

A construction site presents special problems to OSHA enforcement because of the existence of multiple employers (e.g., contractors, subcontractors). Inspections of such employers are not easily separable in distinct worksites.

There are several key points to remember when dealing with construction sites. First, it is well established that a third party can consent to a search of jointly occupied property, as long as the third party has "common authority" over the premises. A general contractor has common authority over the worksite, and therefore, he or she can give a compliance officer permission to investigate a construction site. This authority eliminates the expectation of privacy a subcontractor might otherwise assert.

At the opening conference, however, the general contractor is asked to notify the subcontractors immediately of the inspection and to ask them to assemble in the general contractor's office to discuss the inspection with the compliance officer. A subcontractor who has not been notified at all of the inspection of his premises will not be included in the inspection.

Second, even if a compliance officer does not have permission to enter a worksite, the plain view doctrine allows him to cite employers for violations for any conditions that are visible to someone in public areas.

Third, in early § 5(a)(2) decisions by the Occupational Safety & Health Review Commission, an employer who created a hazardous condition was only responsible for the exposure of his own employees and no others. However, in Brennan v. OSHRC (Underhill Construction Corp.), the court held that an employer who controls an area and creates a hazard may be found in violation of § 5 (a)(2) if its own employees or the employees of another engaged in a common undertaking are exposed. The court also ruled that in order to find a violation, the Secretary only needs to prove that there is access to the zone of danger rather than actual exposure. The Review Commission has followed the principles of Underhill in subsequent decisions.

Fourth, if violations are found, every employer who has employees exposed to the hazards presented by the violations can be cited, even if another employer was contractually responsible for providing the necessary protection. Subcontractors can protect themselves by proving either that they took whatever steps were reasonable under the circumstances to protect their

405

employees against the hazards or that they lacked the expertise to recognize the condition as hazardous. One way in which subcontractors can protect themselves and their employees is by notifying the responsible contractor that created the hazard and the general contractor of violations on the site. This can usually be done during meetings held by the general contractor.

In any event, subcontractors should ask the general contractor and the subcontractor who created the hazards to correct them. Except for minor violations, however, merely complaining to the general contractor is not adequate where alternative means of protection are available. A subcontractor may also try to protect its employees by informing the exposed employees of the presence of the hazard and instructing them to avoid the areas where the hazards exist whenever possible.

Lastly, there is often a question as to whether a general contractor should be held liable for violations created by its subcontractors. The commission has held that, on multi-employer construction sites, the general contractor is responsible for violations of its subcontractors that the general contractor could reasonably be expected to prevent or to detect and abate by reasons of its supervisory capacity over the entire worksite, even though none off its employees are exposed to the hazard.

General contractors normally have the responsibility and the means to ensure that other contractors fulfill their obligations with respect to employee safety where those obligations affect the construction worksite ... Furthermore, the duty of a general contractor is not limited to the protection of its own employees from safety hazards, but extends to the protection of all the employees engaged at the worksite.

Under the Commission's test in Grossman Steel & Aluminum Corp. for determining whether a general contractor should be liable for safety violations by its subcontractors, three factors are considered: (1) degree of supervisory capacity, (2) nature of the safety standard violation, and (3) nature and extent of precautionary measures taken.

HOW CONSTRUCTION INSPECTION CANDIDATES ARE CHOSEN

There are over 2,000 OSHA inspectors stationed in various places across the country. They conduct unannounced inspections of places where people are at work. If they discover conditions or practices that they think are in violation of OSHA requirements-and they nearly always do-citations and penalties against the employer will soon follow.

About half the inspectors are federal employees. The others are state employees. Who administers the OSHA program where you work depends on the state you are in. The federal government enforces OSHA in 29 of the 50 states. State government employees run the program in the other 21 and in two U.S. territories (Puerto Rico and the Virgin Islands).

Scheduling OSHA inspections in the construction industry is difficult. At any given moment, for example, some construction projects are being completed and others are beginning. There is no central list of such projects or their duration. Consequently, OSHA has no way of knowing the locations where construction work is occurring.

OSHA is continually experimenting with ways to select its construction industry inspection targets in order to reach those that are now overlooked, but it has not yet developed a satisfactory method. It currently utilizes a list of the larger construction worksites (not employers) that was developed by a university contractor. OSHA has contracted with the University of Tennessee's Construction Resources Analysis Department (UTenn-CRA) to provide computer-generated inspection lists to its Area Offices. UTenn-CRA receives the information contained in Dodge Reports of Construction Potentials and generates a monthly, randomly selected, construction inspection list for each Area Office based upon:

- Counties located within Area Office boundaries
- Estimated number of worksite inspections to be conducted during the monthly scheduling period (to be determined by the Area Director)

The selection criteria are determined by the Area Director based on local conditions, approved by the Regional Administrator, and provided to UTenn-CRA for entry into the OSHA computer. These selection criteria may be designed to include any class of worksites within the computerized selection process. Some examples of such criteria are

- A minimum dollar value of the construction project
- A minimum size of the project in square feet
- A certain length of time the project is likely to last
- Specific stages of construction projects (in percent complete)
- Specific types of construction projects

GUIDANCE FOR FOCUSED INSPECTIONS IN CONSTRUCTION

The information below will provide basic guidance to safety managers for determining which projects are eligible for focused construction inspections and how OSHA inspectors will be conducting them.

Background

Under previous agency policy all construction inspections were comprehensive in scope, addressing all areas of the workplace and by inference all classes of hazards. This guidance may have caused compliance officers to spend too much time and effort on a few projects looking for all violations and, thus, too little time overall on many projects inspecting for hazards which are most likely to cause fatalities and serious injuries to workers. Previously, a contractor was likely to be cited for hazards that were unrelated to the four leading causes of death that make up 90 percent of all construction fatalities (falls from elevations, 30 percent; struck by, 22 percent; caught in/between, 18 percent; and electrical shock, 17 percent). Although other conditions are important, the time and resources spent to pursue them on a few projects can be better spent pursuing conditions on many projects related to the four hazard areas most likely to cause fatalities or serious injuries. The goal of OSHA's construction inspections is to make a difference in the safety and health of employees at the worksite. To accomplish this, the Compliance Safety Health Officer's (CSHO) time will be more effectively spent inspecting the most hazardous workplace conditions. The CSHO shall conduct comprehensive, resource-intensive inspections only on those projects where there is inadequate contractor commitment to safety and health. It is this group of employers that will receive our full attention.

All construction inspections shall have opening conferences consistent with current agency procedures, and then shall proceed as follows:

- During all inspections, the CSHO shall determine whether there is project coordination by the general contractor, prime contractor, or other such entity that includes:
- An adequate safety and health program/plan that meets the guidelines set forth below
- A designated competent person responsible for and capable of implementing the program
- If the above general contractor, prime contractor, or other such entity meets both of these criteria, then a focused inspection shall be made. When either of these criteria is not met, then the inspection shall proceed in accordance with previously established procedures for comprehensive inspections.
- Inspectors are to take the time necessary to conduct comprehensive inspections based on the conditions of the project and the effectiveness of any safety and health program.
- If the project does not qualify for a focused inspection, then the CSHO is to conduct the same type of inspection that would have been conducted before the focused inspection policy.
- During all safety fatality/catastrophe, complaint, and referral inspections, the CSHO shall inspect the worksite in regard to the fatality/complaint/referral item(s), and then will proceed.

The OSHA Inspector will assess the company's safety and health program and will consider the following:

- The comprehensiveness of the program
- The degree of program implementation
- The designation of competent persons as are required by relevant standards
- How the program is enforced, including management policies and activities, effective employee involvement, and training

Employees will be interviewed during the walk-around inspection to aid in the evaluation of the program.

Examples of safety and health programs can be found in the Safety and Health Program Management Guidelines published January 26, 1989 in the Federal Register (54 FR 3904), in the ANSI A10.33 "Safety and Health Program Requirements for Multi-Employer Projects," and in Owner and Contractor Association model programs that meet the 29 C.F.R. 1926 Subpart C standards.

Focused inspections will concentrate on the project safety and health program and the four leading hazards that account for the most fatalities and serious injuries in the construction industry: falls, electrical hazards, caught in/between hazards (such as trenching), and "struck-by" hazards (such as materials handling equipment and construction vehicles).

During the course of focused inspections, citations will be proposed for the four leading hazards and any other serious hazards observed.

Other-than-serious hazards that are abated immediately, and this abatement is observed by the CSHO, shall not normally be cited.

If during the walk-around inpsection, the CSHO determines that the number of serious and other-than-serious hazards found on the project indicates that the safety and health program is inadequate or is ineffectively implemented, then the inspection shall be comprehensive.

General Guidelines

The Focused Inspections Initiative that became effective October 1, 1994, is a significant departure from how OSHA has previously conducted construction inspections. This Initiative will recognize the efforts of responsible contractors who have implemented effective safety and health programs and will encourage other contractors to adopt similar programs. The number of inspections is no longer driving the construction inspection program. The measure of success of this new policy will be an overall improvement in construction jobsite safety and health.

The Focused Inspections Initiative will enable OSHA to focus on the leading hazards that cause 90 percent of the injuries and deaths.

The leading hazards are

- Falls, (e.g., floors, platforms, roofs)
- struck by, (e.g., falling objects, vehicles)
- Caught in/between (e.g., cave-ins, unguarded machinery, equipment)
- Electrical (e.g., overhead power lines, power tools and cords, outlets, temporary wiring)

Under the Focused Inspection Initiative, the CSHO will determine whether there is project coordination by the general contractor, prime contractor, or other such entity and conduct a brief review of the project's safety and health program to determine whether the project qualifies for a Focused Inspection. In order to qualify, the following conditions must be met:

- The project safety and health program meets the requirements of 29 C.F.R. 1926 Subpart C General Safety and Health Provisions
- There is a designated competent person responsible for and capable of implementing the program

If the project meets the above criteria, an abbreviated walk-around inspection shall be conducted focusing on

- Verification of the safety and health program effectiveness by interviews and observation
- The four leading hazards listed above
- Other serious hazards observed by the CSHO

The CSHO conducting a Focused Inspection is not required to inspect the entire project. Only a representative portion of the project need be inspected.

The CSHO will make the determination as to whether a project's safety and health program is effective, but if conditions observed on the project indicate otherwise, the CSHO shall immediately terminate the Focused Inspection and conduct a comprehensive inspection. The discovery of serious violations during a Focused Inspection need not automatically convert the Focused Inspection into a comprehensive inspection. These decisions will be based on the professional judgment of the CSHO.

The Focused Inspection will be publicized to the maximum extent possible so as to encourage contractors to establish effective safety and health programs and concentrate on the four leading hazards prior to being inspected.

Specific Guidelines

The Focused Inspections Initiative policy applies only to construction safety inspections. Construction health inspections will continue to be conducted in accordance with current agency procedures.

A project determined not to be eligible for a Focused Inspection shall be given a comprehensive inspection with the necessary time and resources to identify and document violations.

A comprehensive inspection shall be conducted when there is no coordination by the general contractor, prime contractor, or other such entity to ensure that all employers provide adequate protection for their employees.

A request for a warrant will not affect the determination as to whether a project will receive a Focused Inspection.

On jobsites where unprogrammed inspections (complaints, fatalities, etc.) are being conducted, the determination as to whether to conduct a Focused Inspection shall be made only after the complaint or fatality has first been addressed.

All contractors and employee representatives will, at some time during the inspection, be informed, why a focused or a comprehensive inspection is being conducted.

Although the walk-around inspection shall focus on the four leading hazards, citations shall be issued for any serious violations found during a Focused Inspection and for any other-than-serious violations that are not immediately abated. Other-than-serious violations that are immediately abated shall not normally be cited nor documented.

Only contractors on projects that qualify for a Focused Inspection will be eligible to receive a full "good faith" adjustment of 25 percent.

TIPS ON WHAT TO DO WHEN AN OSHA INSPECTOR APPEARS AT YOUR WORKSITE

Surviving an OSHA inspection in the construction industry involves three essential elements: Understanding, preparation, and responding. Each is discussed below.

Understanding

The Report of The House Committee that sent the occupational safety and health bill to the floor in 1970 included the minority views of six congressmen who labeled the legislation an "authoritarian, penalty-oriented, 'bull-in-the-china-shop' approach."

Although improvements in the legislation were made before its subsequent enactment, there are no words that could more accurately describe the manner in which that Act has been implemented in recent years by the Occupational Safety and Health Administration (OSHA), the Agency created to enforce it.

Few can argue with the Act's noble purpose of reducing occupational hazards that annually result in the injury and illness of thousands of hardworking employees. But its purpose is not being served by OSHA's cops-and-robbers method of enforcement.

It's doubtful that occupational injuries and illnesses can be significantly reduced by any government regulatory system. Numerous studies have demonstrated that approximately 80 percent of work-related casualties result from deliberate misconduct, negligence, habits of work and human error, causes that are not-and cannot be-controlled by government regulation. For example, the second leading cause of worker fatalities in the U.S. is homicide. Motor vehicle accidents are first. In 1991, both Texas and Massachusetts reported more homicides on the job than motor vehicle accident fatalities.

OSHA inspections do not change a company's worker injury rate. Companies that have been inspected by OSHA and found to have no violations have worker injury/illness rates that are indistinguishable from similar companies that have never experienced an OSHA inspection, as well as others that have been cited for hundreds of OSHA violations. Indeed, OSHA adopted a Voluntary Protection Program (VPP) during the Reagan Administration that exempts certain companies from programmed OSHA inspections. The annual injury/illness rate for the companies that opted for the exemption is substantially lower than comparable companies that are not exempt from OSHA inspection. So much for the notion that OSHA inspections serve a worthwhile purpose.

There are currently more than ~~two thousand~~ 2420 OSHA inspectors-some federal, some state. None of them is familiar with all the machines, systems, operations and work processes they are called upon to inspect. But their superiors evaluate their job performance on the number and severity of the citations and penalties that they issue, so, not surprisingly, they issue a lot of them. As a result, many such citations call for unnecessary changes in work conditions and practices.

The inspectors know that OSHA has not succeeded in its injury/illness reduction mission. They think the fault lies with employers and with their own Agency's failure to put more employers in jail and penalize them more severely. Those view appear in a November 1990 General Accounting Office report, "Inspectors' Opinions on Improving OSHA Effectiveness."

Although there are some exceptions, OSHA inspectors often exhibit an anti-business, pro-union bias. They have been known to accept misinformation and fabrications from employees without any attempt at verification.

OSHA inspectors get help from the AFL-CIO on their personal wages, hours, and working conditions. The Labor Department's management engages in collective bargaining with its employees on those subjects. The OSHA inspectors are represented in that process by an AFL-CIO union.

It's not unusual for OSHA inspections to be scheduled to coincide with a union campaign against a company or for OSHA to cite the company for allegedly violative conditions and practices pointed out to the inspector by a union representative. Frequently those citations have little or nothing to do with employee safety or health.

Employers who understand the foregoing will be better equipped to respond to an OSHA inspection. Many employers, however, conduct themselves during an OSHA inspection as though they believed in the theory that if you're nice to a crocodile, you'll be the last one eaten. Paradoxically, the record shows that the industry giants with the best safety records who lay out the welcome mat when OSHA calls, get hit the hardest. They get the multi-million dollar OSHA fines. The inspectors like to work where they feel welcome, and they believe that big business deserves to be hit big.

Preparation

Most employers will never receive an OSHA inspection, nor will most taxpayers be audited by the IRS. But you could be one of the unlucky ones, so you should prepare accordingly.

There are literally thousands of OSHA regulations. It's impossible to come up with an accurate count because many of them incorporate by reference privately-developed standards that have never been printed in the Federal Register and are not readily accessible. No one has ever read them all.

Many OSHA standards are so vague and ambiguous that they can be interpreted any way an OSHA inspector wants to interpret them. Some inspectors boast that they can find OSHA violations whenever and wherever they choose. No person with any experience in this field will dispute that.

Consequently, there is no way to assure yourself that you have achieved such a state of OSHA compliance that you will not be cited. But there's a lot you can do to keep the adverse consequences of an OSHA inspection to a minimum.

Over the past ten years, OSHA has increasingly focused upon paperwork requirements. There are 138 OSHA regulations that require employers to adopt written programs specifying the means and methods the employer has adopted to satisfy particular OSHA standards. Another 366 OSHA standards require that records be kept, documents maintained, notices be provided, or warnings or instructions be given. Some 400 others require the employer to provide instruction and/or to train employees in the safety and health aspects of their jobs, or to limit work assignments to employees who are "certified," "competent," "designated," or "qualified."

Those are by far the most-cited standards on OSHA's annual hit list. They are very popular with the inspectors. They don't require extensive knowledge of the employer's complex work processes. The inspector doesn't even have to get his hands dirty through a visit to the factory floor or the places where construction work is ongoing.

Those paperwork requirements are buried within the fine print and arcane language of OSHA's vast regulatory tomes.

Few employers are aware of all of OSHA's paperwork requirements. The inspectors know that. They are aware that they can increase their violation totals by simply asking the employer to produce his OSHA-required paperwork for examination. Those questions are regularly asked within minutes after the inspection's opening conference begins.

Improper recordkeeping and the absence of required records or documents are a fertile source of OSHA violations. So too, are company papers and documents that, though not required, can be used to show that the company hasn't fully implemented every suggestion that could conceivably improve employee working conditions. Millions of different documents fall within this category. Included are safety audits, minutes of meetings, capital improvement requests, doctor's reports, memoranda of subordinate company officials, even sales solicitations.

For example, a salesman advises you of a new upgrade for equipment that you use. He provides you with a flyer or prospectus. You don't buy, but you keep the information in your files. Later on, an accident happens and OSHA inspects. The inspector learns that you knew of the availability of the upgrade but didn't obtain it. He concludes that it could have prevented the accident. You are cited for knowingly and willfully exposing your employees to hazard by your indifference to the prospective upgrade.

The same scenario applies when the "smoking gun" is contained in safety-meeting minutes, suggestions made by subordinate supervisors, unions, employees, or contained in books or letters.

OSHA will subpoena your records in an effort to locate documents of that kind whenever there is an accident or industrial calamity. There have even been cases where anonymous tips have caused OSHA to issue subpoenas for literal truck loads of company documents.

Consequently, always include a written response to documents of that kind if you keep them in your files. Explain why the suggestion is impractical, unnecessary, or not feasible. Failure to do so can open you up to a "careless indifference" charge.

The foregoing should not be taken as an argument in favor of paper-shredding. You should always retain the OSHA-required records and written programs that have been referenced above, and you should keep records of everything you do-and every dollar you spend-on employee health and safety. It will come in handy if you're charged with a willful OSHA violation.

Keep records of all employee training and instruction sessions and the issuance of safety rules and equipment to employees. Get signed receipts. Get acknowledgments from job applicants that they are qualified to perform the job they seek. Never forget that a fatality may someday force you to defend yourself in court against charges that the deceased was assigned to work for which he was not qualified or equipped. Some employers have served jail time on such charges.

The vast majority of OSHA violations are based upon information provided to OSHA by the cited employer. If he wasn't told, the inspector would never know, for example, how a particular machine works, how frequently its operated, the chemical composition of particular substances, the amount used in work processes, the work schedules of particular employees, and similar things that are needed to show noncompliance with various OSHA standards. Employers should be aware of that phenomena and decide how much and what kind of information they will voluntarily share with OSHA, as well as the circumstances under which it will be disclosed.

Each employer should also adopt a policy on OSHA inspections. It should cover, at a minimum, whether or not warrantless OSHA searches will be permitted and who will represent the company during the on-site inspection.

An OSHA inspection can have as devastating an effect upon a business as a fire, flood, or similar disaster. Prepare yourself accordingly. Know the different consequences of OSHA inspection without a warrant and those conducted with a warrant, and then adopt a company policy on the issue.

Above all, have a trained, designated representative at each workplace whose job is to represent the company's interest during an OSHA inspection. That person should be thoroughly acquainted with the matters discussed above as well as those that follow.

Responding to the Inspection

No one in management except the company's designated representative should engage in any substantive discussion with an OSHA inspector. When the inspector arrives, have him wait until that representative arrives. Everything that follows are suggestions for your company representative.

Greet the inspector politely. Look at his credentials, copy down his name and address, or get a copy of his business card. Ask him why he wants to inspect your plant and why you were selected for inspection. Make notes of the answers. Save them for future reference. This is important. A number of employers have learned, to their chagrin, that the reasons the inspector gives in person while he is at their plant vary from those given later.

If your company has a policy against warrantless searches, ask him if he has a warrant. If you get a negative response, advise the inspector of company policy. Don't give any reason why the policy exists. Simply state that it's contrary to company policy to let him in. Ordinarily he will leave without further discussion. If, however, he asks for additional information, do not provide it. Tell him that it is contrary to the company's policy to provide the information in the absence of a warrant.

If the inspector shows up with a warrant, get a copy of it. Read it carefully. Keep it with you throughout the entire inspection. It will state the time limits and the ground rules for the inspection. You'll need it if issues arise over what OSHA is authorized to do. If the warrant refers to any other document, insist upon a copy of that document, too.

If your company has no policy against warrantless searches, allow the inspection to begin. Invite the inspector in to your office or some similar place where the ground rules for the inspection can be discussed.

Treat the inspector with respect, but don't grovel or cringe. Keep in mind that it's your business. He is your guest. Neither of you can dictate to the other but, as in all interpersonal relationships, one person's desires may prevail over the other's. Try to have your wishes prevail if possible. For example, express your own views as to where and when various phases of the on-site inspection will be conducted. He may surprise you by honoring your wishes. If you do not express yourself on those matters, of course, his views will prevail.

All OSHA inspections begin with an "opening conference." The inspector has a checklist of topics to ask you. It will take about an hour to cover them. He will ask you questions about your business similar to those that might be asked by a census taker. For example: What kind of business is this? How many employees? Is it a corporation? Who owns it? Do you operate in interstate commerce? You shouldn't have any problem with those questions but, always keep in mind, you have UNLIMITED TIME OUTS. Use them to consult with others whenever a potential problem arises.

The opening conference will also cover the following 14 matters:

1. **Purpose:** the purpose and scope of the inspection, i.e., to investigate an accident or complaint or to conduct a "programmed" wall-to-wall inspection of the entire facility. The OSHA inspector will tell you that. He will also state whether it is a "safety" inspection or a "health" inspection. The latter type of inspection is conducted by an OSHA industrial hygienist and usually includes air sampling to determine if there are any airborne toxic contaminants, and/or measurement of noise levels to see if OSHA noise limitations are being observed. If you want more information than the inspector provides, speak up. Ask anything you please. You are entitled to be provided with full and complete information.

2. **Labor Union:** whether your employees are represented by a union. If so, he will request a meeting with the shop steward or similar union representative. He will want to include the union in the opening conference. You can decline that request. If so, OSHA will generally hold a separate opening conference with the union.

3. **Escort:** He will advise you that both you and an employee representative are entitled to accompany him throughout the physical inspection of your business. You should always do so. If one of your employees (the shop steward, for example) also chooses to do so, you have the right under the OSH Act to require that the employee do it on

his own time. Most on-site OSHA inspections are conducted without employee or union representation.

4. Injury Log: your log of employee injuries and illnesses (and perhaps other records). OSHA will want to look them over to see the kinds of injuries and illnesses listed, their frequency, and whether there is a pattern (such as eye injuries, falls, etc., or a particular job function that is connected to each injury). The OSHA inspector does that so he'll know where to concentrate his attention when he begins the physical inspection. The inspector may also decide to calculate your rate of injury/illness. If so, he'll ask you for an estimate of the hours worked by employees over a year's time. He may also want to verify the accuracy of your injury logs. If so, he'll ask for other records to compare them with.

5. Handouts: The inspector will give you various OSHA pamphlets and informational materials and will tell you how to get additional copies.

6. Sales Pitch: He will usually give a brief promotional pitch, such as OSHA's goal of reducing worker injuries and illnesses and its desire to help you achieve that objective. You should treat that the same way you treat other sales pitches.

7. Other Employers: The inspector will ask if there are people working on the premises who are not employed by your company. If so, he will want to know who they are, what they do, and who they work for. He may want to include other such employers in the opening conference.

8. Trade Secrets: He will ask if you wish to identify certain matters as trade secrets. If you do, he will promise to keep them confidential-but he will inspect them.

9. Hazard Communication: He will ask for a copy of your written hazard communication program together with the list of chemicals that are present on the premises. If you don't have one, you are probably in violation of the OSHA hazard communication standard. Virtually every business in the country is required to have their own written hazard communication program.

10. Lockout/Tagout: If you are covered by the OSHA lockout/tagout standard (and if you have any machines or similar kinds of equipment, you probably are), he will ask to see your written program and he may question you about the details of your lockout/ tagout procedures.

11. Other Records: He may ask you for equipment inspection and maintenance records, certification records, medical surveillance records, results of monitoring conducted (or measurements made) of workplace noise or air contaminants, safety committee minutes, checklists, records of audits or inspections made by insurance companies, consultants or others. You do not have to provide him with everything he requests. When in doubt, call a "time out," go to the telephone and call for advice from your attorney or some knowledgeable source that you trust. Be especially careful about records of internal safety audits or recommendations for improvements in your machines or equipment.

12. Posting: He will want to see if you have the OSHA poster ("Job Safety and Health Protection") on display. It is required to be posted at every place where people work. He may also inquire about the posting of injury/illness log summaries (required only between February 1 and April 30 each year).

13. Safety Programs: He will ask to see any written safety programs you have so he can check into whether they are being observed during his physical inspection of the workplace.

14. Photographs and Videotapes: He will tell you of his intention to use one or both of them during his physical inspection and he may ask if you object. If you do, he'll take a "Time Out" and call back to his office for guidance.

You must be prepared to deal with those fourteen "opening conference" subject matters before OSHA shows up. If you are uncomfortable with any of them, get help now.

You are not required to either give the OSHA inspector copies of, or let him look at, everything he asks about. He is making a request. You have a right to ask why. You also have a right to call "time out" so you can make a telephone call or consult with others on how you should respond to his requests. You also have the right to say "no."

After the opening conference ends, the walk-around inspection (or plant tour) begins. The inspector will look over the workplace, see what work is being done and observe the working conditions and practices. That will generally be followed by a more intensive investigation of things he thinks might constitute OSHA violations. He will probably want to question employees, foremen and supervisors; take photographs; make measurements; do air sampling and things of that nature. You should do likewise. Stay with him at all times. Don't let anyone else take your place. Never let him out of your sight. Take your own notes and photographs. Make your own measurements. Take control. Remember, it's your business. Act accordingly. If you are ever unsure or in doubt about anything, call a "time out" and seek advice.

The inspector will try to get you (and perhaps others in management) to explain what particular employees are doing, how certain machines work, what substances and materials are being used, and how much. It's not simply curiosity. His objective at all times will be to find violations of OSHA requirements. He will watch the employees at work to see if they are exposed to- or have access to-conditions or practices that he considers to be hazardous. That process may take a few hours, a few days, or weeks or months. It depends upon how long he is permitted to stay, what conditions exist, or what he learns from you, your employees, or other company people. Remember that, unless he is there pursuant to a warrant, he is there with your consent. You can withdraw that consent at any time.

He will constantly take notes-on what he sees and hears and what is said to him. Those notes will be the basis of his testimony against you in the event that you dispute the resulting citations and penalties.

Some inspectors have been known to give orders to employers or employees. They have no such authority. OSHA can only act through properly-issued written citations and orders. If the inspector exceeds his authority, you may want to remind him of those limitations. If a friendly reminder doesn't work, you may have to telephone the OSHA Area Director of the office to which the inspector is assigned. That's one reason why you obtained a copy of his business card when he first arrived.

He may point out things that he believes to be OSHA violations. If you agree that they are, you will almost surely be cited and fined. That will happen even if you correct them while he watches. His own job performance evaluation depends upon the issuance of citations and fines—not upon changes you agree to make in your business.

Keep in mind at all times that what you say can—and will—be used against you. The vast majority of OSHA violations are based on what the employer said to the inspector—not what employees said—or what the inspector saw during the inspection. Don't engage in idle talk or chit-chat. Inspectors will sometimes take things out of context and use it against you.

Don't be reluctant to disagree with the inspector whenever he expresses an opinion that a particular condition or practice constitutes an OSHA violation. He is not infallible on any subject. Remember, for a violation to exist there must be a reasonable probability—not a possibility—that the condition or practice could result in injury or illness to an employee. OSHA is not a building code. Its only purpose is to prevent employee injury and illness.

Never automatically accept an OSHA inspector's advice or instruction on anything. Few OSHA inspectors are experts on all OSHA requirements. It would be foolhardy to accept an inspector's word on what OSHA requires without first obtaining verification from a source you know to be reliable and well-informed.

Do not conduct any demonstrations for OSHA's benefit. In an attempt to gather evidence showing a certain non-operating machine to be in violation, OSHA may ask you to start it up so they can see it in operation. Don't do it! Thousands of convictions have been based on nothing more than an employer's willingness to conduct just such a demonstration. You don't have to do it. When you get a request to conduct a demonstration for OSHA's benefit-and this happens often-remember that it is simply that: a request. A simple "no" from you is all you need to handle this potential problem. It is no different from a request from the inspector to borrow your car. You aren't obligated to do so and it can only work to your disadvantage if you do.

The inspector may want to interview some of your employees privately. That is permissible if it doesn't interfere with the employee's work. If it can be done at the employee's workstation at a time when he is not otherwise occupied, stand within view but outside of earshot. Make a note of who the inspector speaks with, as well as the beginning and ending time of the conversation. If the interview cannot be conducted at the employee's workstation without interfering with his work, do not permit the interview at that time. Tell the inspector he is free to obtain the employee's name and telephone number (if the employee will give it) and that he can arrange to conduct the interview at some other time or place. Keep in mind that the OSH Act guarantees you the right to accompany the inspector "during the physical inspection" of your workplace-not just some of the time but all of the time. This is an important right which you should not allow to be abused. Stay with the inspector during every minute he is on your premises. Remember he is your guest-treat him as such and make sure he never abuses his status as a guest.

OSHA inspectors sometimes will want to use your employees as test-subjects or experimentees in order to monitor noise levels or atmospheric conditions in your workplace. To do so, OSHA will not only want to attach measuring devices to their clothing or bodies, OSHA will also want to control the employees' movement and activities during the time such devices are attached in order to insure the integrity of the test results. OSHA does not have the right to do this if the subject employee objects or if it will interfere with work operations. The inspector, however, won't always advise employees of their right to decline. If he doesn't, you should. It is also important to remember that OSHA must show that your employees are exposed to allegedly offensive noise or atmospheric conditions for a sufficient number of hours or days. Otherwise, it won't constitute a violation. Consequently, answering such seemingly innocuous questions as "What hours do employees work in this shop?" or "How often do you use this substance?" could supply OSHA with a vital element of proof in such cases.

Never give estimates. You may be providing false information to OSHA, a criminal offense, or you may be thereby wrongfully convicting your company of an OSHA violation. If you don't have accurate and reliable information when asked, say so. But make sure the inspector understands that the information may exist elsewhere. You can tell him that the company would not have any need for the various experts it employs if you knew everything.

If the inspector wants to interview another management representative-and you decide to permit it-you should also be present during that interview. Sometimes OSHA will try to get contradictory statements from different employer representatives. That can be trouble for your company. Try to prevent it if you can.

Make notes of everything that happens-and when it happens. Many such notes have proved to be worthless because they did not include calendar dates and times of day. Make the same measurements as the inspector. Take the same photographs. If he uses a video camera, do likewise. Get someone to help you with the camera or with your note-taking if needed. A video of the inspector in action can be very helpful. Some inspectors have shown to be less inclined to use videotape when they are aware that they are simultaneously being videotaped. The evidence you gather in this manner will be helpful when defending against a subsequent OSHA citation.

You can be sure that the inspector will testify at that trial as to the accuracy of his notes, the preciseness of his measurements, and the fact that his photographs are truly and fully representative of the situations that existed in your plant during the inspection. If you disagree, you can testify to the contrary and maybe you'll convince the judge. But your testimony would be strengthened if you had made the same measurements, photographs, and notes as did the inspector or had a video of the inspector in action. You have the same rights as the inspector and you would be well advised to exercise them. Some employer representatives bring along a stenographer or a cameraman, or carry a portable dictating machine or a tape recorder during the inspection to make their note-taking easier. That is perfectly permissible.

While accompanying the inspector, don't hesitate to ask him questions. Indeed, it can be very important to you. Make a record of the answers you receive. Find out why he is making that measurement, what he thinks is shown by the picture he took, what he is writing in his notes. Many times OSHA inspectors will take a photograph in such a way that it will show an apparent violation (a "blocked exit," for example). Your photograph should be taken of the entire scene in order to show the true condition. Don't be bashful. Ask the inspector about his pertinent background, knowledge, and experience whenever he shows particular interest in a machine, apparatus, process, or such. He may not answer you fully, but keep track of what he says. It may be invaluable to you if a citation issues. Frequently he will concede that he has never previously seen the type of machine he is claiming to be in violation. Keep a record of any such statement because it is unlikely that he will make such an admission later. If a citation issues and there is a trial, he might then testify that he is one of the world's leading expert that type of machine.

At the end of each inspection day, go over your notes and measurements. Make sure they are complete and accurate. Have them typed if possible. Be sure to record the date you made them—a simple matter which many people overlook. Be certain also to record the name of the OSHA inspector (or inspectors) and, if you are recording conversations which took place, be sure to specify who said what. That includes you—if you said anything to the inspector. When your photographs are developed, label each of them by what they depict and the date and time they were taken. Once you have done it, carefully retain all of it (including your original notes, measurements and tapes, even when you later have them typed). That will be important to you if you are cited. Even if you aren't cited, this material can be helpful in your efforts to understand and comply with OSHA requirements. Should you again be inspected at some future date, it could be very important on that occasion too.

Frequently, the inspector will give you copies of publications or other documents. Don't refuse them even if you already have copies. Keep all of them. Write on each one the date it was given to you and the name of the person who gave it to you. There have been occasions when employers have been given out-of-date publications by OSHA inspectors—not the documents an OSHA inspector may later claim that the employer was given. Because such matters may have importance to the resolution of contested OSHA citations, you should take the few minutes necessary to make the identifying notes on the publications you receive and retain the same for future reference.

Be particularly careful about providing OSHA with company documents. You do not have to give OSHA any records unless a warrant requires you do so. In such cases, the warrant will ordinarily authorize OSHA to inspect certain records. That is exactly what you should permit when faced with such a warrant. Let OSHA look at the particular record. Do not give them a copy. If the OSHA inspector wants to copy the record down by hand, let him do so. It is not to your advantage, however, to give him an extra copy or let him use your photocopying equipment. Should you do so, prepare to defend yourself against a citation based upon the record you supplied. OSHA can subpoena records if you don't voluntarily supply them. That is generally better for you than disclosure upon request because (a) the inspector may decide it's not worth his while to go through the subpoena process, or (b) you can always negotiate the terms of a subpoena once it has issued—and you can sometimes get all or part of it set aside.

Resist the temptation to voluntarily provide OSHA with safety rules, audits, studies, or other things that in your opinion reflect favorably upon your company. In many cases, that material can be used against you. For example, in 1975 the Forging Industry Association retained a university-based team of noise experts to study ways of controlling noise in forge shops. A lengthy study produced a comprehensive report concluding that there were no feasible engineering controls to reduce forge hammer noise to within OSHA limits. Within the report were a few suggestions for lowering noise from other sources. OSHA was supplied with copies of the report because the forging industry thought it would deter OSHA from issuing noise citations. OSHA, however, used it to cite forge shops for willful violations because the forging companies had not done those things the report had simply suggested. Similar things happen all the time. That's one of the reasons why supplying OSHA with documents can cause you trouble.

When the plant-tour phase of the inspection has been completed, OSHA procedures call for a "closing conference" with the employer. This is not to be thought of by the employer as a friendly chat or a chance for you to talk the inspector out of a citation. It is the time when the inspector discusses his findings and looks for admissions from the employer of facts which will help OSHA prove their violation charges. Most admissions against interest are made by employers during the closing conference. You can be sure that the inspector will be looking for such statements and will take note of them. Indeed, the OSHA Field Inspection Reference Manual requires him to do so. It tells the inspector that in filling out his report of the inspection, he should show any supportive information from the closing conference substantiating the employer's general attitude, any general admission of violations, and any agreement about abatement dates. It is not uncommon, for example, for an employer to respond to the inspector's description of a condition observed during the inspection by saying: "Well, we've been trying to get that changed but can't do it." That statement is good evidence both that a substandard condition exists and that you, the employer, have been aware of its existence for some time. That statement will cost your employer plenty. You might be better off to bite your tongue.

The inspector is supposed to use the closing conference to advise you of "apparent violations" he has observed during his inspection. If he uses the word "violation," it is never to your advantage to agree with him. Semantics are very important here. You may concede that a particular "condition" or "practice" exists—but that doesn't necessarily establish that it constitutes an OSHA "violation" (there's a lot of law on this point).

However, when the inspector states that this or that condition or practice is an apparent violation, don't agree, but do ask him (as to each) why he thinks so-and try to find out what, in his opinion, would constitute a feasible means of abatement. Get as many details on this point as you can, but do not argue with him about his response. For example, if he says "I don't know," make a note of it and move on to the next question. If he states that you can correct the excessive noise in your machine shop by buying a can of anti-noise spray at the corner drug store and spraying the machine with it every Tuesday at noon, don't tell him he's crazy. You'll have plenty of time for argument later (after you have contested the resulting citation).

The inspection is the time for acquiring information. Get as much of it as you can-no matter how irresponsible you may think that information to be. Be sure, however, to record the inspector's responses fully. You can do so either by a tape recorder, your own note-taking, or by having a stenographer present. Perhaps you should not use the word "feasible" during your questioning, since it tends to raise red flags in the minds of OSHA people. What you want answered are the details of what he thinks you ought to do in order that the condition or practice he referred to will be corrected or prevented from happening in the future. You want all possible details on the means and methods of abatement; what disciplinary, instruction or employee training program he believes that you should have (including all the details thereof); a full list of all equipment and materials needed; how much they will cost; where they can be acquired; what result they will accomplish; and precisely where they must be installed. Ask those questions. Also ask what safety and health benefits your employees can expect if you do what the inspector wants you to do. Have him explain the "how" and "why" of each position he takes.

Be careful in responding to this question which he will be almost certain to ask: "How much time will you need for abatement?" An unexplained time-estimate by you can be interpreted as both an admission that there is a violation and that your company will be able to abate it within the time period you state. If you don't think the condition mentioned is a violation, say so. Tell him there is no need to abate-or there is no feasible way to do so. Don't worry if he tells you in response that an arbitrarily brief abatement period will be assigned. You'll receive an automatic extension of any such time when you contest the citation.

You are not required to have a closing conference, and if you lack the courage or the skills necessary to avoid unintended admission against your interests, you would probably be wise to tell the inspector that you do not want any closing conference. Don't be fearful that you'll miss out on something if you skip the closing conference. There will be plenty of later opportunities to ask questions and conduct negotiations with OSHA. They will come after citations have been issued and you have contested them. That is usually a good time to get information because you don't have to worry about what you say being used against you. There is also another alternative. You can request that the closing conference be conducted by telephone. OSHA will generally grant such a request.

Once the inspection is concluded either with or without a closing conference, the inspector returns to his office to write his report the inspection. That may take days or weeks. Occasionally, during that process, he may telephone you with a request for additional information or documents. You have the same rights (mentioned above) to say "no" to his request or to delay a response until you have conferred with your own attorney or any one else you rely upon.

The inspector will prepare the citations and proposed penalties after returning to his office. They will be sent to you by certified mail (or personally delivered) a few weeks or a few months after the inspection ends (except that in California, handwritten OSHA citations are sometimes given to the employer before the inspector leaves the worksite). Remember that OSHA must act by means of written citations. An inspector will occasionally forget that fact, and will tell you (or your employees) that this or that must be done. Those verbal orders are not enforceable and you are not under any legal obligation to observe them. A wise policy to follow is "Get it in writing." That applies to every order he gives, or advice or instruction that he offers. If he tells you, it's in the regulations or contained in a particular publication, have him list the specific section, subsection, paragraph, page, and sentence he is relying upon. Don't accept generalities.

The citations and penalty proposals will be in writing. They will be typed on a government form and will be accompanied by a two-page form-letter from the OSHA Area Director, a booklet explaining your rights and responsibilities, and perhaps some other OSHA handouts. You will have 15 working days (approximately three calendar weeks) to contest the citations. You should always do so. It's a very simple process-a one-sentence letter will do the trick. Don't worry about a hearing. You'll have at least two months after you contest to negotiate a settlement. OSHA will be anxious to have that occur. They don't like hearings either. Keep in mind that citations that are not contested are the moral equivalent of guilty pleas. That can come back to haunt you. There are ways to settle contested citations that can avoid or ameliorate potentially dire consequences in the future. Consult with people who are knowledgeable on such matters and follow their advice. It could be very important to your company in defending against law suits, criminal indictments, and future OSHA confrontations.

Frequently, an OSHA inspection will state that a "hazard" exists. If you do not agree with his assertion that a particular condition or practice is hazardous, don't be bashful. Tell him he's wrong whenever you think he is. If you don't, you may create additional problems for yourself. Here's an example:

During the closing conference, the inspector stated that he had found nearly a hundred OSHA violations during his three-day inspection. He asked why there was so many. The company's safety director replied: "I'm only one person. I can't do it all." The inspector treated that reply as an admission that the company knew about the alleged violations. He therefore cited all of them as "willful" with a penalty of $300,000 rather than the $30,000 fine he was considering.

That's an example of how your own words can hurt. Those "willful" citations probably would not have issued if the safety director had replied by asking the inspector for an itemization of his claim, or had simply responded, "That's your opinion."

The goal of OSHA's Focused Inspections policy is concentrating on those projects that do not have effective safety and health programs and limiting OSHA's time spent on projects with effective programs.

To qualify for a Focused Inspection, the project safety and health program will be reviewed and a walk-around inspection will be made of the jobsite to verify that the program is being fully implemented.

During the walk-around inspection, the compliance officer will focus on the four leading hazards that cause 90 percent of deaths and injuries in construction. The leading hazards are

- Falls (e.g., floors, platforms, roofs)
- Struck by (e.g., falling objects, vehicles)
- Caught in/between (e.g., cave-ins, unguarded machinery, equipment)
- Electrical (e.g., overhead power lines, power tools and cords, outlets, temporary wiring)

The compliance officer will interview employees to determine their knowledge of the safety and health program, their awareness of potential jobsite hazards, their training in hazard recognition and their understanding of applicable OSHA standards.

If the project safety and health program is found to be effectively implemented, the compliance officer will terminate the inspection.

If the project does not qualify for a Focused Inspection, the compliance officer will conduct a comprehensive inspection of the entire project.

If you have any questions or concerns related to the OSHA inspection or conditions on the project, you are encouraged to bring them to the immediate attention of the general contractor.

_____

Project/site

_____   _____

DATE   CONTRACTOR/SUBCONTRACTOR

This document should be distributed at the site and given to all Contractors for conducting a walk-around inspection of their work site.

Figure 19.1 Sample Handout for Contractors and Subcontractors

NOTICE TO OSHA

It is the policy of XYZ Construction Company, Inc., to cooperate with any governmental agency seeking to lawfully enforce federal, state, or local laws and regulations pursuant to the safeguards guaranteed to us by the Constitution of the United States.

XYZ Construction Company, Inc. is familiar with federal court decisions and the ruling of the Supreme Court of the United States in Marshall v. Barlow's Inc. that all warrantless searches or inspections of company property, or any portion thereof, conducted pursuant to Section 8(a) of the Occupational Safety and Health Act of 1970 are unconstitutional. Therefore, XYZ Construction Company, Inc. chooses to exercise its constitutional rights and will not permit, nor is any employee of this Company authorized to permit, any search or inspection by any representative of the Occupational Safety and Health Administration unless conducted pursuant to a valid warrant.

XYZ Construction Company, Inc. does not believe that probable cause exists for an OSHA inspection of its property. In the event that the Secretary of Labor believes otherwise and decides to make application for inspection warrant, XYZ Construction Company, Inc. hereby requests that it be given advance notice of such application so it can have an opportunity to oppose the same.

This statement of policy has been reduced to writing and is being delivered in hand to each OSHA representative seeking to make a warrantless search of our property so that OSHA will be officially advised of everything stated herein.

[NOTE: The foregoing notice should be prepared in the name of your company today. Have it ready to deliver to any OSHA inspector upon his arrival on your premises. In the event that your plant is located in a state which has its own OSHA-approved state plan, adapt the notice accordingly.]

Figure 19.2 Sample Notice to OSHA Inspectors on Policy Regarding Warrantless Searches

Notice of Protest

The XYZ Construction Company, Inc., acknowledges the existence of an inspection warrant which appears on its face to authorize an OSHA inspection to be conducted on these premises. The company does not believe that the said warrant was issued pursuant to law and does not believe that probable cause exists for the warrant.

However, since the warrant is facially in proper form, the company believes it could be cited for contempt of court if it declined to allow the inspection to begin as stated in the warrant. The company therefore will allow the inspection UNDER PROTEST. By so doing, we do not waive our right to challenge the validity of the inspection or the authorization for its conduct and specifically preserve the right to do so should any OSHA enforcement proceeding be commenced against this company based on this inspection or should there be any other reason to litigate this matter.

This Notice of Protest was delivered in hand on the premises of this company to _____ , a representative of OSHA, by _____ , a representative of this company, on _____, 20 _____ at _____ o'clock.

[NOTE: The foregoing Notice of Protest should be prepared in the name of your company and delivered to any OSHA inspector who has a warrant to conduct an inspection of your business. In the event that your plant is located in a state which has its own OSHA-approved state plan, adapt this Notice of Protest accordingly.]

Figure 19.3 Sample Notice of Protest to OSHA Inspectors Who Have a Warrant

Glossary

A

Additional criteria: This includes the following conditions when they are work-related:

- Any needle-stick injury or cut from a sharp object that is contaminated with another person's blood or other potentially infectious material
- Any case requiring an employee to be medically removed under the requirements of an OSHA health standard
- Tuberculosis infection, as evidenced by a positive skin test or diagnosis by a physician or other licensed healthcare professional after exposure to a known case of active tuberculosis.

Anchorage: A secure point of attachment for lifelines, lanyards, or deceleration devices capable of withstanding the anticipated forces applied during a fall.

Annual summary: This consists of a summary of the occupational injury and illness totals for the year from the OSHA No. 300.

Annual survey: Each year, the Bureau of Labor Statistics (BLS) conducts an annual survey of occupational injuries and illnesses to produce national statistics. The OSHA injury and illness records maintained by employers in their establishments serve as the basis for this survey.

All other occupational illnesses: This is a term used on the OSHA 300 Log to record illnesses. Examples would include anthrax, brucellosis, infectious hepatitis, malignant and benign tumors, food poisoning, histoplasmosis, and coccidioidomycosis.

Barricade: An obstruction to deter the passage of persons or vehicles.

B

Body belt: A work-positioning safety belt, designed to fit around a workers' waist and used in conjunction with a lanyard, lifeline, or rebar assembly. Body belts (single or double D-ring) are designed to restrain a person in a hazardous work position and to reduce the possibility of falls. They should not be used when a fall potential exists, but for positioning only.

Body support: A belt or harness consisting of single or multiple straps that are arranged and assembled for the purpose of providing body support both during and after a fall arrest. The body support is designed to distribute arresting forces over the body (e.g., full body harness).

433

Bureau of Labor Statistics (BLS): This is the agency of the Labor Department that compiles statistics on total injuries and illnesses. These figures are then given to OSHA to use in planning their programmed inspections. The Bureau of Labor Statistics is the agency responsible for administering and maintaining the OSHA recordkeeping system.

C

Certification: The process of preparing the Log and Summary of Occupational Injuries and Illnesses, OSHA Form 300-A. A responsible person certifies that the information is true and complete by signing this form. A company executive is required to certify the summary.

Code of Federal Regulations (CFR): The official register of the U.S. government that contains all final OSHA regulations. The OSHA standards are part of Title 29, the section of the CFR assigned to labor regulations.

Competent person: An individual knowledgeable about fall protection equipment and systems, including the manufacturer's recommendations and instructions for their proper erection, use, inspection, and maintenance. This person is capable of identifying existing and potential fall hazards and has the authority to take prompt corrective action to eliminate those hazards.

Compliance Safety Health Officer (CSHO): An OSHA inspector. He or she inspects a facility for violations of OSHA standards.

Connecting means: A device, lanyard, or lifeline used to connect the body support to the anchorage in such a way as to provide protected movement during an elevated work task.

D

Days away from work: This contains the requirements for recording work-related injuries and illnesses that result in days away from work and for counting the total number of days away associated with a given case. This requires the employer to record an injury or illness that involves one or more days away from work by placing a check mark on the OSHA 300 Log in the space reserved for days away from work in the column reserved for that purpose. The employer is not to count the day of the injury or illness as a day away, but to begin counting days on the following day.

The employer must record the number of calendar days that the employee was unable to work as a result of the injury or illness, regardless of whether or not the employee was scheduled to work on those days. Weekend days, holidays, vacation days, or other days off are included in the total number of days recorded if the employee would not have been able to work on those days because of a work-related injury or illness. The employer may "cap" the total days away at 180 calendar days.

Death: The employer must record an injury or illness that results in death by entering a check mark on the OSHA 300 Log in the space for fatal cases. Employers are also directed to report work-related fatalities to OSHA within eight hours.

Decibel: A unit for measuring the relative loudness of sounds. The noise standard establishes a 90-decibel limit on employee noise exposure when averaged over an eight-hour workday, and requires hearing protection at and above the 85-decibel level.

E

Establishment: An establishment is a single physical location where business is conducted or where services or industrial operations are performed. For activities where employees do not work at a single physical location, such as construction; transportation; communications, electric, gas, and sanitary services; and similar operations, the establishment is represented by main or branch offices, terminals, or stations that either supervises such activities or are the base from which personnel carry out these activities.

Employee: A person who works for another in return for financial or other compensation.

Employee representative: Anyone designated by the employee for the purpose of gaining access to the employer's log of occupational injuries and illnesses.

Employer: Any person engaged in a business affecting commerce who has employees.

Exposure records: Records that must be kept of measurement of any exposures to air contaminants or noise in the workplace. These records must be kept for 30 years beyond the term of employment. Employers must also provide access to employee's medical records.

F

Fall arrest system: Includes the proper anchorage, body support (belt/harness), and connecting means (lanyards and lifelines) interconnected and rigged to arrest a free fall.

Fall hazard: Occurs during any construction activity that exposes an employee to an unprotected fall which may result in injury.

Fall prevention: Any means used to reasonably prevent exposure to an elevated fall hazard, either by eliminating work at elevation or by using aerial lifts, scaffolds, floors, guardrails or by isolating an area.

Fall protection: Involves using fall arrest equipment and systems to minimize the effects of a fall once it has occurred.

Fall protection work plan: A written plan in which the employer identifies all areas on the job site where a fall hazard exists. The plan describes the method(s) of fall protection necessary to protect employees and includes safe work practices required during the installation, use, inspection, and removal of the fall protection method selected.

Fall restraint system: An approved device and any necessary components that function together to restrain an employee in such a manner as to prevent that employee from the exposure of falling to a lower level. When standard guardrails are selected, compliance with applicable regulations governing their construction and use shall be followed.

Fatality: A death resulting from an accident.

Fatality/Catastrophe inspection: A type of OSHA inspection in which an employee has been killed on the job; there have been multiple employee hospitalizations; or there has been a newspaper, radio, or TV report of an accident, explosion, spill or calamity of some kind. An OSHA inspector will probably show up shortly thereafter.

Federal Register: The official source of information and notification on OSHA's proposed rulemaking, standards, regulations, and other official matters, including amendments, corrections, insertions, or deletions.

First aid: Includes only the following treatments (any treatment not included in this list are not considered first aid for recordkeeping purposes):

- Using a nonprescription medication at nonprescription strength
- Administering tetanus immunizations
- Cleaning, flushing or soaking wounds on the surface of the skin
- Using wound coverings such as bandages, gauze pads, or butterfly bandages
- Using hot or cold therapy
- Using any non-rigid means of support, such as elastic bandages, wraps, non-rigid back belts, etc.
- Using temporary immobilization devices while transporting an accident victim
- Drilling of a fingernail or toenail to relieve pressure, or draining fluid from a blister;
- Using eye patches
- Removing foreign bodies from the eye using only irrigation or a cotton swab
- Removing splinters or foreign material from areas other than the eye by irrigation, tweezers, cotton swabs or other simple means
- Using finger guards
- Using massages
- Drinking fluids for relief of heat stress

First report of injury: A workers' compensation form which may qualify as a substitute for the supplementary record, OSHA No. 301.

Full body harness: A body support configured of connected straps to distribute a fall arresting force over at least the thighs, shoulders, and pelvis. The harness provides a D-ring for attaching a lanyard, lifeline, or deceleration devices.

G

General industry: The OSHA standards in 29 CFR 1910 that apply to most businesses and industry.

H

Horizontal lifeline: Provides an attachment for the worker's lanyard or other fall arrest device to protect him while moving horizontally and to control dangerous swing falls. It may be a cable or wire rope that is installed horizontally and that serves as an anchoring line rigged between two or more fixed anchorages on the same level. Horizontal lifelines must be positioned above waist high on a worker and all horizontal lifelines and their installation should be approved and supervised by a qualified person.

Hospitalization: The confinement of an employee to a hospital because of on-the-job injuries or illnesses.

I

Incidence rates: An incidence rate is the number of recordable injuries and illnesses occurring among a given number of full-time workers (usually 100 full time workers) over a given period of time (usually one year).

L

Lanyard: The connecting means (rope, webbing) used to attach a body belt or harness to a lifeline or an anchorage point. Lanyards are usually two-, four-, or six-feet long and come with or without a shock-absorber.

Leading edge: The advancing edge of a floor, decking, or form work which changes location as additional sections are placed. Leading edges not actively under construction are considered to be "unprotected sides and edges," and appropriate methods of fall prevention shall be required to protect exposed workers.

Log (OSHA Form 300): The OSHA recordkeeping form used to list injuries and illnesses and to note the extent of each case.

Loss of consciousness: The term is defined as any time a worker becomes unconscious as a result of a workplace exposure to chemicals, heat, an oxygen deficient environment, a blow to the head, or some other workplace hazard that causes loss of consciousness. Unconsciousness means a complete loss of consciousness and not a sense of disorientation, "feeling whoozy," or a diminished level of awareness. If the employee is unconscious for any length of time, no matter how brief, the case must be recorded on the OSHA 300 Log.

Lost workday: Any day an employee would have worked if an occupational illness or injury had not occurred. These are cases which involve days away from work, or days of restricted work activity, or both.

Low-hazard industries: Selected industries in retail trade, finance, insurance, real estate, and services which are regularly exempt from OSHA recordkeeping. To be included in this exemption, an industry must fall within an SIC not targeted for general schedule inspections and must have an average lost workday case injury rate for a designated three-year measurement period at or below 75 percent of the U.S. private sector average rate.

M

Medical records: Employee medical and employment questionnaires, the results of examinations and tests, medical opinions, descriptions of treatments, and records of prescriptions. These records must be kept for the duration of employment plus 30 years.

Medical treatment: Medical treatment means the management and care of a patient to combat disease or disorder. For recordkeeping purposes, it does *not* include (a) visits to a physician or other licensed healthcare professional solely for observation or counseling; (b) diagnostic procedures such as x-rays and blood tests, including the administration of prescription medications used solely for diagnostic purposes (e.g., eyedrops to dilate pupils); or (c) any treatment contained on the list of first-aid treatments.

O

Occupational illness: Any abnormal condition or disorder, other than one resulting from an occupational injury, caused by exposure to factors associated with employment. It includes acute and chronic illnesses or disease which may be caused by inhalation, absorption, or ingestion of or direct contact with a hazardous substance.

Occupational injury: Any injury such as a cut, fracture, sprain, amputation that results from a work accident or from a single instantaneous exposure in the work environment. Conditions resulting from animal bites, such as insect or snake bites, and from one-time exposure to chemicals are considered to be injuries. Needlestick injuries and cuts from sharp objects that are contaminated with another person's blood or other potentially infectious material must also be recorded.

Occupational Safety and Health Act (OSH Act): The OSH Act of 1970 is the statute enacted by Congress that sets forth the law that is carried out by OSHA. The purpose of the Act is "to ensure so far as possible every working man and woman in the nation safe and healthful working conditions and to preserve our human resources." The Act strives for maximum effectiveness by covering virtually every person or business organization that employs people.

Occupational Safety and Health Administration (OSHA): An agency of the U.S. Department of Labor with safety and health regulatory and enforcement authority for most United States industries and businesses.

OSHA Form 300, (Log of Work-Related Injuries and Illnesses): This is the form which replaces the old Form 200 and can now be printed on legal size paper as opposed to its former large format. This is where employers are required to record work-related deaths, injuries, illnesses, and days away from work. Employers are also forbidden from entering employees' names on this form for some sensitive illness or injuries including sexual assaults, HIV infections, and mental illness.

OSHA Form 300-A, (Summary of Work-Related Injuries and Illnesses): This is a new form aimed at easing employers duties to calculate incidence rates which must be posted every year at each establishment by February 1 and remain in place until April 30. To make sure that management is aware of the safety status quo, the summary must be certified by a company executive, the owner, an officer of the corporation, or the highest ranking company official working at that site or his or her supervisor.

OSHA Form 301, (Injury and Illness Accident Report): This form, which replaces the old Form 101, includes more specific information about how an accident occurred. Employees, former employees, and their representatives must have access to the information on this form so that they can review reports of their own injuries and illnesses. Employee representatives are barred from seeing personal information recorded on it.

P

Poisoning: This is a term used on the OSHA 300 Log. Examples would include poisoning by lead, mercury, cadmium, arsenic, or other metals; poisoning by carbon monoxide, hydrogen sulfide, or other gases; poisoning by benzol, carbon tetrachloride, or other organic solvents; poisoning by insecticide sprays such as parathion and lead arsenate; and poisoning by other chemicals such as formaldehyde, plastics, and resins.

Posting: The is the annual summary of occupational injuries and illnesses (OSHA Form 300-A) that must be posted at each establishment by February 1 and remain in place until April 30 to provide employees with the record of their establishment's injury and illness experience for the previous calendar year.

Physician or other licensed healthcare professional: A physician or other licensed healthcare professional is an individual whose legally permitted scope of practice (i.e., license registration or certification) allows him or her to independently perform, or be delegated the responsibility to perform, the activities described by this regulation.

Q

Qualified person: A person who, by reason of education, experience, or training is familiar with the operation to be performed and the hazards involved. The design of fall arrest systems must be engineered by a qualified person.

R

Recordable injuries and illnesses: Occupational deaths, loss of consciousness, restriction of work or motion, transfer to another job, or medical treatment (other than first aid) that must be recorded on the OSHA 300 Log.

Regularly exempt employers: Employers regularly exempt from OSHA recordkeeping include all employers with no more than ten full- or part-time employees at any one time in the previous calendar year; and all employers in retail trade, finance, insurance, real estate, and services industries; (i.e., SICs 52-89, except building materials and garden supplies, SIC 52; general merchandise and food stores, SICs 53 and 54; hotels and other lodging places, SIC 70; repair services, SICs 75 and 76; amusement and recreation services, SIC 79; and health services, SIC 80). Note: Some state safety and health laws may require these employers to keep OSHA records.

Respiratory condition: This is a term used on the OSHA 300 Log to record respiratory conditions. Examples would include pneumonitis; pharyngitis; rhinitis, or acute congestion due to chemicals, dusts, gases, or fumes; farmer's lung.

Restriction of work or motion: Occurs when the employee, because of the result of a job-related injury or illness, is physically or mentally unable to perform all or any part of his or her normal assignment during all or any part of the workday or shift.

Restricted work or transfer to another job: The employer must record the number of calendar days that the employee was restricted or transferred. Restricted work occurs when the employer keeps the employee from performing one or more of the routine functions of his or her job or from working the full workday. Restricted work also occurs when a physician or other healthcare professional recommends that the employee not perform one or more of the routine functions of his or her job or not work the full workday. The employer may stop counting days if the employee's job has been permanently modified in a way that eliminates the routing functions the employee has been restricted from performing. For example, a permanent modification would include reassigning an employee with a respiratory allergy to a job where such allergies are not present or adding a mechanical assist to a job that formerly required manual lifting.

Rope grab: A fall arresting device that provides employees protection while moving in the vertical direction (such as climbing). Rope grabs are designed to move up or down a vertical lifeline which is suspended from a fixed overhead anchorage point. The vertical lifeline is independent of the work platform and is attached to a harness by a rope grab and lanyard. In the event of a fall, the rope grab locks onto the lifeline to arrest the fall. The use of a rope grab device is ideal for fall protection during work from two-point suspension scaffolds.

Routine functions: An employee's routine functions that he or she regularly performs at least once per week.

S

Safety monitor system: A system used in conjunction with a warning line system. A competent person is assigned, as his sole duty, to monitor the proximity of workers to fall hazards when working between the warning line system and the unprotected sides and edges of a work surface.

Safety nets: Used to provide passive fall protection under and around an elevated work area.

Self-retracting (retractable) lifeline: A deceleration device containing a drum-wound line which may be slowly extracted from or retracted onto the drum under slight tension during normal employee movement, and which after onset of a fall, automatically locks the drum and arrests the fall. This device limits the fall to approximately 18 inches and is used during climbing operations or with horizontal lifeline systems.

Shock absorbing lanyard: A flexible line of webbing, cable, or rope used to secure a body belt or harness to a lifeline or anchorage point that has an integral shock absorber. The shock absorbing effect minimizes the forces distributed to the employee and anchorage points.

Significant work-related injuries and illnesses: Employers must record work-related injuries and illnesses that are significant or are diagnosed by a physician or other licensed healthcare professional. Record any work-related case involving cancer, chronic irreversible disease, a fractured or cracked bone, or a punctured eardrum. The employer must record such cases within seven days of finding out.

Signals: Moving signs provided by workers, such as flagmen, or by devices, such as flashing lights, to warn of possible or existing hazards.

Signs: The warnings of a hazard temporarily or permanently affixed or placed at locations where hazards exist.

Skin disorders: This is a term used on the OSHA 300 Log to record skin disorders. Examples would include contact dermatitis, eczema, or rash caused by primary irritants and sensitizers or poisonous plants; oil acne; chrome ulcers; chemical burns; and inflammations.

Small employers: Employers with no more than ten full- and/or part-time employees among all the establishments of their firm at any one time during the previous calendar year.

Source of injury or illness: This is the object, substance, exposure, or bodily motion that directly produced or inflicted the disabling condition cited. Examples are a heavy box, a toxic substance, fire or flame, and bodily motion of injured or ill worker.

Standard Industrial Classification (SIC): A classification system developed by the Office of Management and Budget, Executive Office of the President, for use in the classification of establishments by type of activity in which engaged. Each establishment is assigned an industry code for its major activity which is determined by the product manufactured or service rendered. Establishments may be classified in 2-, 3-, or 4-digit industries according to the degree of information available. They are used for statistical and research purposes. The US Department of Labor keeps track of employee injuries and illness statistics based on SIC codes. OSHA then uses those figures to program their inspections.

State agency: State agency administering the OSHA recordkeeping and reporting system. Many states cooperate directly with BLS in administering the OSHA recordkeeping and reporting programs. Some states have their own safety and health laws which may impose additional obligations.

State plan states: There are 23 states and two territories that are known as state plan states. A state can operate its own OSHA program if it obtains the Secretary of Labor's approval to do so. Employers in these states are subject only to regulation by their state OSHA agency. Federal OSHA authorities play virtually no enforcement role in these states. These 25 jurisdictions are also eligible for 50 percent federal funding of their occupational safety and health activities. The Department of Labor only requires that every state plan be at least as effective as federal OSHA standards in providing safe and healthful employment.

Supplementary record (OSHA Form 301): The form (or equivalent) on which additional information is recorded for each injury and illness entered on the log.

T

Tags: Temporary signs, usually attached to a piece of equipment or part of a structure, to warn of existing or immediate hazards.

U

Unprotected sides and edges: Any side or edge of a form, deck, floor, or structure where there is no protection from a falling hazard.

W

Warning line system: A barrier erected on the working surface to warn employees they are approaching an unprotected fall hazard.

Work environment: This consists of the employer's premises and other locations where employees are engaged in work-related activities or are present as a condition of their employment. The work environment includes not only physical locations, but also the equipment or materials used by the employee during the course of his or her work.

Workers' compensation systems: This is the state system that provides medical benefits and/or indemnity compensation to victims of work-related injuries and illnesses.

Index

A

Access and egress 105, 173, 264, 275

Access to employee exposure and medical records 69

Accident prevention 69–70, 83, 206, 208–209

Accident prevention signs and tags 83, 206

Administrative controls 7, 250, 353, 382–383, 403

Administrative subpoena 64

AFL-CIO 3, 412

Air monitoring 105–108, 175–176, 353

Alarm system 76, 296

Alaska 2, 10, 206

Anchorage 96, 103–104, 252, 258–262, 264, 267, 269, 433–435, 437, 441

Annual survey of occupational injuries and illnesses 13, 433

arc welding 86, 168

Arc welding and cutting 86, 168

Arizona 2, 206

Asbestos 9, 112, 120, 145, 160–161, 367–378, 399

Atomic Energy Act of 1954 2

Audiogram 54, 56–59, 188, 191

Audiometric testing 54–59, 77, 189–192

Awkward postures 215–216, 232

B

Barricades 68, 83–84, 113, 197, 200–202, 209–210, 266, 273–278, 433

Batteries 87, 91, 100, 336, 341, 343–344

Battery charging 87, 91, 336, 341

Benzene 31, 36, 49, 120

Blaster qualifications 112, 179

Blasting and use of explosives 112, 178

BLS (Bureau of Labor Statistics) 13–14, 16–17, 22, 25, 42–45, 58, 153, 157, 216, 235, 433–434, 442

BLS survey 17, 44

B (continued)

Bureau of Labor Statistics (BLS) 13–14, 16–17, 22, 42–45, 58, 60, 153, 157, 216, 235, 433–434, 442

Bureau of Mines 160, 294, 298

C

Cadmium 31, 36, 49, 120, 439

California 2, 118, 206, 427

Carpal Tunnel Syndrome (CTS) 214, 216–219

Caution signs 88, 207

CDC (Centers for Disease Control) 358

Centers for Disease Control (CDC) 358

Certified 32, 56–57, 62, 114, 149, 178, 303, 305, 413, 427, 439

Chains 97, 251

Code of Federal Regulations 28, 149, 158, 434

Collective bargaining 7–8, 412

Competence 54, 57, 153, 343

Competent person 69, 99, 101–102, 105, 107–108, 111, 114, 168, 170–178, 182, 266, 273–275, 278–282, 288, 290–291, 293, 295–297, 303, 305, 369, 408–409, 434, 441

Compliance audit(s) 389

Compliance officer 61, 64, 405, 429

Compliance officer checklist 64

Compliance Safety Health Officer (CSHO) 407–410, 434

Compressed air 68, 85, 104–105, 113, 177, 239, 357, 372, 403

Concrete 68, 84, 101–103, 115, 122, 174, 183–184, 201, 258–259, 261–264, 268, 300–301

Concrete and masonry construction 68, 101–102, 174

Congress 1–2, 4–6, 8, 14–15, 67, 205, 212, 439

Constitution 2, 430

Construction inspection 70, 119, 406–407, 409

Construction Safety Act 67, 183

Construction training requirements 158, 182

Container labeling 123, 140, 142

Contractor requirements 69

Crane inspection(s) 122, 302, 332

Crane inspection records 332

Cranes and hoists 109–100

CSHO (Compliance Safety Health Officer) 7, 407–410, 434

CTS (Carpal Tunnel Syndrome) 214, 216–219

D

Danger signs 179, 206, 403

Days away from work 23, 28–30, 35, 41, 44, 46, 52–54, 213, 434, 438–439

De Quervain's disease 217

Death 4, 16, 23, 29–30, 39, 46, 50–53, 62, 87, 211, 255, 270–271, 308, 335, 364, 399, 407, 435–436

Demolition 68, 111–112, 118, 122, 178, 293–296, 347

Demolition operations 111, 178, 293

Department of Agriculture 2

Department of Labor 2, 25, 67, 99, 121, 172, 235, 346, 439, 442

Diagnosis of a significant injury or illness 23–24

Directional signs 208

Dusts 77, 107–108, 160, 239, 339, 440

E

Education 68, 71, 150–151, 153, 157–159, 191, 248, 264, 266, 305, 440

Egress 69, 72, 75, 105–106, 159, 173, 264–265, 275

Emergencies 75–76, 110, 124, 247, 354, 368, 371

Emergency planning and response 388–389

Employee Emergency Action Plan 388

Employee information and training 80, 125, 161–163, 362–363, 370

Employee involvement 19, 25, 187, 408

Employee privacy 24

Energized substations 115, 181

Equipment 80–101

Equipment selection 239, 274

Ergonomic injury 213

Ergonomic inspections 212

Ergonomics 211–213, 215–216, 219–235

Ergonomics checklist 230–235

Ergonomics program 212, 220–229

Establishments 14, 17, 22, 25, 28, 39–40, 43–45, 64, 433, 442

Excavation(s) 68, 100–101, 105, 108, 113, 122, 173–174, 181, 183, 255, 273–291

Excavation, Trenching, and Shoring Program 273

Exit markings 75

Exit sign(s) 207

Exposure monitoring 351, 364, 369, 374

Exposure records 72, 74–75, 363, 374, 435

Eye and face protection 80, 241, 251

F

Fall arrest system(s) 94–96, 98, 171, 257–261, 263, 265–266, 435, 440

Fall hazard 95, 98, 119, 255, 257–258, 261–262, 264–266, 269–270, 435–436, 443

Fall hazard locations 255, 262

Fall protection 68, 94–98, 103–104, 111–112, 119, 121, 171, 182, 255, 258–260, 264, 266, 268–271, 273, 278, 434, 436, 441

Fall Protection Program 259, 264, 268–271, 273, 278

Fatality 15–16, 22, 25–26, 29–30, 63, 408, 410, 414, 436

Fire protection 9, 68–69, 82, 91, 109, 112, 121, 164, 296, 338, 341, 381, 386

First aid 13, 15, 19, 23–24, 47–48, 53, 68, 70–71, 76–77, 100, 124, 126, 135, 159–160, 214, 294, 400, 436, 440

Flagger control 200

Floor and wall openings 111–112

Floor openings 5, 111, 255, 271

Foot protection 80, 245

Forceful exertions 216

Forceful lifting 215

Forklift safety 333, 338–346

Forklift Safety Program 339–344

Formaldehyde 31, 36, 49, 121, 439

Frequently asked questions on hazard communication 136–148

Frequently asked questions on reporting occupational hearing loss cases 58

Fuel handling and storage 340

Fumes 35, 77, 86, 90–91, 107–108, 160, 247, 252, 298, 341, 347, 440

G

Gas welding and cutting 85, 166

Gases 31, 35–36, 77, 82–83, 86, 90–91, 106–108, 110, 159–160, 175, 241, 294, 340–341, 379, 439–440

Gases, vapors, fumes, dusts, and mists 77, 160

Gassy operations 106–108

General Duty Clause 5(a)(1) 10, 211–212, 216

Grounding of equipment 89

Guarding 85, 88–92, 139, 170, 314

Guardrail protection 257, 266

H

Hand protection 246, 252

Hand signaling 200, 209

Hand signals 99, 301, 303–304

Hawaii 2, 206

Hazard assessment 239

Hazard Communication Standards (HCS) 78–80, 123, 125, 137–143, 144–147, 158, 162, 372, 389, 417

Hazard determination 123, 141, 144

Hazardous chemical(s) 9, 77–80, 137–144, 147, 149, 161, 163, 294, 379–380, 385, 388, 393–395

Hazardous substance(s) 68, 120–121, 123, 125–126, 157, 276, 438

HCS (Hazard Communication Standards) 78–80, 123, 125, 137–143, 144–147, 158, 162, 372, 389, 417

Head protection 80, 242–244, 251

Health effects of lead 348

Healthcare professional 24, 30, 38, 47, 49, 53–54, 58, 60, 64, 73–74, 433, 438, 440–441

Hearing Conservation Program 77, 185, 193

Hearing loss 54–60, 183, 185, 188–191

Hearing loss disorders 57

Hearing loss recording criteria 54

Hearing protection program 58

Hearing protectors 55, 183, 188–191

Heat cramps 400

Heat exhaustion 31, 36, 399–401

Heat rashes 401

Heat stress 31, 36, 48, 399, 401–404, 436

Heat stress questionnaire 404

Heat stroke 399–400, 403

Heating 82, 86, 88, 115, 168, 347, 380

Hepatitis 31, 36, 47, 49, 433

Hot work 107–109, 176, 387, 401, 403

Hot work permit 387

Housekeeping 69, 161, 181, 257–258, 348, 357, 368, 372

Hydrogen sulfide 31, 36, 107, 439

Hygiene 159, 161, 357, 359, 363

Hygiene areas and practices 357

I

Illness 8, 13–17, 22–26, 28–31, 35–36, 38–39, 41–44, 46–54, 56, 59–61, 64, 73, 82, 159, 177, 213–216, 385, 411–412, 417, 420, 433–435, 438–440, 442

Illumination 69, 77–78, 105, 109, 175, 278

Incident investigation 388

Indiana 2, 206

Infirmary 159, 294

Injuries and illnesses 13–17, 19, 22–26, 28–30, 32, 38, 40–41, 45–46, 48, 50, 52–53, 60–61, 64, 149–150, 157, 191, 212–215, 235, 385, 411, 417, 433–435, 437–441, 443

Injury 8, 13–17, 22–26, 28–31, 35–36, 38–39, 41–44, 46–54, 56, 60–61, 64, 81–82, 105, 113, 149, 152, 155, 157, 159, 212–217, 219, 238, 242–243, 245, 251–252, 255–256, 259–261, 265–267, 269–271, 294, 313, 334–335, 338, 385, 399, 411–412, 417, 420, 433–436, 438–440, 442

Injury/illness records 8

Ionizing radiation 31, 36, 77–78, 160

Iowa 2, 206

J

Job Hazard Analysis 152–153, 157, 303–304

K

Kentucky 2, 204, 206

L

Labor Department 1, 412, 434

Labor-management relations 6–8

Ladders 68, 111–112, 118–119, 122, 182, 255, 257, 275, 320

Lanyards 81, 96–97, 115, 252–253, 258–259, 261–263, 269, 433, 435, 437

Lead 6–7, 11, 31, 36, 49, 77–78, 112, 162, 219, 308, 347–366, 403, 439

Lead Exposure Control Program 348

Lifelines 81, 96–97, 252–253, 257–259, 261–263, 269, 277, 294, 320, 433, 435, 437

Loss of consciousness 13, 15, 23, 48, 52–53, 399, 438, 440

Lost workdays 15, 43, 45

Low hazard industry exemption 17

M

Maryland 206

Masonry 5, 68, 101–102, 111, 122, 174

Masonry construction 68, 101–102, 174

Material handling equipment 99, 117, 122, 173, 341

Material Safety Data Sheet (MSDS) 9, 61, 78–80, 83, 124–125, 137–148, 157–158, 162, 211, 213–217, 219, 362, 380

Mechanical demolition 111, 178

Medical access orders 64

Medical records 62, 65, 69, 72–73, 75, 364, 435, 438

Medical removal 7, 11, 49, 162, 354, 358–361, 363

Medical removal protection 7, 162, 358, 360–361, 363

Medical services and first aid 76–77, 159, 294

Medical surveillance 49, 74, 162, 348, 358, 360–361, 363–364, 370, 372, 374, 403, 417

Medical treatment 13, 15, 19, 23–24, 38, 47, 52–54, 74, 212, 400, 438, 440

Medical treatment beyond first aid 53

Mental illness 24, 28, 49, 439

Methenediamileine 49

Methylene chloride 49, 121

Michigan 3, 10, 204, 206

Minnesota 3, 206

Mists 77, 107–108, 160, 247

Mobile equipment 181, 276, 296

Monitoring noise 186–187

MSD (Musculoskeletal Disorder) 213–216

MSDS (Material Safety Data Sheet) 9, 61, 78–80, 83, 124–125, 137–148, 157–158, 162, 211, 213–217, 219, 362, 380

Multi-employer job sites 5

Multiple hospitalizations 25, 63

Musculoskeletal Disorders (MSD) 61, 211, 213–214, 216

N

National Institute of Occupational Safety and Health (NIOSH) 74–75, 110, 216, 247, 348, 353–354, 363, 370

National Labor Relations Act 7–8

Needlesticks 49

Nevada 3, 206

New Mexico 3, 206

Newly covered industries 17

NIOSH (National Institute of Occupational Safety and Health) 74–75, 110, 216, 247, 348, 353–354, 363, 370

Nixon, Richard 1–2

Noise control 183

Noise measuring 188

Noise Standard 3, 7, 11, 54–59, 77, 183, 435

Nonionizing radiation 31

Non-routine tasks 126, 140

North Carolina 3, 204, 206

O

One or more days away from work 46, 52, 434

Oregon 3, 206

OSH Act 2, 8, 10, 13–14, 16, 26, 67, 205, 416, 421, 439

OSHA Area Office 63

OSHA Form 300A 32, 41

OSHA Form 301 26, 38, 439, 442

OSHA inspection(s) 8–9, 11, 60, 186, 212, 405–406, 411–412, 414, 416–417, 427, 429–431, 436

OSHA inspector(s) 5, 7, 60, 62–63, 70–71, 139, 147, 186, 212, 406–408, 411–412, 414, 416–418, 420–421, 423–424, 430–431, 434, 436

OSHA regulation(s) 8, 142, 237, 244, 299, 412–413, 434

OSHA representative 14, 72, 430

OSHA requirements 4, 8, 97, 146, 198–199, 274, 307, 309–310, 312–323, 338, 379, 388, 418, 420, 423

OSHA Standard(s) 3, 5–7, 9–10, 14, 26, 49, 67, 69, 75, 78, 84, 123, 136, 139, 141, 143–144, 149–152, 158, 184–185, 191, 198, 205, 211–212, 244, 248, 266, 268, 338–339, 347–348, 351, 355, 357, 379–382, 385, 387–389, 412–414, 429, 434, 437, 442

OSHA violation(s) 6, 8–9, 411, 414, 418–421, 425, 428

Overhead lines 115, 181

Overloading 119, 262, 306, 308, 321–322

Oxygen 48, 107–109, 159, 217, 276, 400, 438

P

Partially exempt industries 17–18

PEL (Permissible Exposure Level) 184–185, 347–348, 350–357, 362, 365–369, 371–373

Penalties 1, 9, 62–63, 144, 147, 406, 411, 419, 427–428

Permissible Exposure Level (PEL) 184–185, 347–348, 350–357, 362, 365–369, 371–373

Personal Protective Equipment (PPE) 69, 71, 74, 78, 80–82, 85, 92, 105, 126, 138, 161, 175, 185, 237–242, 249–253, 269, 274–275, 355, 360, 373, 383

Physician 23–24, 38, 46–50, 53–54, 56–58, 60–61, 64, 73, 159, 162, 177, 190, 294, 354, 358–360, 363–364, 372–374, 400, 433, 438, 440–441

Poison(s) 159, 362

Police and fire contact 295

Posting 9, 24–25, 83–84, 107, 149, 278, 417, 440

Powered industrial truck hazards 333–346

Power operated hand tools 85

Power transmission and distribution 68, 115, 122, 180

Powered industrial truck 173, 333–346

Powerline contact 306–307

PPE (Personal Protection Equipment) 69, 71, 74, 78, 80–82, 85, 92, 105, 126, 138, 161, 175, 185, 237–242, 249–253, 269, 274–275, 355, 360, 373, 383

Preconstruction planning 303–304

Process Hazard Analysis 380–383, 386, 389

Process safety information 380, 383, 385, 387–389

Process Safety Management 77–78, 149, 163, 379–380, 389–392

Programmed inspections 7, 434

Protective work clothing and equipment 81, 137, 161, 238, 348, 352, 355–356, 363, 369–371, 403

Public health service 9

Puerto Rico 3, 206, 406

Q

Qualified 69, 88, 96, 99, 102, 104, 110, 113, 149, 153–154, 158, 160, 169–175, 177–181, 199, 242, 259–260, 262, 265–266, 294, 296–297, 300, 369, 382, 413–414, 437, 440

R

Raynaud's Syndrome 217–218

Recording injuries to soft tissues 24

Recordkeeping 13–17, 19, 25–26, 28, 39, 47, 50, 52, 54–56, 60–63, 65, 70–72, 274, 363, 373–374, 413, 434, 436, 438, 440, 442

Recordkeeping forms 26

Recordkeeping inspection 63, 65

Repetitive motion 211, 215–218

Reporting occupational hearing loss cases 54

Respiratory conditions 31, 35, 440

Respiratory protection 81, 136, 164, 247–249, 276, 347, 350, 353, 355, 359, 370–371

Restricted work 23–24, 30, 46, 52–54, 438, 440–441

Restriction of motion 15

Restriction of normal job functions 23

Retention of records 31, 39, 75

Rope grabs 258, 261, 263, 441

Ropes 96–97, 300, 302

S

Safe blasting procedures 293, 296

Safety and Health Program(s) 13–14, 45, 70, 80–82, 149, 151–152, 213, 408–410, 429

Safety-related work practices 87–89

Scaffolding 68, 93, 169

Secretary of Labor 1–2, 4, 10–11, 13, 15, 67, 107, 178, 215, 430, 442

Sexual assault 24, 49

Sharps 19, 50, 63–64

SIC Code 17, 32, 64, 157

Signal(s) 68, 76, 83–84, 99, 107, 113, 179, 184, 197, 200, 202, 209, 248, 257, 269–297, 301, 303–304, 316, 334–335, 340, 442

Significant criteria 48

Signs 22–24, 32, 53, 68, 73–75, 82–84, 88–89, 97, 106–107, 109–110, 114, 124, 141, 149, 162, 178–179, 197, 199–202, 205–209, 244, 274, 282, 286, 293–294, 297, 334, 348, 358, 362, 370, 399–401, 403, 442–443

Signs and tags 83–84, 206

Signs, signals, and barricades 68, 83–84, 197, 209

Skin disorders 30, 35, 442

Small employer exemption 17

Stairway(s) 68, 82, 112, 118–119, 122, 182, 257

Standard threshold shift (STS) 54–59, 190–191

State OSHA plans 2

State plan standards 206

State Right-to-Know Laws 80

STS (Standard threshold shift) 54–59, 190–191

Subcontractor 5–6, 69, 230, 405–406, 429

Subpart C—General Safety and Health Provisions 61, 67–68, 323, 369, 408–409

Subpart D—Occupational Health and Environmental Controls 67, 76–77

Subpart F—Fire Protection and Prevention 68, 82, 109, 121

Subpart G—Signs, Signals, & Barricades 68, 83–84, 197–198, 206

Subpart H—Materials, Handling, Storage, Use, and Disposal 68, 78, 84

Subpart I—Tools (Hard and Power) 68, 85–86

Subpart J—Welding and Cutting 68, 85–86, 121

Subpart K—Electrical 68, 86–87, 91–92, 121

Subpart L—Scaffolding 68, 93

Subpart M—Fall Protection 68, 85, 94, 112, 121

Subpart N—Cranes, Derricks, Hoists, Elevators, and Conveyors 68, 98–99, 173, 313

Subpart O—Motor Vehicles, Motorized Equipment, and Moving Operations 68, 99–100

Subpart P—Excavations 68, 100–101, 122, 282, 284–286

Subpart Q—Concrete & Masonry Construction 68, 101, 122

Subpart R—Steel Erection 68, 103

Subpart S—Underground Construction 68, 104–105

Subpart U—Blasting and Use of Explosives 68, 112–113

Subpart V—Power Transmission and Distribution 68, 115–116

Subpart W—Rollover Protective Structures and Overhead Protection 68, 100, 117–118

Subpart X—Stairways & Ladders 68, 112, 118–119, 122

Subpart Z—Toxic and Hazardous Substances 67–68, 78, 120–121

Surface transportation of explosives 112, 179

T

Tags 83–84, 90, 180, 206, 208–209, 338, 443

Tendinitis 214, 216–217, 219

Tenosynovitis 214, 217

Termination 15, 75, 199

Termination of employment 15, 75

Threshold quantities (TQ) 379–380

Time-weighted average (TWA) 55, 184–185, 188, 347, 350, 355, 365–369

Toilets 144

Tools 38, 45, 68, 85, 89, 91–92, 101, 109, 115–117, 140, 153–154, 157, 165–166, 183–184, 208, 211, 215, 217–219, 231, 238, 240, 242, 245, 271, 275, 314, 341, 401, 409, 429

Tools and equipment 85, 92

Toxic and hazardous substances 68, 120–121

Toxic substance(s) 31, 36, 72–73, 348, 359, 442

TQ (Threshold quantities) 379–380

Trade secret(s) 73–74, 124, 389, 417

Training and education 68, 71, 150–151, 153, 157, 159, 191

Transfer 13, 15, 19, 23–25, 29–30, 35, 38, 46, 52–53, 74, 109, 140, 154, 294, 364, 374, 440

Transfer to another job 13, 15, 23, 46, 53, 440

Trenching 101, 173, 273, 408

Trigger finger 217

Turntable failure 306, 321

TWA (Time weighted average) 55, 184–185, 188, 347, 350, 355, 365–369

Two-blocking 306, 310–312, 314

Type A soil(s) 283, 287, 290

Type B soil(s) 284, 287

Type C soil(s) 281, 284, 287, 290

U

U.S. Bureau of Mines 160, 294

U.S. Department of Labor 25, 67, 99, 121, 172, 235, 439

U.S. Public Health Service 9

Underground construction 68, 104–106, 108–109, 122, 175–176

Unguarded moving parts 306, 313

Unsafe hooks 306, 314

V

Vapors 35, 77, 90, 107–108, 160, 241, 247, 252, 339

W

Wall openings 111–112

Warning line system 97, 441, 443

Waste disposal 136, 161, 372

Water accumulation 173, 276

Welding 31, 36, 68, 85–86, 108–109, 118, 121, 127, 166–168, 176, 238, 262, 347, 387

Welding, cutting, and heating 86, 168

Wire ropes 300, 302

Woodworking machinery 165

Work classification 367–368

Work relationship 16, 57

Written fall protection plan 95, 268

Government Institutes Mini-Catalog

PC #	**ENVIRONMENTAL TITLES**	Pub Date	Price*
949	ABCs of Environmental Regulation, Second Edition	2002	$79
812	Achieving Environmental Excellence	2003	$79
898	Book of Lists for Regulated Hazardous Substances, Tenth Edition	2001	$99
846	Clean Water Handbook, Third Edition	2003	$125
822	Environmental Compliance Auditing & Management Systems (ECAMS)	2001	$1,395
897	EH&S CFR Training Requirements, Fifth Edition	2001	$125
4300	EH&S Daily Federal Register Notification Service (e-mail)	2003	$150
890	Environmental Biotreatment: Technologies for Air, Water, Soil, and Wastes	2002	$149
4045	Environmental, Health & Safety CFRs on CD-ROM, single issue	2003	$450
825	Environmental, Health and Safety Audits, 8th Edition	2001	$115
4900	Environmental Law Expert, CD-ROM	2003	$695
955	Environmental Law Handbook, Seventeenth Edition	2003	$99
688	Environmental Health & Safety Dictionary, Seventh Edition	2000	$95
956	Environmental Statutes, 2003 Edition	2003	$125
957	Fundamentals of Environmental Sampling	2003	$79
689	Fundamentals of Site Remediation	2000	$85
907	Hazardous Materials Transportation Training, Student's Manual	2002	$89
958	Information Technology Solutions for EH&S Professionals	2003	$115
823	Integrating Environmental, Health, and Safety Systems	2001	$495
588	International Environmental Auditing	1998	$179
819	ISO 14001: Positioning Your Organization for Environmental Success	2001	$115
936	Managing Your Hazardous Wastes	2002	$89
830	RCRA CFRs Made Easy	2002	$115
841	Risk Management Planning Handbook, Second Edition	2002	$115
816	Stormwater Discharge Management	2003	$89
946	Wastewater and Biosolids Treatment Technologies	2002	$129

PC #	**SAFETY and HEALTH TITLES**	Pub Date	Price*
697	Applied Statistics in Occupational Safety and Health	2000	$105
893	Beyond Safety Accountability	2001	$79
894	Building Successful Safety Teams	2001	$79
843	Emergency Preparedness for Facilities	2003	$125
904	Ergonomics: A Risk Manager's Guide	2002	$159
949	Excavation Safety	2003	$85
663	Forklift Safety, Second Edition	1999	$85
814	Fundamentals of Occupational Safety & Health, Third Edition	2003	$79
818	Job Hazard Analysis	2001	$79
888	Keys to Behavior Based Safety	2001	$85
662	Machine Guarding Handbook	1999	$75
838	Managing Chemical Safety	2002	$95
889	Managing Electrical Safety	2001	$69
4800	OSHA Compliance Master CD-ROM	2003	$375
668	Safety Made Easy, Second Edition	1999	$75
947	Safety Metrics	2003	$85
815	So, You're the Safety Director, Third Edition	2003	$89

Government Institutes
4 Research Place, Suite 200 • Rockville, MD 20850-3226
Tel. (301) 921-2323 • FAX (301) 921-0264
Email: giinfo@govinst.com • www.govinst.com

Please call our customer service department at (301) 921-2323 for a free publications catalog.

CFRs now available online. Call (301) 921-2323 for info.

*All prices are subject to change. Please call for current prices and availablity.

Government Institutes Order Form

4 Research Place, Suite 200 • Rockville, MD 20850-3226
Tel (301) 921-2323 • Fax (301) 921-0264
Internet: http://www.govinst.com • E-mail: giinfo@govinst.com

4 EASY WAYS TO ORDER

1. Tel: **(301) 921-2323**
Have your credit card ready when you call.

2. Fax: **(301) 921-0264**
Fax this completed order form with your company purchase order or credit card information.

3. Mail: **Government Institutes Division**
ABS Group Inc.
P.O. Box 846304
Dallas, TX 75284-6304   USA

Mail this completed order form with a check, company purchase order, or credit card information.

4. Online: **Visit http://www.govinst.com**

PAYMENT OPTIONS

❏ **Check** *(payable in US dollars to **ABS Group Inc. Government Institutes Division**)*

❏ **Purchase Order** *(This order form must be attached to your company P.O. Note: All International orders must be prepaid.)*

❏ **Credit Card**   ❏ **VISA**   ❏ **Master Card**   ❏ **American Express**

Exp. ____ /____

Credit Card No. _____

Signature _____

(Government Institutes' Federal I.D.# is 13-2695912)

CUSTOMER INFORMATION

Ship To: (Please attach your purchase order)

Name _____
GI Account # (*7 digits on mailing label*) _____
Company/Institution _____
Address _____
(Please supply street address for UPS shipping)
_____
City _____ State/Province _____
Zip/Postal Code _____ Country _____
Tel ( ) _____
Fax ( ) _____
E-mail Address _____

Bill To: (if different from ship-to address)

Name _____
Title/Position _____
Company/Institution _____
Address _____
(Please supply street address for UPS shipping)
_____
City _____ State/Province _____
Zip/Postal Code _____ Country _____
Tel ( ) _____
Fax ( ) _____
E-mail Address _____

Qty.	Product Code	Title	Price

30 DAY MONEY-BACK GUARANTEE

If you're not completely satisfied with any product, return it undamaged within 30 days for a full and immediate refund on the price of the product.

Subtotal _____
MD Residents add 5% Sales Tax _____
Shipping and Handling (see box below) _____
Total Payment Enclosed _____

SOURCE CODE: BP03

Shipping and Handling	Sales Tax
Within U.S: 1-4 products: $6/product 5 or more: $4/product **Outside U.S:** Add $15 for each item (Global)	Maryland .... 5% Texas .... 8.25% Virginia 4.5%